Developments in Carbohydrate Chemistry

Edited by

Richard J. Alexander
Penwest Foods Co.
Cedar Rapids, Iowa

Henry F. Zobel
ABCV Starch
Darien, Illinois

The American Association of Cereal Chemists
St. Paul, Minnesota, USA

Cover design adapted from an illustration of the three-dimensional structure of B-type starch by A. Imberty and S. Perez in *Biopolymers* 27:1205-1221. Copyright © 1988 John Wiley & Sons, Inc. Reprinted by permission.

This book has been reproduced directly from computer-generated copy submitted in final form to the American Association of Cereal Chemists by the editors of this volume. No editing or proofreading has been done by the Association.

Library of Congress Catalog Card Number: 92-71691
International Standard Book Number: 0-913250-76-7

Reference in this volume to a company or product name by personnel of the U.S. Department of Agriculture or anyone else is intended for explicit description only and does not imply approval or recommendation of the product to the exclusion of others that may be suitable.

Printed in the United States of America on acid-free paper

American Association of Cereal Chemists
3340 Pilot Knob Road
St. Paul, Minnesota 55121-2097, USA

CONTENTS

APPLICATIONS

CONTRIBUTORS

R. J. ALEXANDER, Penwest Foods Co., Cedar Rapids, Iowa

C. G. BILIADERIS, Department of Food Science, University of Manitoba, Winnipeg, Canada

A. CORONA, Department of Chemical Engineering, Worcester Polytechnic Institute, Worcester, Massachusetts

M. J. GIDLEY, Unilever Research, Sharnbrook, Bedford, England

J. M. HARPER, Office of Vice President for Research, Colorado State University, Fort Collins, Colorado

R. E. HEBEDA, Enzyme Biosystems, Ltd., Arlington Heights, Illinois

K. B. HICKS, USDA, Eastern Regional Research Center, Philadelphia,Pennsylvania

A. T. HOTCHKISS, Jr., USDA, Eastern Regional Research Center, Philadelphia, Pennsylvania

K. W. KIRBY, Penford Products Co., Cedar Rapids, Iowa

J. LELIÈVRE, Food Science Department, Acadia University, Wolfville, Nova Scotia, Canada (present address: Research and Development, Ault Foods Ltd., London, Ontario, Canada)

J. F. ROBYT, Department of Biochemistry, Iowa State University, Ames, Iowa

J. E. ROLLINGS, Department of Chemical Engineering, Worcester Polytechnic Instititute, Worcester, Massachusetts

C. R. SULLIVAN, Department of Chemical Engineering,Worcester Polytechnic Institute, Worcester, Massachusetts

W.M. TEAGUE, Enzyme Biosystems, Ltd., Arlington Heights, Illinois

D. B. THOMPSON, Department of Food Science, Pennsylvania State University, University Park, Pennsylvania

P. J. WOOD, Centre for Food and Animal Research, Agriculture Canada, Ottawa, Ontario, Canada

H. F. ZOBEL, ABCV Starch, Darien, Illinois

PREFACE

At the time of the Annual AACC meeting in 1992 the Carbohydrate Division will celebrate its 25th Anniversary. Such an event is particularly meaningful to the membership for several reasons: for many members, memories of their active involvement in the organization; the interpersonal relationships the group has fostered; and especially the technical/scientific programs the group has sponsored. It is to the latter that this book is primarily dedicated.

The Carbohydrate Division has been conducting technical symposia since it was founded in 1967. For the year presented, most of these symposia have centered around some specific aspect of carbohydrate chemistry. These have included starch hydrolysis (and resulting products), functionality of carbohydrates, starch structure, methods of carbohydrate analyses, modified food starches and their uses, and nutritional aspects of carbohydrates.

It was decided that an appropriate activity to commemorate the anniversary would be a special symposium on carbohydrate chemistry presented at the 1992 meeting and made available in book form at the time of the meeting. This symposium (book) will concentrate on current advances and theories concerning carbohydrates, and will incorporate updated portions of several of the earlier symposia.

In this way, it is hoped that the Carbohydrate Divison can make a permanent contribution to the field of carbohydrate chemistry. With this book we will be able to provide carbohydrate chemists, cereal chemists, food technologists, and related technical and scientific personnel with an update on the status of carbohydrates. In doing so, we will be helpful in creating potential new products and technology for the immediate future and for years to come.

Richard J. Alexander

Henry F. Zobel

Developments in Carbohydrate Chemistry

STARCH GRANULE STRUCTURE

Henry F. Zobel

A B C V Starch
Darien, IL 60559

INTRODUCTION

Like many other complex materials our knowledge about starch granule structure has developed slowly and sometimes indirectly. New information on starch is often the result of using new methods of characterization and investigation. In recognition of the 25th anniversary of the founding of the Carbohydrate Division of AACC, progress in defining granule structure and related properties will be reviewed. Shifts in research emphasis from 1967 to 1992 will be noted and, where possible, the significance of findings will be indicated.

HISTORICAL

We should first, however, place the findings of our generation in perspective by acknowledging prior insights into structure. Leeuwenhoek in 1719, for example, used his newly invented microscope to examine starches. He reported on the size and shape of granules, their swelling in hot-water, the attendant diffusion of solubles, and the resultant sac-like residue (Radley, 1968).

In 1858, Nägeli introduced micellar theory to describe starch and cellulose properties, postulating that molecular aggregates (micelles) rather than individual molecules could account for observed properties. With time, a blocklett picture emerged such as that shown in **Figure 1A** where particles were held together with an amorphous material responsible for water uptake and swelling. Once it was realized that molecules were many times longer than those ascribed to particle lengths, the model became outmoded and led to the continuous structure model (**Fig.1B**). The next advance was the micellar network or fringe micelle theory (**Fig. 1C**) which was developed to account for the gel properties of gelatins (Gerngross *et al.*, 1930). This first model is represented by **Figure 1C**. The concept was adapted to both starch and cellulose and received widespread attention in the process. For cellulose, a pronounced fibrillar molecular orientation was incorporated into the fringe models (Sterling, 1968; Howsmon and Sisson, 1954).

In many ways, the lineage of structural studies on starch can be traced to the research and writings of K H. Meyer and co-workers (Meyer, 1942a, 1942b, 1952). Meyer viewed granules as containing

1

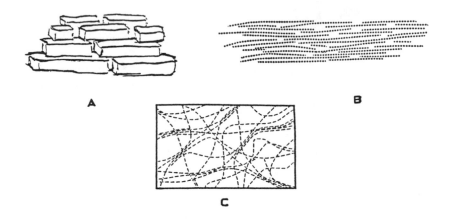

Figure 1. Blocklett (A), continuous (B), and fringe micelle (C) models for natural polymers.

Figure 2. Model for native (Left) and swollen (Right) starch granules; adapted from proposals by K. H. Meyer.

radially oriented needle-like crystals to explain their insolubility and optical properties. To explain swelling and elastic properties, he first observed that isolated "true" spherulites disintegrated under pressure, but granules did not. Therefore, granules must be held together by an elastic non-crystalline or amorphous element. Specifically, a system in which crystallites are built from giant branched molecules of amylopectin and united by molecular threads. To Meyer, "fringe micelles" were formed with branches of different amylopectin molecules gathered together in crystalline bundles, which themselves were linked together by those parts of the amylopectin molecule not included in the crystal structure (Meyer, 1952). His first reference to such a model may have been at a conference in Madrid in 1932 (Meyer, 1942b). **Figure 2** reflects the granule features described above and the several schematics published by Meyer for native and swollen (partially gelatinized) granules. In accord with developments to be discussed here, the model remains acceptable.

2

Meyer, a high molecular weight advocate for characterizing natural polymers, often held views contrary to that of his colleagues. He viewed molecules to be chain-like arrangements of building units united by primary valence forces; micellar structures were formed through secondary valences. Meyer's overall conclusions can be summarized as follows: molecular groups, and not the elemental molecules as such, determine the technically significant properties of starch pastes and solutions. He further suggested that only by consideration of such systems could one understand correctly the nature of starch substances.

In the starch literature, the Meyer fringe micelle model often has been used to illustrate native and swollen granules but with little discussion of the underlying concept. One adaptation of the model was to account for the low rigidity of gels after the starch had been degraded by enzyme action. Even though starch crystallinity increased, the semicrystalline gel network was weakened by enzymic attack in the amorphous regions of the gel (Zobel et al., 1959). The same reasoning was used by the authors to account for the effects of enzymes added to retard bread firming. A renewed appreciation of the fringe micelle as a model for starch has resulted from the writings of Slade and Levine 1987a, 1987b, and 1988. In utilizing such a model, tacit recognition is given to the roughly 60-90% amorphous content of granules and the role this fraction has in determining starch properties.

With the work on natural polymers in place, the fringe micelle model was adapted to depict the morphology of semicrystalline synthetic polymer structures (Flory, 1953, Wunderlich, 1973, Billmeyer, 1984). Although widely used, the model is not universal; sometimes even judged as totally inadequate for explaining polymer crystallization (Geil, 1963, Meares, 1965).

Within this limited sampling of early starch structure concepts, we find remarkable resemblances to our views of today. For this, we must accord high marks to those early workers, who used simple tools and techniques, but applied keen powers of observations and intellect to their work.

STARCH CRYSTAL STRUCTURES

Native starches has been known to be partly crystalline since about 1920. Subsequently, Katz established the designations A, B, and C that are still used for native and retrograded starch showing crystalline structures. The presence or absence of crystalline order in starches is often a basic factor underlying their properties. Accordingly, studies have persisted to determine the size, molecular conformation and chain packing that describe the basic repeating element (unit cell) of a starch crystallite (Zobel, 1988a,b).

Figures 3-5 show 25 years of progress in these endeavors beginning with the B-structure (Fig. 3) published by Rundle et al., 1944. Their proposed cell was orthorhombic (three perpendicular axes) with dimensions of a = 16.0 Å, b = 10.6 Å (fiber axis) and c = 9.2 Å. The cell contained 8 glucose residues, about 10 molecules of water and single chains in a conformation subsequently shown to require distorted bond angles and lengths.

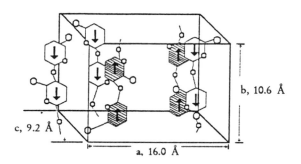

Figure 3. Rundle, Daasch, and French structure for B- starch.

The idea of double stranded rather than single helices for native starch polymorphs was first presented by Kainuma and French, 1972. The proposal addressed concerns about chain packing and a cell density (French, 1972b) that did not require inordinate amounts of water. Feasibility for helices being formed by both amylose and amylopectin was demonstrated using space-filling molecular models (French, 1972a). These models clearly indicated that α-(1,6) linkages were ideal points for promoting double helix formation (Zobel, 1988b). In the absence of computer modeling facilities, such models remain an alternative to help visualize starch molecules and their potential for crystallization, particularly amylopectin.

For structural studies, the advent and development of computer modeling vastly enhanced our capabilities for investigating starch (French and Murphy, 1977; Pérez and Vergelati, 1987; French and Brady, 1990). Figures 4 and 5 show computer-generated double helices resulting from an extensive examination of possible helical conformations and packing arrangements. In these Figures, two parallel helices are shown in their best fit (lowest energy) arrangement.

4

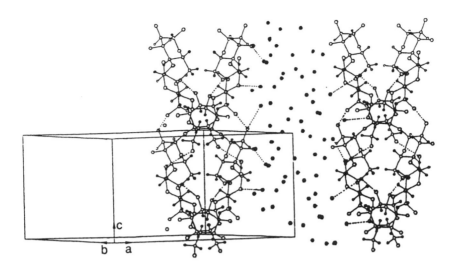

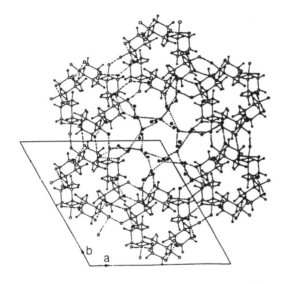

Figure 4. Crystalline structure of B-starch.
(Top) Two strands of double helices represented in three-dimensions along their fiber axis. Water molecules ☉ and interconnecting hydrogen bonds (•••) also are depicted.
(Bottom) An a,b plane projection showing the unit cell and nearby double helices. Water molecules ☉ are located in the center channel (Reproduced from Imberty and Pérez, 1988 and by permission of John Wiley & Sons. Copyright © 1988 John Wiley & Sons, Inc.)

Wu and Sarko, 1978a,b published B- and A-structures with double helices using data gathered from x-ray fiber patterns taken of oriented amylose films and computer modeling. Their data indicated a best fit for unit cells with two right-handed, parallel stranded, antiparallel packed helices (12 residues). Cell dimensions of B were: a = b = 18.5 Å, and c = 10.40 Å (helix axis). For A, cell size was: a = 11.90 Å, b = 17.70 Å, and c = 10.52 Å. Cell packing for B and A was hexagonal and orthogonal, and each contained 36 and 8 molecules of water, respectively (Sarko and Wu, 1978).

In 1987, Pérez and co-workers began to publish new three-dimensional structures for both native starch polymorphs (Imberty et $al.$, 1987a,b; Imberty et $al.$, 1988a,b). Structure development incorporated electron diffraction data from micron-sized single crystals, x-ray powder diffraction data, and fiber pattern data from Wu and Sarko.

Lattice parameters for B are the same as those given above, chains are packed in an hexagonal array where $\alpha = \beta = 90°$ and $\gamma = 120°$. The unit cell has two left-handed, parallel-stranded double helices that are arrayed in parallel: there are 12 glucose residues and 36 water molecules (27%). Half the water is tightly bound to the double helices and the other half is centered around a 6-fold screw axis parallel to the c axis. The repeating unit is α-(1,4) linked maltose with 3_1 symmetry generating a double helix with six residues per turn of each chain; the repeat distance is 2c or 21Å. Calculated cell density is 1.41 compared to 1.45, determined. **Figure 4** shows the B crystal structure with the double helices aligned along the fiber direction (c-axis). Another view is that of the a,b crystal axis plane in which helices are perpendicular to the paper; the additional helices illustrate the location of a central column of water. In both views, the isolated dots θ represent water molecules. The dotted lines show hydrogen bonding between adjacent helices and between water molecules.

The A unit cell (**Fig.5**) is represented in the same fashion as B with a three dimensional projection of the double helices along the c-axis and one onto the a,b plane. Like the B-structure, the unit cell for A contains 12 residues located in two left-handed chains; however, there are only 4 water molecules θ between helices. Crystallization occurs in a monoclinic lattice with a = 21.24 Å, b = 11.71 Å, c = 10.69 Å, and C = 123.5°. Helices are in the B2 space group. Translations along the axes permit the "crests" and "troughs" of adjacent helices to form densely packed structures (see also **Fig.4**). Interstrand stabilization is provided by $O_{(2)}....O_{(6)}$ hydrogen bonds and van der Waal's forces. The helix inner channel is 3.5 Å, too small for water inclusion; helix outside dimension is 10.3 Å. The individual helical repeat of 2c = 21.38 Å is generated by a two fold screw axis which means the asymmetric repeat unit for A is a maltotriose residue. Calculated cell density for A is 1.48 and close to the observed density, 1.51. Solid state ^{13}C nuclear magnetic resonance (NMR) spectra confirms the asymmetry assignments for both the A- and B- structures (Imberty *et al.*, 1988a).

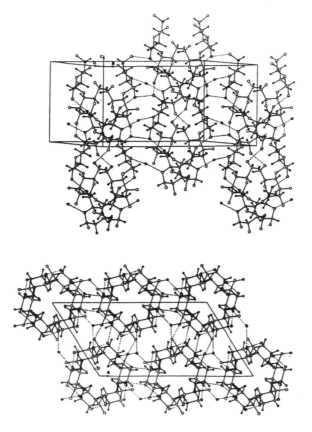

Figure 5. Crystalline structure of A-starch.
(Top) Double helices displayed along their fiber axis.
(Bottom) An a,b plane projection showing the unit cell, water molecules (⊖) and hydrogen bonds (··). (Reproduced, with permission, from Imberty *et al.*, 1987. Copyright 1987 American Chemical Society.)

In summary, the three dimensional organization of B- and A-structures differ, but their molecular conformations are practically identical. Both structures have double helices, the same chain chirality, parallel-strandness of helices and parallel registry in their packing. French,

1984 cites the latter two characteristics as likely from the viewpoint of granule biosynthesis. Model structures developed for amylopectin show parallel arrays of short branches suited to forming double helices and crystallites through their aggregation (Robin *et al.*, 1974). It is less clear how long amylose molecules form extensive double helices beyond their intertwining at the end of two chains.

Left-handed chirality places B- and A-chains in agreement with their precursor materials (Sarko and Marchaessault, 1967; Sarko and Biloski, l980). Fibers showing a native polymorph, for example, are generated from crystalline amylose acetate which is left-handed and single helical in conformation. Fibers are stretched for crystal orientation and then deacetylated with alkali, while under tension. The resulting alkali amylose fibers are then carefully humidified to achieve conversion to B or A. Other evidence for the chirality of amylosic chains has been obtained by optical rotation and by nuclear magnetic resonance (NMR) spectroscopy studies (Imberty *et al.*, 1988a). In addition, left-handed helices are reported to be somewhat more stable than their right-handed counterparts in aqueous solutions (Brant, 1980).

Since A has an amylosic helix in its center rather than a column of water, a more stable structure is suggested. Stability is evidenced by crystalline A- and B- spherulites melting at 90° and 77°C, respectively, when measured at a 0.9 volume fraction of water (Whittam *et al.*, 1990).

This difference in cell make-up between B and A would seem to account for several well-known effects in starches. Like, corn (A) loses less crystallinity than potato starch (B) when both are oven-dried. Likewise, displacement of the water column with a starch helix would explain the results of heat/moisture treatments on potato starch. Thus, potato starch can be transformed from B to A when heated under conditions of about 27% moisture, 100°C, and high humidity (Sair, 1967; Donovan *et al.*, 1983). Conversion is an example of a solid-state transformation in which potato starch becomes more cereal-like in its properties. On the other hand, A to B transformations cannot take place without the destruction of A and recrystallization (Zobel, 1988a). A stepwise description of the B to A transition has been proposed from computer modeling studies (Pérez *et al.*, 1990; Imberty *et al.*, 1991).

DOUBLE vs. SINGLE HELICES

Elegant proof for amylosic chains forming left-handed double helices is presented by Hinrichs *et al.*, 1987. This single crystal study describes a double helix in a polyiodide complex of p-nitrophenyl-α-maltohexaose at atomic levels of resolution. The resulting precise structural parameters are expected to help interpret fiber diffraction data on native polymorphs.

Both double and single chain helices have been established in crystalline starch material using x-ray diffraction. A critical limitation of x-ray methods, however, is that discrete diffraction lines become increasingly diffuse as crystallite size decreases. At a point, crystalline structure that would be indicative of helices can no longer be identified. To some extent, this gap in analysis is bridged by developments in high-resolution carbon 13 (^{13}C) NMR in which cross-polarization and magic-angle spinning are used to study solid samples (Gidley and Bociek, 1985, 1988). M. J. Gidley's chapter, this volume, reviews these developments as well as other NMR techniques with respect to starch structure and functionality.

Briefly, NMR spectra of solids can be used to identify order/non-order on a molecular basis in terms of double helix/single chain conformations (Gidley and Bociek, 1985). For exactness, it should be pointed out that single chain here should not be understood to mean the tight, six residue helix prescribed for V-structures by x-ray data. In this case, the starch chains are probably best described as having an extended (shallow) helix. Using ^{13}C-NMR spectra, Jane *et al.*, 1985 identified compact and extended helical conformations in aqueous amylodextrin and amylose solutions with and without added complexing agents, respectively.

With development of the ^{13}C methodology, structure/property relations can be sought beyond crystallinity to which primary consideration has been given. A major future achievement would be to distinguish helices, either in or out of crystalline registry, by NMR alone. Current methodology requires both NMR and x-ray data (Gidley and Bociek, 1985; Gidley and Cooke, 1991).

Introduction of the double helix concept is cited as a pivotal event in the history of starch chemistry. (For a concise summary of views about granule structure when double helices were first being considered, the book by Banks and Greenwood, 1975 is recommended.) While double helix findings have been applied mainly to crystal structure determinations, other possible structure/property relationships would include the following.

1. Double helix compactness can be a factor in granule density; the result a practical energy package.

2. If double helices are regarded as one molecular chain "complexed" with another, their presence could account for the resistance to chemical and enzymatic degradation shown by starches (Kainuma and French, 1972; Jane and Robyt, 1984; Matsunaga and Kainuma, 1986).

3. NMR measurements on 4 granular starches indicated double helix contents ranging from 38-53% (Gidley and Bociek, 1985). Since a 3.5 Å inner helix channel precludes absorption of water (Imberty et al., 1987a), a similar exclusion of impurities is possibly a factor in the high purity of isolated native starches.

4. Given the resistance of high amylose starches to gelatinization, the total amount of double helix may merit consideration rather than crystallinity alone. For such a starch, Gidley and Bociek, 1985 measured a 38:62 ratio of double (ordered) versus single (disordered) helices. The crystallinity of these starches, however, is generally on the low side of a 15-25% range (Zobel, 1988b). The authors cite 25%, which means crystallinity could account for 65% of granular order; at 15%, the percentage is only 39%.

5. Single V-amylose helices, amorphous or crystalline, however, can be very sensitive to water (Schoch, 1942) unless complexed with an appropriate adjunct (Gray and Schoch, 1962; Krog, 1971; Zobel, 1988a). Dispersion of "labile" single helical chains supports Gidley and Bociek, 1985 findings that associate double and single helical chains with ordered and non-ordered conformational states, respectively.

6. Gelation and retrogradation can be interpreted as the result of double helices forming a network of physically cross-linked molecules. As initial juncture points grow into helical segments and then aggregate into B-crystallites, gels or retrograded materials become more rigid and difficult to disperse (Sterling, 1960; Matsukura et al., 1983; Gidley, 1989; Thompson, 1992, this volume).

On the other hand, introduction of complexing agents into starch preparations disrupts double helix conformations by forming "stable" single chain V-conformation helices. Accordingly, control can be realized over processes of crystallization or retrogradation (Matsunaga and Kainuma, 1986; Zobel, 1988a).

7. In contradistinction to gelation, gelatinization must involve the unraveling of double helices. A relevant consideration then would be the relative stability of isolated double helices versus those in crystalline lattices. Is the disruption of helices a part of, a precursor to, or post crystal melting event? For the gelatinization process, how can the role of various elements of granular organization be distinguished?

One approach has been that of Gidley and Cooke, 1991 who followed the gelatinization of a waxy maize starch by birefringence, crystallinity, gelatinization enthalpy, and molecular order (see #4) measurements. In this way, the authors showed that loss of birefringence not only preceded other changes but also proceeded at a faster rate. Initial loss of birefringence and granule swelling were perceived at the same time while the latter continued throughout the heating process. While the other measures lagged birefringence loss, they also indicated losses of structure beyond complete loss of birefringence. Although differences between crystalline and molecular order losses were not resolved, insight was provided about levels of organized structure present during gelatinization. Their findings paralleled DSC data showing the persistence of structure at temperatures higher than those at peak Brabender viscosities (Zobel, 1984). In addition, the data gives quantitative support to views on the effects of amorphous and crystalline regions on starch swelling and gelatinization.

8. Are double helices a factor in explaining anomalies in starch behavior? For example: (a.) The results of Colwell et al., 1969 suggested that starch crystallization became less important at temperatures over $21°$ C in firming of gels (or bread staling). (b.) High temperature "melting" peaks on the order of $140-180°C$ are reported for low moisture (9-25%) starches (Donovan, 1979; Colonna and Mercier, 1985; Biliaderis et al., 1986; and Wittwer et al., 1987). X-ray patterns taken of comparable samples, however, are not amorphous but rather show evidence of crystallinity (Zobel et al., 1988). (c.) On the other hand, an annealed rice showing a sharp $75°C$ DSC melting peak was also shown to be amorphous. The authors concluded that the transition was due to changes other than those detectable by x-ray (Nakazawa et al., 1984.

How effective, for example, is the short-range "order" in double helices, which are not in a crystalline registry? Colonna and Mercier, 1985 schematically depict a double helix to coil transition to illustrate gelatinization, but with little discussion. Likewise, Wittwer et al., 1987 simply comment that a high temperature endotherm represents the melting enthalpy of starch helices. An obvious challenge is to be able to detect helix to random coil transitions under a variety of circumstances.

Research in this area is reported by Cooke and Gidley, 1991 who used DSC, NMR, and x-ray to follow the gelatinization of a starch under excess water conditions. They report a concurrent loss of molecular and crystalline order for several cereal starches and tapioca but not potato starch. Under their conditions, results were interpreted as indicating that enthalpy of gelatinization primarily reflects loss of double helical order. Gelatinization data obtained on a variety of starches using multiple techniques should provide new insights into gelatinization as well as starch structure.

Prior references to single helices of the V-type requires the following brief review. In general, the V conformation is a result of amylose being complexed with substances such as aliphatic fatty acids, surfactants, emulsifiers, n-alcohols, glycerol, dimethyl sulfoxide (Zobel, 1988a). While indirect methods have implicated the linear outer branches of amylopectin with complex formation, direct proof has also been obtained (Biliaderis and Vaughan, 1987; Gudmundsson and Eliasson, 1990). Crystallographic studies show that 6 glucopyranose residues repeat in about 8 Å generating helices in a close-packed array (French and Murphy, 1977). Single chain structures and their respective x-ray patterns are designated as V-anhydrous, hydrate, or DMSO as the case may be. Extruded starches may show a regular V or an E structure (Mercier et al., 1979). The latter appears to be an expanded V.

Our concern here, however, is limited mainly to the presence of V-structures in granular starches. As a rule, common industrial starches do not show V in the sense of three dimensionally ordered structures. Some of the exceptions to this rule are amylomaize starches with amylose content of 49%, or above, and certain genetic mutants of maize. For example, V-structures can be found in native starches grown under the control of recessive genes such as the amylose extender (*ae*), dull (*du*), and sugary (*su*) genotypes in single or multiple combinations. **Figure 6** compares starches showing a regular B-pattern and both B- and V-patterns, respectively.

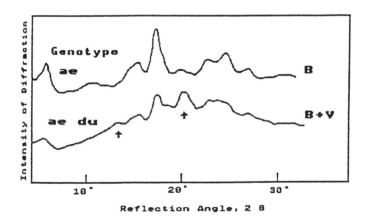

Figure 6. X-ray diffractograms of maize genotypes showing either B- or B- plus V-type patterns.

Within the criteria mentioned, it is not uncommon to find starches grown under control of the *ae* gene to show mixed patterns. Singly or in combination, *su*$_2$ and *su*$_2$*du* have shown B + V, A + V, and C + V patterns (Zobel, 1988a). The reason for the appearance of a V-structure in native starches is not completely understood. Amylopectin average chain length can be a factor much like its apparent role in determining native A, B, or C structures (Hizukuri *et al.*, 1983).

Natural occurring V-structures have not been observed in starches with less than 30% amylose; polar lipids also are needed. Solid state ^{13}C NMR studies, however, indicate that the preferred state for the amorphous phase in granules is a V-type conformation (Gidley and Bociek, 1988). Accordingly, such a conformational predisposition should facilitate the formation of V-structures in starches under the conditions described in the next two paragraphs.

When amylose and polar lipids are present in native starches, V-structures can result from gelatinization, both during heating and upon cooling (Sterling, 1960; Zobel, 1988a). Similarly, the development of V-structure has been identified in parboiled rice (Priestly, 1976) and in extruded starches containing polar lipids (Mercier *et al.*, 1979). For details on extrusion processing of starch see the chapter by Harper, 1992, this volume.

To modify starch properties while essentially retaining native structure, V can be introduced into granules by heat, moisture, and time treatment conditions such as: 18-45% moisture, 90-130° C, and holding times of 1-16h (Fukui and Nikuni, 1969; Zobel, 1988a).

STARCH GELATINIZATION, METHODOLOGY

Gelatinization is the first step in many cases of starch utilization whereby native granular structure is either partially or completely disrupted. In effect, an order/disorder transition takes place. During the 25 years of our interest, commonly used methods for following gelatinization include; loss of granule birefringence, consistency/viscosity changes, and granule swelling and solubility behavior. From birefringence observations, definite structure inferences can be made about molecular chain orientation but less so about crystalline order due to crystallite size. From granule swelling and solubility curves, as a function of temperature, Leach *et al.*, 1959 interpreted the data as indicating two levels of bonding forces within a granule structure.

More sophisticated methodology for determining starch gelatinization, along with supplying structural information, includes: x-ray diffraction (Zobel *et al.*, 1988), NMR spectroscopy (Jaska, 1971;

Hennig et al., 1976), light scattering (Marchant and Blanshard, 1978), electron spin resonance (ESR) (Windle, 1985) and electron microscopy (Liu et al., 1990).

Within our time frame, two determinant factors stimulated research into gelatinization per se as well as its basic relationship with granular and molecular structures. These two factors were the application to starch of: first, differential scanning calorimetry (DSC) and secondly, Flory's melting theory for polymers (Flory, 1953; Mandelkern, 1964).

DSC thermograms show first order endotherms indicating that crystallite melting is a major factor in starch gelatinization (Donovan, 1979; Zobel, 1984). DSC provides a facile method for measuring the heat energy required for gelatinization. As such, processes utilizing starch are readily assessed from this standpoint. The sealed pans, if viewed as miniature reactors, can be opened and examined. In this way, a variety of gel textures and associated thermal conditions can be determined to establish potential conditions for manufacture. A DSC study by Liu and Lelièvre, 1992 extended these possibilities by exploring the gelatinization of wheat and rice granular starch blends. DSC is a way to measure the temperature and energy needed to reverse starch retrogradation; or, to measure its level in a starch product (White et al., 1989; Chang and Lui, 1991). Biliaderis, 1990 reviews DSC for retrogradation measurements. Furthermore, DSC analysis provides a method for characterizing the amorphous phase of starch by glass transition measurements.

Representative literature citations on glass transitions in starch include: Slade and Levine, 1987a, 1987b, 1988; Zeleznak and Hoseney, 1987; and Biliaderis, 1990, 1991. A subsequent chapter in this volume by Lelièvre discusses glass transitions as well as thermal analysis of aqueous starch systems in general.

Flory's theory, expressed by the Flory Huggins (F-H) equation (see below), has been used extensively in synthetic polymer studies but less so with natural polymers. The latter category includes work with: a.) collagen, Witnauer and Fee, 1957; Flory and Garrett, 1958; b.) starch/Nägeli dextrins, Zobel et al., 1965; Lelièvre, 1973; Donovan, 1979; Donovan and Mapes, 1980; Biliaderis et al., 1980; Lelièvre, this volume; c.) heat/moisture treated starches, Donovan et al., 1983; and d.) short chain amyloses, Whittam et al., 1990. Lelièvre, 1976 used the Flory-Huggins equation to successfully model the effect of sugars on starch gelatinization. Evans and Haisman, 1982 modified the equation to predict the effect of different solutes on gelatinization temperature.

The arguments advanced by Flory were based on thermodynamic considerations and equilibrium melting conditions. The latter conditions, however, are generally compromised by realistic time constraints to achieve equilibrium. Regardless, thermal data developed under conditions less than ideal have been confirmed by methods in which equilibrium conditions were assured (Mandelkern, 1964).

According to Flory, a rigorous test for polymer melting being a first order transition is to determine the effect on heat of fusion and on melt temperature by mixing a second component (diluent) with the polymer. A smooth response to diluent action effectually substantiates existence of distinct crystalline and amorphous phases. Experimentally, melting point depression depends on the volume fraction of diluent (plasticizer or solvent) in the blend and on its interaction with the amorphous fraction of the polymer. The consequence of an added diluent is much like the well-known effect of melting point depression observed in monomeric binary mixtures. At a critical diluent concentration, melt temperatures generally remain unchanged upon further dilution (Mandelkern, 1964). For water-starch systems, this point is about 60% water (basis, g of water/g of dry starch). For the chemist or technician, the F-H equation provided the first rational explanation of starch gelatinization (melting) temperatures and the role of water.

The melt temperature (T_m), in degrees Kelvin, of a polymer-diluent mixture is related to diluent concentration by the Flory-Huggins equation. Where, T_m^0 is the melting point of the pure polymer, ΔH is the enthalpy of fusion of the polymer repeating unit, V_u and V_1 are the molar volumes of the repeating unit and diluent, respectively, R is the gas constant, v_1 is the volume fraction of diluent, and χ_1 is a parameter for polymer-diluent interaction.

$$\frac{1}{T_m} - \frac{1}{T_m^0} = \frac{R}{\Delta H_u} \frac{V_u}{V_1} (v_1 - \chi_1 v_1^2)$$

Analysis of starch, like that of collagen, has the factor of polymer hydration that needs consideration. For starch this means allowing for the volume of water held by the crystallites in calculating diluent volume fractions. In addition, determined T_m^0 values will represent the melting of the starch-diluent complex and not that of the dry starch. Thus, determinations will not be independent of diluent.

In using this equation, T_m is assumed to approximate equilibrium melting conditions where all traces of crystallinity are lost by a theoretical macroscopic crystal. Deviations from equilibrium conditions can limit the theoretical basis for determined thermal data. However, a Flory-Huggins type analysis has value as a way to systematically compare starches and to aid in practical applications. Non-equilibrium aspects of starch melting are discussed by Shiotsubo and Takahashi, 1984; Blanshard, 1987; Slade and Levine, 1988; and Biliaderis, 1990. The 1984 article reports that non-equilibrium conditions prevail at heating rates above $0.5°K$ min^{-1}, which is much lower than in conventional DSC practice.

Annealing is one way to assure that melt temperatures are at, or near, equilibrium. Of interest here is the result of starch annealing by Larsson and Eliasson, 1991. To avoid gelatinization problems, the authors used an intermediate water/starch ratio (1:1) to anneal maize, waxy-maize, wheat, and potato starches for 24 h at 50° C. As a result, gelatinization ΔH (enthalpy) did not change, single and narrowed peaks were detected, and peak temperatures were higher by 4 to 10 degrees. The temperature at which the last traces of crystallinity disappeared (T_c) is also of interest since both native and annealed starches gave practically the same value. What this suggests is that past use of T_c for non-annealed starches in the Flory expression may have had more merit than realized up to this time. Annealed starches with the above properties appear ideally suited for analysis using the F-H expression.

The polymer-diluent interaction parameter (χ_1) and the interaction energy (B) are related by the expression: $B = \chi_1 RT/V_1$. These measures provide a way to evaluate added ingredients, plasticizers, or solvents in terms of their interaction with the starch. Example applications are reported by: Witnauer and Fee, 1957; Flory and Garrett, 1958; Donovan, 1979; Lelièvre, 1976; and Evans and Haisman, 1982. Notwithstanding its limitations, this approach is suggested as a semi-empirical way to quantify interactions between components in baking studies (Blanshard, 1986).

STARCH GELATINIZATION, MECHANICS

For the 25 year period of our review, the article published in 1979 by Donovan deserves special attention. In this work, thermograms showing single and biphasic melting endotherms were shown by potato starch when water:starch ratios were varied. To date, all starches examined in this way appear to behave in the same manner. For potato, Donovan observed a single peak (G) in excess water, above about 155% water (basis, g water/g dry starch). At 115% water (db), a shoulder was apparent on the leading edge of the thermogram. The shoulder developed into a single peak (M) and melted at increasing temperatures with decreasing water content. At about 50% water (db), the G peak was virtually absent. Typical endotherm tracings with these features are shown in **Figure 7**.

One legacy of this work was the introduction of G (gelatinization) and M (melting) to distinguish these two endotherms although both indicate melting behavior. For this reason, the scheme of Biliaderis, 1990 is preferable in that these two endotherms are labeled M_1 and M_2 with subsequent melt peaks labeled M_3 and M_4.

Another application of the G and M designations is to consider the terms descriptive of the physical nature of the heated starch. Namely, the starch is more likely to be "gel-like" in appearance as well as in properties under conditions where a G peak is observed. On the other hand, under M conditions, the resulting product is more "melt-like" in appearance and in properties.

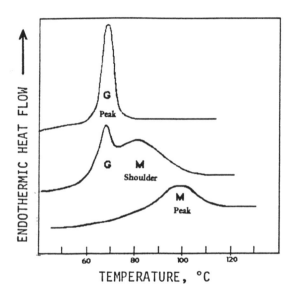

Figure 7. Tracings of differential scanning calorimeter thermograms illustrating G (gelatinization), M (shoulder), and M (peak) endotherms.

Regardless of terminology, efforts to understand the reasons for G and M endotherms have provided insight into granule structure as well as gelatinization. For example, x-ray patterns were recorded concurrently with the gelatinization of potato starch at 69, 59, 52, 38 and 30 % water (basis, total sample wt). These mixtures were chosen to match the excess, intermediate, and low water levels used by Donovan, 1979. At each water level, a steady loss of crystallinity was observed during gelatinization over a temperature range dependent upon water level. Under DSC conditions where G or M occurred either alone, or together, only melting was noted. This finding ruled out any possible recrystallization taking place between the G and M endotherms which would have been evidence for non-equilibrium melting. From the standpoint of granular structure, the data reflected a crystallite population with varying degrees of internal order (Zobel et al., 1988).

A significant study on Nägeli dextrins helped define the role of the amorphous phase during gelatinization and the G endotherm. These materials are prepared by a gradual erosion of granular starches with acid to hydrolyze the amorphous phase. Samples included native potato, and potato dextrins isolated after 3-days, and 8-weeks of hydrolysis. In excess water, DSC tracings showed G endotherms peaking at about 64,° 69°, and 73°C, respectively. Respective peak half-widths were narrow, narrow, and very broad (Donovan and Mapes, 1980).

Under reduced water conditions, the native starch showed both G (65°C) and M (70°C shoulder) endotherms. The 3-day sample also showed G and M (shoulder) endotherms but at 67° and 80°C, respectively. The 8-week Nägeli dextrin gave a single broad endotherm (50° to 105°C) with a peak temperature of 90°C. Specific enthalpy of gelatinization was virtually the same for the three samples.

With no change in enthalpy, no loss or gain in crystalline structure appeared to have taken place. In the limited water analysis, the absence of a G endotherm from the 8-week dextrin dictated the results as a melt transition. Because the two phases had been markedly decoupled, transition temperatures were high and the endotherm broadened. The endotherm was interpreted as revealing the intrinsic stability, and heterogeneity in size, and perfection of crystalline regions in granular starches. For the native and 3-day samples, their lower transition temperatures and narrower peaks supported the view that swelling in amorphous regions destabilized crystalline regions.

Disappearance of the G endotherms with loss of the amorphous phase was clear evidence of their interdependence and the inherently low stability of some of the crystallites. A related finding is that of Zeleznak and Hoseney, 1986 on the melting of crystalline isolated amylopectin samples. One endotherm was observed in excess water and two under reduced water conditions. In effect, granular structure was eliminated as a factor in limiting water to account for biphasic endotherms. To obtain the G endotherm, apparently both relatively high levels of water and an amorphous phase are needed.

Donovan's original hypothesis (1979) stated that the gelatinization process (G endotherm) is one of disorder being aided by the swelling action of water in the amorphous phase. In the literature, this process has been referred to as water-mediated, and as solvation or hydration-assisted melting. Donovan further described the gelatinization phase transition as the disordering of individual chains being separated from ordered regions with the possibility that crystallites might not be left for melting at a higher temperature. A process he picturesquely described as "stripping". He also alluded to the unfolding and hydration of helices as a result of their being separated from crystallites. Subsequently, Donovan, 1982 described the process as one in which crystallites were being "pulled apart as increased thermal energy and swelling pressures overcame the internal binding forces of the crystallites".

In this context, we need to look back at the earlier discussion of helix stability and its short-range order (points 6 and 7 under double vs. single helices). If it is determined that individual chains are operative, will it be a hydration-assisted unfolding of double helices? Russell, 1987 suggests this might be so, attributing the G endotherm to the disordering of double helices and of short-range order.

An interesting corollary proposal is one that crystallinity may primarily provide density rather than structural integrity to granules, a role assigned to double helices (Cooke and Gidley, 1991).

STARCH: CHANGES IN PERSPECTIVE

If one compares the major starch topics of 1967 with those of the present, noticeable contrasts can be made. Examples of topical shifts are as follows:

Amylose vs. Amylopectin

The prominence given to amylose in starch studies is understandable since retrogradation and its consequence are relatively easy to identify and follow. In addition, enzymic methods for structure determinations are fairly straightforward procedures. In a way, the properties expected of a linear molecule like amylose made it difficult to accept the development of crystallinity in the branched amylopectin fraction. Crystallinity data indicated otherwise with native waxy starch (100% amylopectin) showing 40% compared to high amylose at about 15%. Crystallization of isolated amylopectins has been reported by Ring, 1985 and Hoseney et al., 1986. An interesting exercise by Imberty and Pérez, 1989 was to computer model the α-(1,6) branch point. The result was finding one geometric arrangement where limited disruption of the original double helix would occur. Furthermore, indications were that a branch point could facilitate crystallite formation. This is a striking finding since experience with synthetic polymers generally has been that branch points interfere with crystallinity (Mandelkern, 1964).

Amylose can be leached from starches leaving the amylopectin as well as granule crystallinity largely intact. This and findings such as those from reaction studies (Steeneken and Smith, 1991) have led to the conclusion that amylose is mainly in the amorphous phase (Lineback, 1984; Manners, 1989). Conversely, amylopectin is the main component of the crystalline fraction. This opinion begs the question of mixed crystal formation which remains open except for subjective arguments based on differences between potato and corn starches (Zobel, 1988b).

Since the major fraction in most starches is amylopectin, its fine structure and structure relationships to starch utility are increasingly being investigated (Robin et al.,1974; Kobayashi et al.,1986; Sanders et al., 1990; Yuan et al., 1992). The methodology employed relies heavily on the use of specific enzymes for degradation of the amylopectin. Also required are ways to separate degradation products and to monitor their separation. Progressive elucidation of amylopectin fine structure is detailed by Whelan, 1976; Manners, 1979; Manners, 1985; and Hizukuri and Maehara, 1990, 1991. Investigation of amylopectin structure is a case in point where progress is dependent on methodology developments; in this instance, both enzymic and instrumental.

The development of separation and detection methodology in order to obtain macromolecular information on starch polymers is reviewed by Sullivan et al., this volume. The authors stress the heterogeneous nature of starch and the need to obtain detailed rather than average property measurements. The joint acquisition of macromolecular and amylopectin fine structure data should key future developments in starch products and end use applications.

For many years, the prevailing model for amylopectin, or some version of it, was one formulated by Meyer and Bernfeld in 1940. A representative schematic of their model is shown in **Figure 8**. Often referred to as "bush-" or "tree-" like, the model was the first to show multiple branching. In the last 20 years, however, the "cluster" model has received general acceptance. Shown in **Figure 9**, the model also has been described as a "racemose" or "grape" structure. Both Nikuni, 1969, and French, 1972a are credited with its development. While their basic concept remains, modifications have been proposed since the model was introduced (Nikuni, 1978; Lineback, 1984; Manners, 1989; Hizukuri, 1986).

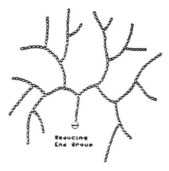

Figure 8. Schematic of K. H. Meyer's amylopectin structure.

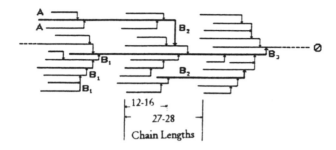

Figure 9. Hizukuri model for amylopectin structure (from Hizukuri, 1986 and by permission of Elsevier Science Publishers).

The ratio of A- to B-chains, a measure of multiplicity in branching, is a major element in defining amylopectin structure. A-chains are linked via an $\alpha,(1\text{-}6)$ bond at its reducing end while B-chains are linked in the same manner but may also have more than one A-chain attached. C-chains function like B-chains and carry the one functional reducing end group in the molecule. Manners, 1989 indicates a range in ratios of between 1.1 to 1.5 with potato amylopectin at 1.2 (av of 2). Hizukuri, 1986 assigns a ratio of 0.8 to potato amylopectin. Since techniques and weighting of results can give different results, absolute values have yet to be obtained.

Other measures used to differentiate fine structure between amylopectins include exterior and inner chain lengths, average chain lengths, and profiles of chain distributions. **Figure 10** shows a comparison of chain profiles for potato and waxy rice amylopectins (Hizukuri, 1986). Chain profiles require complete debranching of the amylopectin with subsequent fractionation. Profiles will depend on choice of enzyme, separation method, and whether calculations are on a weight or number basis. As profile development continues and data bases are enlarged, however, one prospective result is a predictive method for amylopectin and starch properties. For example, consider the waxy rice profile (**Figure 10**) which differs markedly from that of potato as well as other profiles determined by Hizukui, 1986, 1990. Waxy rice has unusual, and unexplainable, freeze thaw stability (Schoch, 1967). An interesting consideration is whether or not the chain profile provides a clue to this stability.

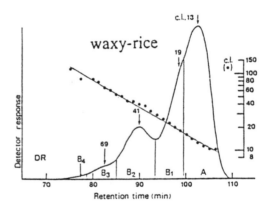

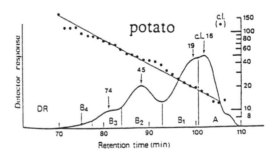

Figure 10. Molecular distribution profiles for waxy rice and potato amylopectins by gel-permeation HPLC (from Hizukuri, 1986 and by permission of Elsevier Science Publishers).

The model in **Figure 9** is that of Hizukuri, 1986. Accordingly, the role of B1, B2, and B3 chains are emphasized whereby 1, 2, or 3 clusters are connected by these B-chains, respectively. For A-chains the average chain length is 14 which is similar to values cited by French, 1972a and Robin et al., 1974. The suggested chain length difference between clusters is 27-28, the difference of B3-B2.

Based on a cluster model, selected starch properties are possibly accounted for as follows:

a.) The tying together of clusters by a common molecule conforms with amylopectin viscosities being higher than those for the larger glycogen molecule.

b.) Gelatinized potato starch can exhibit highly swollen resilient "sacs" that resist dispersion; this characteristic behavior could be the effect of a relative abundance of connecting B- chains.

c.) If A-chains are taken as a factor in crystallite size, their calculated length is about 49Å (14 x 3.5Å, the repeat distance for 6 glucose residues in 21Å). Crystallites of this order in size are also indicated in the literature (Sterling and Pangborn, 1960, who also critique methods giving 150Å sizes; Robin, et al., 1974; Zobel, 1988b). Supporting evidence for these dimensions, as well as for the cluster model, is provided by studies using electron microscopy. Kassenbeck, 1978 and Yamaguchi et al., 1979 report periodicities on the order of 50-70Å in starches

d.) A corresponding calculation using the earlier value of 28 chain lengths for cluster separation gives a distance of 98Å. In this sense, the model closely agrees with low-angle x-ray diffraction indicating, a 97-99Å repeat distance in potato starch (Sterling, 1962; Zobel, 1988b). This spacing is attributed to a periodic change in electron density such as the center to center distance between amorphous areas with an intervening crystallite cluster. If crystallite size is considered (see c), granule composition would be estimated then as half crystalline and half amorphous in structure.

Heat / Moisture Treatment vs. Annealing

Heat/moisture (H/M) treatment of starch commonly refers to processes by which tuber or root starches become more cereal like in properties; often characterized by a B to A transformation in x-ray structure. The conditions for such changes were given earlier in the section on crystal structures. While A-type starches were once thought to be unaffected, H/M treatments have been shown to modify structure as well as properties (Osman, 1967; Fukui et al., 1969; Lund, 1984; Zobel, 1988a). These treatments are basically annealing processes conducted under conditions of limited moisture and fairly high temperatures.   Annealing per se also includes conditions of excess water and temperatures   often on the order of 50°C.

First, in this section, we should note that publication of Sair's article on "heat/moisture   treatment   of starch" and the founding of the Carbohydrate Division both took place in 1967.   Publication followed Sair's talk at the 51st annual meeting of the American Association of Cereal Chemists. To a large extent, participation came about through the persuasive efforts of Tom Schoch who cited interest in the subject. This was an easy task since a favorite topic amongst starch chemists at that time was B to A transformations.   The subject is still a viable one (Donovan et al., 1983).

Secondly, the literature records an interesting shift in the perception of heat-treating   moistened or slurried starches.   For example, "treatment" was often the operative word with analogies to similar processes, and property changes, in synthetic polymers (Banks and Greenwood, 1975).   In 1978, articles by Marchant and Blanshard, and by Ahmed and Lelièvre, used annealing in a descriptive sense. Yost and Hoseney, 1986 may have been one of the first to cite annealing in the title of an article about starch.

Increased   gelatinization   temperature   is a common result of annealing starch. Gelatinization also is generally more sharply defined which seemingly indicates better homogeneity in crystallite size and/or perfection.   Thus while these effects imply an increase in heat of gelatinization (ΔH) it is not always the case, and at best, sometimes no change occurs. If ΔH values decrease, clear evidence is provided that gelatinization   occurred · during treatment (Gough and Pybus, 1971; Nakazawa et al., 1984; Krueger et al., 1987a; Krueger et al., 1987b; Knutson, 1990; Larsson and Eliasson, 1991).   Expectation of better crystallinity is not always confirmed by x-ray analysis; results may even be contradictory (Nakazawa et al., 1984).   On the other hand, Ahmed and Lelièvre, 1978 showed a parallel increase in crystallinity and ΔH. If x-ray data can be obtained, however, its comparison with other data should be helpful in understanding the effects of annealing.

It should be pointed out that DSC and x-ray diffraction do not necessarily measure the same levels of order in materials. (See earlier section on double/single helices, point 7.) It is further apparent that the effects of annealing are not completely understood, particularly at a molecular level. In other words, for a given sample, it is often difficult to find a consistent scheme for all the data points.   Marchant and Blanshard, 1978 broadly describe annealing as a conditioning in which granules assume a more stable configuration.   In the process, a general realignment of the polymer chains is viewed as taking place within the noncrystalline regions of the granule as well as in crystallites.

24

In the context of this paper, this description could include development of double helical order apart from any crystalline structure. In addition, standard methods may not detect some of the changes taking place.

The effects of starch annealing due to processing conditions have been investigated for wheat (Lorenz et al., 1978), rice (Lund, 1984), and maize, (Krueger et al., 1987b). Wet-milling processes were demonstrated to have an annealing effect on wheat and corn starches (Lorenz, Krueger). Larsson and Eliasson, 1991 investigated maize, waxy-maize, wheat and potato, finding for wheat starch that T_0 increased under holding times as low as 10 min at $50\,^{\circ}C$ (1:1, ratio of starch:water). The effects noted on these starches include those mentioned above as well as changes in water absorption, viscosity, solubility, swelling, and leaching of amylose.

The most that can be said is that water acts to plasticize the amorphous regions in starch to facilitate annealing. At high water levels, any correlation with glass transition temperatures (Tg) is difficult to establish since they approach $0\,^{\circ}C$. With moisture contents in the 13 to 20% range and Tg's on the order of $60\text{-}30\,^{\circ}C$ (Zeleznak and Hoseney, 1987), relationships between glass transition and annealing temperatures could be significant. To anneal starches for application studies, the use of 50%, or less, water contents is preferred over excess water conditions where it is more difficult to avoid gelatinization.

Crystalline vs. Amorphous

Much like the example of amylose, effects due to starch crystallization have been relatively easy to recognize and to measure. Once crystalline regions are formed, their reversal is dependent on molecular type: amylose, 100 to $120\,^{\circ}C$; amylopectin, as low as $50\,^{\circ}C$ for samples having a low degree of order. When formed in gels, crystalline regions act as physical cross links in a gel network.

In reviewing crystallinity, the x-ray polymorphic forms have been described in terms of structure, their function in granules, changes that occur with annealing, loss due to melting (gelatinization), and recrystallization upon retrogradation. Through macro-crystalline domains, crystallinity also contributes to granule ultrastructure.

These structures are often revealed by etching granules, or granule fragments, with acid or enzymes followed by microscopic examination. Most studies require the use of transmission or scanning electron microscopy but large growth rings can be seen with an optical microscope. Observed structures show periodic density fluctuations that vary in size and shape, depending on the extent to which amorphous regions and/or imperfect crystallites have been disrupted.

Ultrastructure domains of from about 50-200 Å in width are commonly observed and lamella or fibrillar structures can be 500 Å or longer in length. In some cases, residual structures resemble the blockletts shown in **Figure 1a**. Examples of ultrastructural features are illustrated, or cited, by the following authors: Sterling, 1965; Kassenbeck, 1978; Yamaguchi *et al.*, 1979; French, 1984; Lineback, 1984; Zobel, 1988b; and Imberty *et al.*, 1991. Comprehensive reviews on this subject are those of Hood and Liboff, 1982 and Gallant and Boucher, 1986.

The amorphous granule fraction increasingly is being regarded in attempts to understand and to modify starch properties. Consequently, the relevance of the amorphous fraction has been considered for each topic under discussion. In addition, granule water sorption in relation to the amorphous fraction warrants review.

Water in the crystalline fraction, for example, is limited to the amount held in the crystal lattice. Granules, however, absorb relatively high levels of water when slurried in water or held at high relative humidities. Hellman *et al.*, 1952 used the latter method to minimize water trapped between granules. Water absorption was measured at 100% humidity by carefully maintaining sample and surroundings at the same temperature. Final moisture levels were: maize, 39.9%; potato, 50.9%; tapioca, 42.9%; and waxy maize, 51.4%. (basis, g water/g dry starch). The water contents represent points where additional (excess) water would be primarily outside the granule, in a separate phase. At this saturation point, granules are highly plasticized due to water sorption by the amorphous fraction and are amenable to swelling (Hellman *et al.*) and gelatinization.

An interesting aspect of these water contents is their relation to the transition from G to M type endotherms (Evans and Haisman, 1982). When excess water prevails, heating leads to "gelatinization" as evidenced by a G peak. The picture is one of water being pulled from outside the granule to hydrate the starch and causing swelling (**Fig. 2**). Highly swollen maize granules reportedly can hold as much as 2500% water (Meyer, 1942b). When granule water content is lower than these saturation levels, the M peak develops.

STARCH GRANULE: MOLECULAR DISPOSITION

A survey of normal maize starch properties suggests that, within the granule, amylose is an entity mainly separated from the amylopectin fraction. Some reasons for this are as follows: a.) it easily leaches from granules with hot water, b.) iodine, glycerol, and other agents readily form V structures, c.) rapid setback occurs and resulting gels are firm like isolated amylose, and d.) the starch readily solubilizes in dimethyl sulfoxide (Zobel, 1988b).

The suggestion for such a molecular arrangement within granules comes from the finding that maize starches develop V crystalline structures when annealed or heat/moisture treated. This formation strongly argues for the naturally occurring fatty acids and the amylose fraction to be in close proximity within the granule. The situation is analogous to that of seed crystals needing only the proper conditions for development. Visually, the sequence can be illustrated by absorbing iodine vapors into a dry granule and then adding water. The expected blue color appears when water is added, plasticizing the amylose and permitting it to wrap around the poly-iodide chain. When such a sample is x-rayed, a crystalline V structure is detected.

The alternative for V to appear in these cases would include migration through the granule of a large amylose molecule and/or long-chain fatty acids. A second limiting factor would be to locate sufficient material in one place for crystallite formation. Such schemes appear highly unlikely to occur in order to obtain V crystalline order.

The schematic in **Figure 11** shows a phase separation between amylose and amylopectin. This does not preclude some mixing between amylose and amylopectin which is shown to occur on a small scale. The naturally occurring fatty acids are shown in close association with the amylose.

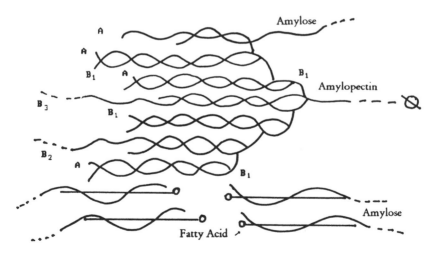

Figure 11. Starch crystallite model showing proposed disposition of amylopectin, amylose, and fatty acids in maize granules.

The drawing is intended to show a more extended form for the amylose compared to that found in a compact single helical complex or that of a double helix (see earlier discussion). The amylopectin is depicted as double helices that are further arrayed into crystalline structures. The scheme intentionally represents the amylopectin structure proposed by Hizukuri, 1986 as much as possible. Hence, the B1, B2, and B3 nomenclature as well as A. Helical lengths are drawn as being largely regulated by the length of A chains. In turn, cluster length is on the order of 50 Å to conform with the discussion earlier of the amylopectin model, paragraphs c and d.

Blanshard, 1987 presents a similar model but with less emphasis on amylopectin details. Of interest, is that while both models were developed independently, granule lipids were a major factor in both cases. Here, it was the formation of complexes as outlined above. For Blanshard, it was the astute detection of a 150 Å spacing that was attributed to radially arranged lipids and modeled in association with the amylose fraction. The radial arrangement for amylose in both models fits the scheme of Kassenbeck, 1978 and the dichroism shown by iodine stained granules (French, 1984).

CONCLUSIONS

In reviewing the literature for this manuscript, it is easy to conclude that the first 25 years for the Carbohydrate Division of the AACC has been one of significant advances in knowledge about starch and its granular structure. Past starch research (1940-1987) has been appropriately categorized by periods dominated by chemical discipline such as organic, biochemical, and physical (Blanshard, 1987). With these skills in place, interdisciplinary efforts are foreseen as needed to meet future challenges. For example, to relate functional properties with granule structure, to better understand granule biosynthesis, and to ultimately create granules to order.

LITERATURE CITED

Ahmed, M., and Lelièvre, J., 1978. Effect of various drying procedures on the crystallinity of starch from wheat grains. Starch 30:78.

Banks, W., and Greenwood, C.T., 1975. Starch and its Components. John Wiley & Sons, New York.

Biliaderis, C.G., Maurice, T.J., and Vose, J.R. 1980. Starch gelatinization phenomena studied by differential scanning calorimetry. J. Food Sci. 45:1669.

Biliaderis, C.G., Page, C. M., Maurice, T.J., and Juliano, B.O. 1986. Thermal characterization of rice starches: A polymeric approach to phase transitions of granular starch. J. Agr. Food Chem. 34:6.

Biliaderis, C.G., and Vaughan, D.J., 1987. Electron spin resonance studies of starch-water-probe interactions. Carbohydr. Polym. 7:51.

Biliaderis, C.G. 1990. Thermal analysis of food carbohydrates. Page 169 in: Thermal Analysis of Foods. V.R.Harwalkar and C.-Y. Ma, eds. Elsevier Applied Science, London.

Biliaderis, C.G. 1991. The structure and interactions of starch with food constituents. Can. J. Physiol. Pharmacol. 69:60.

Billmeyer, F.W., 1984. Textbook of Polymer Science 3rd Ed. Wiley-Interscience, New York.

Blanshard, J.M.V., 1986. The significance of the structure and function of the starch granule in baked products. Page 1 in: Chemistry and Physics of Baking. J.M.V. Blanshard, P.J. Frazier, and T. Galliard, eds. The Royal Society of Chemistry, London.

Blanshard, J.M.V., 1987. Starch granule structure and function: a physicochemical approach. Page 116 in: Starch Properties and Potential. T. Galliard, ed. John Wiley & Sons, New York.

Brant, D.A. 1980. Conformation and behavior of polysaccharides in solution. Page 447 in: The Biochemistry of Plants, Vol. 3:11. Academic Press, Inc., New York.

Chang, S-M and Liu, L-C. 1991. Retrogradation of rice starches studied by differential scanning calorimetry and influence of sugars, NaCl and lipids. J. Food Sci. 56:564.

Colonna, P. and Mercier, C. 1985. Gelatinization and melting of maize and pea starches with normal and high-amylose genotypes. Phytochemistry, 24:1667.

Colwell, K.H., Axford, D.W.E., Chamberlain, N., and Elton, G.A.H.1969. Effect of storage temperature on the ageing of concentrated wheat starch gels. J. Sci. Fd. Agric. 20:550.

Cooke, D. and Gidley, M.J., 1991. Loss of crystalline and molecular order during starch gelatinization: origin of the enthalpic translation. Carbohydr. Res., (in press).

Donovan, J.W., 1979. Phase transitions of the starch-water system. Biopolymers 18:263.

Donovan, J.W., and Mapes, C.J., 1980. Multiple phase transitions of starches and Nägeli amylodextrins. Starch 32:1990.

Donovan, J.W., 1982. Personal communication.

Donovan, J.W., Lorenz, K., and Kulp, K., 1983. Differential scanning calorimetry of heat-moisture treated wheat and potato starches. Cereal Chem. 60:381.

Evans, I.D., and Haisman, D.R., 1982. The effect of solutes on the gelatinization temperature range of potato starch. Starch 34:224.

Flory, P.J., 1953. Principles of Polymer Chemistry. Cornell University Press, Ithaca, New York.

Flory, P.J. and Garrett, R.R., 1958. Melting point depression in fibrous collagen by diluent addition. J. Amer. Chem. Soc. 80:4836.

French, A.D., and Murphy, V.G., 1977. Computer modeling in the study of starch. Cereal Foods World 22:61

French, A.D., and Brady, J.W., 1990. Computer Modeling of Carbohydrate Molecules. American Chem. Soc., Washington, D.C.

French, D., 1972a. Fine structure of starch and its relationship to the organization of starch granules. J. Jap. Soc. Starch Sci. 19:8.

French, D., 1972b. Personal communication.

French, D., 1984. Organization of starch granules. Page 183 in: Starch: Chemistry and Technology. R.L. Whistler, J.N. BeMiller, and E.F. Paschall, eds. Academic Press, Inc., Orlando.

Fukui, T., and Nikuni, Z., 1969. Heat-moisture treatment of cereal starch observed by x-ray diffraction. Agri. Biol. Chem. 33:460.

Gallant, D.J., and Bouchet, B., 1986. Ultrastructure of maize starch granules. A review. Food Microstructure 5:141.

Geil, P.H., 1963. Polymer Single Crystals. Interscience Publishers, Inc., New York.

Gerngross, O., Herrmann, K., and Abitz, W., 1930. Structure formations in gelatin gels. Z. Physik. Chem. (B) 10:371.

Gidley, M.J., and Bociek, S.M., 1985. Molecular organization in starches: A[13]C CP/MAS NMR study. J. Am. Chem. Soc. 107:7040.

Gidley, M.J., and Bociek, S.M., 1988. ^{13}C CP/MAS NMR studies of amylose inclusion complexes, cyclodextrins, and the amorphous phase of starch granules: relationships between glycosidic linkage conformation and ^{13}C chemical shifts. J. Am. Chem. Soc. 110:3820.

Gidley, M.J., 1989. Molecular mechanisms underlying amylose aggregation and gelation. Macromolecules 22:351.

Gidley, M.J., and Cooke, D., 1991. Aspects of molecular organization and ultrastructure in starch granules. Biochem. Soc. 19:551.

Gough, B.M., and Pybus, J.N., 1971. Effect on the gelatinization temperature of wheat starch granules of prolonged treatment with water at 50 °C. Starch 23:210.

Gray, V.M., and Schoch, T.J., 1962. Effects of surfactants and fatty adjuncts on the swelling and solubilization of granular starches. Starch 7:239.

Gudmundsson M., and Eliasson, A.-C., 1990. Retrogradation of amylopectin and the effects of amylose and added surfactants/emulsifiers. Carbohydr. Polym. 13:295.

Hellman, N.N., Boesch, T.F., and Melvin, E.H., 1952. Starch granule swelling in water vapor sorption. J. Am. Chem. Soc. 74:348.

Hennig, V.H.J., Lechert, H., and Goemann, W., 1976. Examination of the swelling of starch by pulsed NMR-method. Starch 28:10.

Hinrichs, W., Buttner, G., Steifa, M., Betzel, Ch., Zabel, V., Pfannemüller, B., and Saenger, W., 1987. An amylose antiparallel double helix at atomic resolution. Science 238:205.

Hizukuri, S., Kanekko, T., and Takeda, Y., 1983. Measurement of the chain length of amylopectin and its relevance to the crystalline polymorphism of starch granules. Biochim. Biophys. Acta 760:188.

Hizukuri, S., 1986. Polymodal distribution of the chain lengths of amylopectins, and its significance. Carbohydr. Res. 147:342

Hizukuri, S., and Maehara, Y., 1990. Fine structure of wheat amylopectin: the mode of A and B chain binding. Carbohydr. Res. 206:145.

Hizukuri, S., and Maehara, Y., 1991. Distribution of the binding of A chains to a B chain in amylopectins. Page 212 in: Biotechnology of Amylodextrin Oligosaccharides. R.B. Friedmann, ed. American Chemical Society, Washington, D.C.

Hood, L.F., and Liboff, M., 1982. Starch ultrastructure. Page 341 in: New Frontiers In Food Microstructure. D.B. Bechtel, ed. American Association of Cereal Chemists, Inc., St. Paul, Minnesota.

Hoseney, R.C., Zeleznak, K.J., and Yost, D.A., 1986. A note on the gelatinization of starch. Starch 38:407.

Howsmon, J.A., and Sisson, W.A., 1954. Submicroscopic structure of cellulose fibers. Page 231 in: Cellulose and Cellulose Derivatives. E. Ott, H.M. Spurlin, and M.W. Grafflin, eds. Interscience Publishers, Inc., New York.

Imberty, A., Chanzy, H., Pérez, S., Buléon, A., and Tran, V., 1987a. Three-dimensional structure analysis of the crystalline moiety of A-starch. Food Hydrocolloids 1:455.

Imberty, A., Chanzy, H., and Pérez, S., 1987b. New three-dimensional structure for A-type starch. Macromolecules 20:2634.

Imberty, A., Chanzy, H., Pérez, S., Buléon, A., and Tran, V., 1988a. The double-helical nature of the crystalline part of A-starch. J. Mol.

Biol. 201:365.

Imberty, A., and Pérez, S., 1988b.   A revisit to the three-dimensional structure of B-type starch. Biopolymers 27:1205.

Imberty, A., and Pérez, S., 1989.   Conformational analysis and molecular modelling of the branching point of amylopectin.   Int. J. Biol. Macromol. 11:177.

Imberty, A., Buléon, A., Tran, V., and Pérez, S., 1991.   Recent advances in knowledge of starch structure. Starch 43:375.

Jane, J.-L., and Robyt, J.F., 1984.   Structure studies of amylose-V complexes and retrograded amylose by action of alpha amylases, and a new method for preparing amylodextrins.   Carbohydr. Res. 132:105.

Jane, J.-L., and Robyt, J.F., 1985. ^{13}C-N.M.R. study of the conformation of helical complexes of amylodextrin and of amylose in solution. Carbohydr. Res. 140:21.

Jaska, E., 1971.   Starch gelatinization as detected by proton magnetic resonance.   Cereal Chem. 48:437.

Kainuma, K., and French, D., 1972.   Naegeli amylodextrin and its relationship to starch granule structure. II. Role of water in crystallization of B-starch. Biopolymers 11:2241.

Kassenbeck, P., 1978.   Contribution to the knowledge on distribution of amylose and amylopectin in starch granules. Starch 30:40.

Knutson, C.A., 1990.   Annealing of maize starches at elevated temperatures.   Cereal Chem. 67:376.

Kobayashi, S., Schwartz, S.J., and Lineback, D.R., 1986. Comparison of the structures of amylopectins from different wheat varieties. Cereal Chem. 63:71.

Krog, N., 1971.   Amylose complexing effect of food grade emulsifiers. Starch 23:206.

Krueger, B.R., Walker, C.E., Knutson, C.A., and Inglett, G.E., 1987a. Differential scanning calorimetry of raw and annealed starch from normal and mutant maize genotypes.   Cereal Chem. 54:187.

Krueger, B.R., Knutson, C.A., Inglett, G.E., and Walker, C.E., 1987b. A differential scanning calorimetry study on the effect of annealing on gelatinization behavior of corn starch. J. Food Sci. 52:

Larsson, I., and Eliasson, A-C., 1991.   Annealing of starch at an intermediate water content. Starch 43:227.

Leach, H.W., McCowen, L.D., and Schoch, T.J., 1959.   Structure of the starch granule I. swelling and solubility patterns of various starches. Cereal Chem. 36: 534.

Lelièvre, J., 1973. Starch gelatinization. J. Applied Polym. Sci. 18:293.

Lelièvre, J., 1976.   Theory of gelatinization in a starch-water-solute system. Polymer 17:854.

Lineback, D.R., 1984. The starch granule organization and properties. Bakers Digest, March 13, 1984:16.

Liu, H., and Lelièvre, J., 1992. A differential scanning calorimetry study of melting transitions in aqueous suspensions containing blends of wheat and rice starch. Carbohydr. Polym. 17:145.

Liu, J., and Zhoa, S., 1990. Scanning electron microscope study of gelatinization of starch granules in excess water. Starch 42:96.

Lorenz, K., and Kulp, K., 1978. Steeping of wheat at various temperatures - effects on physicochemical characteristics of the starch. Starch 30:333.

Lund, D., 1984. Influence of time, temperature, moisture, ingredients, and processing conditions on starch gelatinization. CRC Critical Reviews in Food Sci. and Nutrition 20:249.

Mandelkern, L., 1964. Crystallization of Polymers. Mcgraw-Hill, New York.

Manners, D.J., 1979. The enzymic degradation of starches. Page 5 in: Polysaccharides in Food. J.M.V. Blanshard and J.R. Mitchell, eds., Butterworths, London.

Manners, D.J., 1985. Some aspects of the structure of starch. Cereal Foods World 30:461.

Manners, D.J., 1989. Recent developments in our understanding of amylopectin structure. Carbohy. Polym. 11:87.

Marchant, J.L., and Blanshard, J.M.V., 1978. Studies of the dynamics of the gelatinization of starch granules employing a small angle light scattering system. Starch 30:257.

Matsukura, U., Matsunaga, A., and Kainuma, K., 1983. Structural studies on retrograded normal and waxy corn starches. J. Jpn. Soc. Starch Sci. 30:l06.

Matsunaga, A., and Kainuma, K., 1986. Studies on the retrogradation of starch in starchy foods, Part 3. Starch 38:1.

Meares, P., 1965. Polymers Structure and Bulk Properties. D. Van Nostrand Co. Ltd., London.

Mercier, C., Charbonniere, R., Gallant, D., and Guilbot, A., 1979. Structural modification of various starches by extrusion cooking with a twin-screw French extruder. Page 153 in: Polysaccharides in Food, J.M.V. Blanshard, and J.R. Mitchell, eds. Butterworths, London.

Meyer, K.H., 1942a. Recent developments in starch chemistry. Advances in Colloid Sci. l:143.

Meyer, K.H., 1942b. Natural and Synthetic High Polymers. Interscience Publishers, Inc., New York.

Meyer, K.H., 1952. The past and present of starch chemistry. Experientia 8:405.

Nakazawa, F., Noguchi, S., Takahashi, J., and Takada, M., 1984. Thermal equilibrium state of starch-water mixture studied by differential scanning calorimetry. Agric. Biol. Chem. 48:2647.

Nikuni, Z., 1969. A structure for amylopectin. Chori Kagaku 2:6.

Nikuni, Z., 1978. Granule and amylopectin structures. Starch 30:105.

Osman, E.M., 1967. Starch in the food industry. Page 163 in: Starch: Chemistry and Technology. R.L. Whistler and E.F. Paschall, eds. Academic Press, New York.

Pérez, S., and Vergelati, C., 1987. Solid state and solution features of amylose and amylosic fragments. Polymer Bulletin 17:141.

Pérez, S. Imberty, A., and Scaringe, R.P., 1990. Modeling of infractions of polysaccharide chains. Page 281 in: Computer Modeling of Carbohydrate Molecules. A.D. French and J.W. Brady, eds. American Chemical Society, Washington, DC.

Priestley, R.J., 1976. Studies on parboiled rice. Food Chem. 1:5.

Radley, J.A., 1968. Starch and its Derivatives. Chapman and Hall Ltd., London.

Ring, S.G., 1985. Observations on the crystallization of amylopectin from aqueous solution. Int. J. Biol. Macromol. 7:253.

Robin, J.P., Mercier, C., Charbonniere, R., and Guilbot, A., 1974. Lintnerized starches. Gel filtration and enzymatic studies of insoluble residues from prolonged acid treatment of potato starch. Cereal Chem. 51:389.

Rundle, R.E., Daasch, L.W., and French, D., 1944. The structure of the "B" modification of starch from film and fiber diffraction diagrams. J. Am. Chem. Soc. 66:130.

Russell, P.L., 1987. Gelatinization of starches of different amylose/amylopectin content. A study by differential scanning calorimetry. J. Cereal Sci. 6:133.

Sair, L., 1967. Heat-moisture treatment of starch. Cereal Chem. 44:8.

Sanders, E.B., Thompson, D.B., and Boyer, C.D., 1990. Thermal behavior during gelatinization and amylopectin fine structure for selected maize genotypes as expressed in four inbred lines. Cereal Chem. 67:594.

Sarko, A., and Marchessault, R.H., 1967. The crystalline structure of amylose triacetate I. A stereochemical approach. J. Am. Chem. Soc. 89:6454.

Sarko, A., and Wu, H.-C.H,1978. The crystal structures of A-, B- and C-polymorphs of amylose and starch. Starch 30:73.

Sarko, A., Biloski, A., 1980. Crystal structure of the KOH-amylose complex. Carbohy. Res. 79:11.

Schoch, T.J., 1942. Fractionation of starch by selective precipitation with butanol. J. Am. Chem. Soc. 64:2957.

Schoch, T.J., 1967. Properties and uses of rice starch. Page 79 in: Starch: Chemistry and Technology. R.L. Whistler and E.F. Paschall, eds. Academic Press, New York.

Shiotsubo, T., and Takahashi, K., 1984. Differential thermal analysis of potato starch gelatinization. Agric. Biol. Chem. 48:9.

Slade, L., and Levine, H., 1987a. Structural stability of intermediate moisture foods. Page 115 in: Food Structure. J.M.V. Blanshard and J.R. Mitchell, eds. Butterworths, London.

Slade, L., and Levine, H., 1987b. Recent advances in starch retrogradation. Page 387 in: Recent Developments in Industrial Polysaccharides. S.S. Stivala, V. Crescenzi, and I.C.M. Dea, eds. Gordon and Breach Science, New York.

Slade, L., and Levine, H., 1988. Non-equilibrium melting of native granular starch: Part I. temperature location of the glass transition associated with gelatinization of A-type cereal starches. Carbohydr. Polym. 8:183.

Steeneken, P.A.M., and Smith, E., 1991. Topochemical effects in the methylation of starch. Carboyhydr. Res. 209:239.

Sterling, C., and Pangborn, J., 1960. Fine structure of potato starch. Amer. J. Botany 47:577.

Sterling, C., 1960. Molecular association of starch at high temperature. Starch 3:78.

Sterling, C., 1962. A low-angle spacing in starch. J. Polym. Sci. 56:S10.

Sterling, C., 1965. The submicroscopic structure of the starch grain - an analysis. Food Technol. 19:97.

Sterling, C., 1968. The structure of the starch grain. Page 139 in: Starch and its Derivatives. Radley, J.A. Chapman and Hall Ltd., London.

Whelan, W.J., 1976. New thoughts on the structures of starch and glycogen. J. Jap. Soc. Starch Sci. 23:101.

White, P.J., Abbas, I.R., and Johnson, L.A., 1989. Freeze-thaw stability and refrigerated-storage retrogradation of starches. Starch 41:176.

Whittam, M.A., Noel, T.R., and Ring, S.G., 1990. Melting behaviour of A- and B-type crystalline starch. Int. J. Biol. Macromol. 12:359.

Windle, J.J., 1985. An ESR spin probe study of potato starch gelatinization. Starch 37:121.

Witnauer, L.P. and Fee, J.G., 1957. The effect of diluents on the melting of crystalline regions in cowhides. J. Polym. Sci. 26:141.

Wittwer, F., and Tomka, I., 1987. Polymer composition for injection molding. U.S. Patent 4,673,438, 1987.

Wu, H-C,H., and Sarko, A., 1978a. The double-helical molecular structure of crystalline B-amylose. Carbohydr. Res. 61:7

Wu, H-C.H., and Sarko, A., 1978b. The double-helical molecular structure of crystalline A-amylose. Carbohydr. Res. 61:27.

Wunderlich, B., 1973. Macromolecular Physics, Vol. 1, Academic Press, New York.

Yamaguchi, M., Kainuma, K., and French, D., 1979. Electron microscopic observations of waxy maize starch. J. Ultrastructure Res. 69:249.

Yost, D.A., and Hoseney, R.C., 1986. Annealing and glass transition of starch. Starch 38:289.

Yuan, R.C., Thompson, D.B., and Boyer, C.D, 1992. Fine structure of amylopectin in relation to gelatinization and retrogradation behavior of maize starches from three wx-containing genotypes in two inbred lines. Cereal Chem., in press.

Zeleznak, K.J., and Hoseney, R.C., 1986. The role of water in the retrogradation of wheat starch gels and bread crumb. Cereal Chem. 63:407.

Zeleznak, K.J., and Hoseney, R.C., 1987. The glass transition in starch. Cereal Chem. 64:121.

Zobel, H.F., and Senti, F.R., 1959. The bread staling problem. x-ray diffraction studies on breads containing a cross-linked starch and a heat-stable amylase. Cereal Chem. 36:441.

Zobel, H.F., Senti, F.R., and Brown, D.S., 1965. Studies on starch gelatinization by differential thermal analysis. Abstracts 50th Annual Meeting, American Assoc. Cereal Chemists, 77.

Zobel, H.F., 1984. Gelatinization of starch and mechanical properties of starch pastes. Page 285 in: Starch: Chemistry and Technology. R.L. Whistler, J.N. BeMiller, and E.F. Paschall, eds. Academic Press, Inc., Orlando.

Zobel, H.F., 1988a. Starch crystal transformations and their industrial importance. Starch 40:1.

Zobel, H.F., 1988b. Molecules to granules: A comprehensive starch review. Starch 40: 44.

Zobel, H.F., Young, S.N., and Rocca, L.A., 1988. Starch gelatinization: An x-ray diffraction study. Cereal Chem. 65:443.

EXTRUSION PROCESSING OF STARCH

Judson M. Harper
Colorado State University
Fort Collins, CO 80525

INTRODUCTION

Extrusion has become an increasingly important method of processing starch based products. Initially, ram or piston extruders were used to form long strands of spaghetti from a premixed dough of durham semolina and water. These early batch extruders were called macaroni presses, a term that is still applied to the continuous single screw forming extruders used to form the myriad of macaroni shapes available today. Although the piston is a form of batch-type extruder, its use is limited to specialized or small-scale applications and will not be considered further.

Extrusion describes a process where softened or plasticized materials are forced by pressure through a die or opening to create a specific shape. Food extrusion equipment continuously works, heats, and pressurizes a wide variety of food ingredients. When the heat input to the food ingredients is limited, as is the case for macaroni, no chemical change or cooking occurs and the extruder just provides the means to shape the product. Conversely, cooking extrusion describes processes where the ingredients are heated and sheared under pressure causing significant chemical, physical and functional changes prior to their being shaped by the die at the discharge of the equipment.

The combination of cooking and shaping has made the food extruder an important processing alternative in the production of starch based products (Harper, 1989a). Examples of extruded expanded snack food items include curls (collets), onion rings, chips, pinwheels and balls made from corn, rice, wheat, potato flour and starch (Huber and Rokey, 1990). Extruded ready-to-eat (RTE) breakfast cereals (Miller, 1988 and 1990) constitute another large segment. Other extruded starch based foods are croutons, breadings, baby foods, drink bases, and the modified and cooked starches used as an ingredient in the formulation of many other foods (Dziezak, 1989).

Extrusion cooking of pet foods and animal feeds has become increasingly important and significant (Miller, 1985). In this case, the extruder cooks the major starch based ingredients in combination with other materials in order to increase palatability, digestibility, and availability of protein, energy, vitamins and minerals for a completely balanced diet. Finally, the process has found new and unique food processing applications. For example, the use of extrusion cooking to enhance enzyme saccharification of starch has been studied by Linko *et al*. (1984).

This chapter will describe the operating principles of single and twin screw cooking extruders. It will use that description to provide the basis for increased understanding of how extrusion cooking differs fundamentally from the high moisture cooking procedures typically used to process starch based ingredients. Next, it will focus on how the cooking extrusion process alters the molecular structure and nature of starch. The impact of extrusion on the molecular structure of starch affects the physical and functional characteristics of the food. Finally, a brief discussion of how changes in extrusion processing conditions affect finished products will complete the review.

COOKING EXTRUSION

Extruder Operation

Cooking extruders consist of a stationary barrel in which one or two flighted screws rotate. Figure 1 shows a cross-sectional drawing of single screw extruder's principal features.

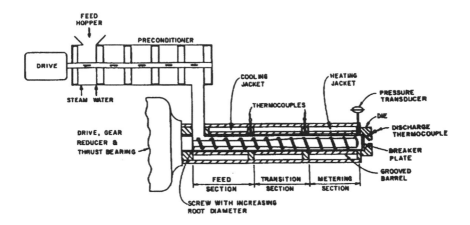

Fig. 1. Cross-sectional view of a single screw cooking extruder equipped with a preconditioner.

The flights on the rotating screw(s) transport
relatively dry (10-40% wb moisture) ingredients down
the barrel. Because of the restricted discharge at
the outlet of the barrel caused by the die, the open
volume of the channels created by the flights wound
around the screw become filled with the food ingredients. To facilitate the transport process of the
ingredients down the barrel, mechanical energy is
applied by a large motor turning the screw(s) which
overcomes fluid friction associated with flow of the
very viscous product. In the extrusion process,
mechanical energy is converted to heat, which rapidly
raises the temperature of the mixture in the barrel.
A very small fraction of the added mechanical energy
goes to increase the pressure of the food ingredients
and force them through the shaping die. The elevated
pressure (25-200 bars) within the extruder allows the
internal temperature to rise above the normal boiling
point of water without the production of steam or loss
of moisture. Given appropriate conditions, the granular free flowing starch based food ingredients are
converted into a homogenous plasticized mass with a
temperature of 100-200°C in less than 2 minutes.
 In the extrusion cooking process, the screw

performs four functions: conveying, mixing, heating and pressurizing. The initial sections of the screw (*i.e.*, those nearest the ingredient inlet) perform only the conveying function, moving the ingredients to be processed into the later screw sections. It is these sections which mix, work and heat the ingredients as they move toward the die.

To increase the amount of mixing, working and consequently heating of the product, the character of the screw is changed to increase shear. In the case of single screw extruders, the root diameter of the screw is often increased in a transition section, which effectively decreases the flight height and the channel area available for flow. This creates a compressive effect and increased mechanical working of the dough. Another effective technique involves reducing the conveying potential of the screw by cutting or notching the flights, allowing protruding mixing pins inserted through the barrel wall to mate with these openings in the flights. Reducing the pitch of the screw's flights is still another method of increasing working.

Twin screw extruders use co-rotating, intermeshing lobed mixing elements, such as those shown in Figure 2, to work and knead the product in process. They aid mixing and dissipate a significant fraction of the mechanical energy added in the extrusion process.

Fig. 2. Lobed kneading elements used in twin screw extrusion to enhance mixing and mechanical energy dissipation.

The screw is constructed and operated so that the lobed screw sections are completely filled with the food ingredients to assure maximum mixing and heating (Chang and Halek, 1991). Twin screw extruder configurations often employ mixing/working sections separated by conveying sections. This allows more precise control of the heating and cooking processes occurring within the extruder.

In twin screw extruders with long barrels, pressure on the ingredients in process may be reduced by venting after heating but before exiting the die (Figure 3). Reduction in pressure at such a vent causes some of the product's moisture to flash off as steam. This obviously removes significant heat, thus reducing the temperature of the dough remaining in the extruder barrel.

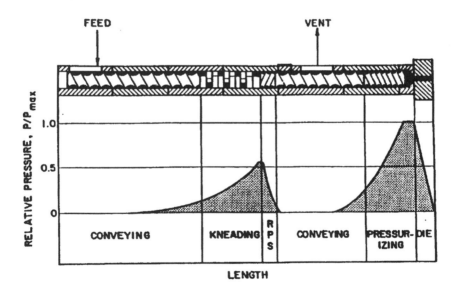

Fig. 3. Illustrative screw and pressure profile for a twin screw food extruder. RPS is reverse pitch screw element which creates a restriction and filling of upstream screw channel.

Versatility in temperature and shear control is the primary reason twin screw extruders are increasingly popular for the cooking and processing of low moisture starch based products. Harper (1989a) has

41

compared other aspects of single and twin screw
extruders in addition to their versatility.

Flow Characteristics

Starch based low moisture melts formed in the
barrel during the extrusion cooking process are very
viscous. The apparent viscosity of these materials is
dependent upon the shear rate experienced during their
passage through the extruder, moisture and temperature, and the extent to which the starch molecules
have lost their native granular structure during the
extrusion process. Harper *et al.* (1971) described the
apparent viscosity (η) and the pseudoplastic behavior
of extrusion-processed starch dough with a modified
power law model

$$\eta = K_0 \dot{\gamma}^{n-1} (\frac{\Delta E}{RT}) \exp(-C \cdot MC)$$

where K_0 (consistency index), n (flow behavior index),
$\Delta E/R$ (activation energy/gas constant), and C
(constant) were determined experimentally. Vergnes
and Villemaire (1987) have found the consistency and
flow behavior indices for starch melts to be a
function of temperature (T), moisture content (MC),
and specific mechanical energy (SME), defined as
mechanical energy input from the screw (W) divided by
the mass flow rate ($\dot{m}$). A comprehensive summary of
starch based dough rheology under extrusion conditions
has been provided by Colonna *et al.* (1989) and Harper
(1989b). The values of n reported for starch based
melts varied between 0.2-0.7, with an average around
0.5.

To estimate the apparent viscosity of melts,
knowledge of the shear rate in various regions of the
extruder is required. Shear rate in the screw channel
is characterized by

$$\dot{\gamma} = \frac{dV}{dh} = \frac{D\omega}{2H}, \ \sec^{-1}$$

42

where dV/dh is the change in dough velocity with respect to channel height, $D\omega/2$ is the maximum tip speed of the screw, and H is the flight height. The equation shows the direct relationship between increasing screw speed and $\dot{\gamma}$.

High shear rates can occur in the die ($\dot{\gamma}_d$) opening and are approximated by

$$\dot{\gamma}_d = \frac{3n+1}{4n} \left[\frac{32\dot{m}}{\pi D_d^3 \rho}\right], \text{ sec}^{-1}$$

where D_d is the die diameter and ρ the density of the dough. Note that shear rate in the die increases rapidly when the diameter of the die opening is reduced or the mass flow rate through a single die opening increases.

The very viscous nature of the starch based doughs and the relatively slow speed of the screw and low flow rate through the die results in laminar rather than turbulent flow throughout the process. During laminar flow, molecules flow in straight lines with variations in flow velocity between adjacent layers of fluid being characterized by the shear rate ($\dot{\gamma}$). This is a significant factor given the large asymmetric molecular shapes of both amylose and amylopectin.

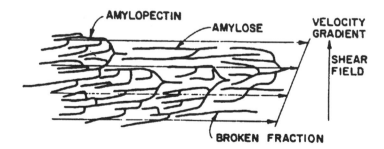

Fig. 4. Starch molecules aligned and being stressed by laminar flow field within a food extruder. (Data from Harper 1986. Used by permission.)

Figure 4 shows a bulky amylopectin molecule attempting alignment in the laminar flow field. This alignment can result in strain on the starch molecule (in high shear areas) which, in turn, can cause mechanical depolymerization and fracturing of granular elements.

Expansion

To overcome flow restrictions in the screw channel and at the die, the flights of the screw increase the pressure on the food material being processed. The pressure on the melt forces it through the die and then drops to atmospheric when the product emerges. Spontaneous puffing of the emerging shaped product occurs when some of the high temperature moisture within the hot product escapes as steam, and stored energy from within the elastic polymeric melt is released (Guy and Horne, 1988). The loss of moisture and heat with the steam causes a rapid temperature drop of the order of 60°C and effectively solidifies the expanded cellular product almost immediately after it exits the die.

Energy Input

Energy input to the extrusion process occurs from three sources:
1. Dissipation of mechanical energy (W) turning the screw filled with viscous fluid;
2. Heat transfer (q) through the barrel wall from heating/cooling jackets or heating bands; and,
3. Latent heat from condensing steam ($\dot{m}_s\lambda$), injected directly into the ingredients in the preconditioner or through the barrel wall.

The total energy E per unit mass is defined as:

$$E = \frac{(W + q + \dot{m}_s\lambda)}{\dot{m}}$$

where high temperature extrusion requires a proportionally larger E. Under normal circumstances, mechanical energy is the largest contributor to E. Higher SME ($W/\dot{m}$) is encountered with lower moisture extrusion and causes the greatest amount of damage to the starch molecule during the extrusion process. To reduce starch damage, more of the heating of the higher moisture ingredients is done with direct steam injection and/or heat transfer through the barrel wall. However, heat transfer through the barrel wall is often limited in large extruders because of the relatively small barrel surface in relation to the large quantity of food being processed and the poor mixing in the screw channel because of the laminar flow environment.

TRANSFORMATION OF STARCH DURING EXTRUSION

The processing of starch by extrusion cooking involves the cooking and modification of the raw starch to provide a uniquely textured, expanded/puffed and shaped product. Cooking extrusion differs from other starch processing methods because the process is performed at low moistures, for short times, at elevated temperatures and pressures in a relatively high shear environment. The extent of alteration and the variety of characteristics of extrusion cooked, starch based food products is rather remarkable due to these unique processing conditions.

Classically, gelatinization describes the loss of starch granules' crystallinity in the presence of heat and excess quantities of water. The process begins with the native starch granules absorbing water and swelling. As more heat is added, the water uptake disrupts the ordered structure within the granule. Initially, amylose diffuses preferentially out of the swollen granules to form a gel which eventually supports the collapsed granules consisting mostly of amylopectin. Gelatinization occurs over a relatively narrow range of temperatures characteristic of the botanic variety of starch and the amount of water available. Very little or no molecular cleavage or depolymerization occurs in the conventional gelatinization process and the viscosity of the resulting gel is maximal for a given concentration of starch.

During extrusion cooking, native starch granules also lose their crystallinity as measured by X-ray

45

diffraction or light scattering techniques; however, the process is substantially different from that described above. Colonna *et al.* (1989) have reviewed the literature on the transformation of starch during the cooking extrusion process. In the extruder, granular starch is first progressively compressed and transformed into a dense solid material. As a consequence of shearing and heating to high temperatures, there is a progressive loss of crystallinity as it melts to form a hot amorphous mass. At 18% moisture, Mercier *et al.* (1979) found that extrusion temperatures between 50-65°C resulted in the deformation of granules while temperatures above 200°C resulted in significant loss of identity (*i.e.*, disruption) of the granules. Substantial granule damage was observed by Guy and Horne (1988) and Della Valle *et al.* (1989) when the SME during extrusion was 500-600 kJ/kg, while flattened and deformed granules occurred in the range of 250-360 kJ/kg. Starch granules may remain intact under high moisture and low temperature extrusion conditions (Richman and Smith, 1985) but, with increasing severity of treatment, they lose their crystalline structure.

During the extrusion process, damage or depolymerization of starch molecules occurs along with the loss of granular crystallinity. The extent of damage and molecular depolymerization have been estimated by size exclusion, intrinsic viscosity, and HPLC techniques (Diosady, 1986). Most of the depolymerization resulted in large fragments rather than producing low molecular weight oligosaccharides. Gomez and Aguilera (1983 and 1984) describe how increasing shear during the conveying and heating of the starch during extrusion causes breakdown or dextrinization to the starch molecule. Diosady *et al.* (1985) postulated a mechanical degradation model for extrusion which included product moisture, temperature and screw speed as variables. Extrusion of low moisture ingredients at high shear resulting from high speed screws or mixing elements had the greatest effect on the starch molecule itself. This molecular degradation of extruded starch has been demonstrated by Mercier (1977), Meuser *et al.* (1982), Davidson *et al.* (1984), Colonna *et al.* (1984), Diosady *et al.*

(1985), and Wen *et al.* (1990) by the use of gel permeation chromatography or intrinsic viscosity measurements. Colonna *et al.* (1989) concluded in their review that the experimental evidence was in favor of random chain cleavage, with amylopectin being somewhat more susceptible than amylose to mechanical dextrinization.

Mercier (1980) described the effect of extrusion temperature and moisture on the X-ray diffraction patterns of the resulting products. Extrusion at lower temperatures (135°C) and higher moistures (22%), produced a V-type structure having a unique peak at 9°54', which is similar to that given by the butanol-amylose complex. At a higher extrusion temperature of 185°C and 13% moisture, an "extruded" or E-type structure resulted which was characterized by an X-ray diffraction peak at 9°03'. Reconditioning products with E-type structure to moistures above 30% caused a reversion to the V-type structure. These differences and the interconvertability of the structures were felt to be related to the helical structure of the amylose in the extruded samples since samples continuing only amylopectin showed no such X-ray diffraction patterns.

Complexing of amylose with lipids and monoglycerides during extrusion has been confirmed by Mercier *et al.* (1980) with manioc starch for C_{12} or longer fatty acids. Meuser *et al.* (1987) found that the less saturated fatty acid chains were complexed to a lesser extent, and the limit of lipid complexed was 3%. The amylose and lipid complex was found to be hydrolyzed more slowly by enzymes and was described as resistant starch. He also showed that high levels of lipid in extruded ingredients reduce SME, the rapid increase in temperature and the mechanical starch damage associated with low moisture starch extrusion without the presence of lipid.

PROPERTIES OF EXTRUDED STARCH

There are a number of commercially significant properties of extruded starch products that can be varied and controlled by changing the extrusion process itself. Examples of such properties include the water solubility/absorption characteristics of extruded starch, paste viscosity, expansion or density of extruded products, and mechanical properties of extruded foams.

Water Solubility/Absorption

The water solubility and absorption of extruded starch impacts the functional properties of the material for use as a food ingredient. Application as thickening and moisture control, the base of a ready-to-eat cereal which must maintain piece integrity and crispiness in milk, or to produce a gravy on a pet food to be rehydrated with water are dependent upon these characteristics. Two indices have been used to characterize starch's solubility and water absorption capacity. The water absorption index (WAI) is determined by the method of Anderson *et al.* (1969), and defined as the weight of gel formed per gram of dry sample. The water solubility index (WSI) is the percentage of the dry matter in the supernatant resulting from the centrifugation of the water/sample mixture previously used to determine WAI.

Vigorous extrusion conditions characterized by elevated SME and high melt temperatures behind the die result in greater mechanical dextrinization of the starch molecule. The extruded material is more water soluble and less able to hold water or form a gel. The inverse relationship between WAI and WSI is the result of the way the indices are defined. Figure 5 shows the relationship between SME and extruded product temperature and WSI found by Meuser *et al.* (1987), and is characteristic of results obtained on both single and twin screw extruders. Both SME and product temperature will increase as feed moisture to the extruder decreases, a result of the increased viscosity of the resulting melt. The increase in WSI with a decrease in extrusion moisture has been observed by Anderson *et al.* (1969), Mercier and Feillet (1975), and Paton and Spratt (1984). Extruders with a high shear screw will also favor starch dextrinization during processing and increased solubility of the resulting product (Kim, 1984).

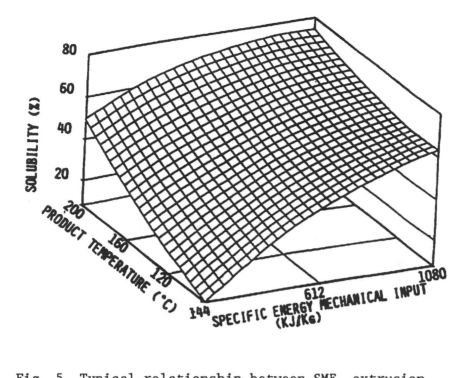

Fig. 5. Typical relationship between SME, extrusion
temperature and WSI.   (Data from Meuser *et al.*, 1987.
Used by permission.)

The amylose-amylopectin ratio in extrusion
ingredients appear to affect the solubility of the
product following extrusion.   Mercier and Feillet
(1975) found a decrease in WSI with increases in
amylose content.   Solubility also decreased when
complexing between starch and lipid occurred [Mercier
et al. (1980), Colonna and Mercier (1983), Schweizer
et al. (1986)].   WAI for extruded starch samples (3-10
g gel/g dry starch) is less than that for samples
which were gelatinized conventionally at high moisture
before being drum or spray dried (10-20 g/g) (Gomez
and Aguilera, 1984).   Increasing thermal treatment in
the extruder has a negative impact on WAI.   Further,
amylose appears to require an extrusion temperature in
excess of 200°C to significantly increase WAI.

Paste Viscosity

The pasting characteristics of extruded starches are important to many of their functional uses as food ingredients. Actual viscosity is important when extruded starches become the basis for infant foods where caloric density is critical (Jansen et al., 1981) or drinks where consistency is an important acceptability consideration. Because of mechanical depolymerization during the process, extrusion cooked starch must be added at higher concentrations to achieve a specific viscosity. The Brabender ViscoamyloGraph is used most frequently to measure the pasting characteristics of a hydrated sample over a range of temperatures. Other types of rheometers, including rotating cylinder, coaxial cylinders or couette, have been used to measure the rheology of starch pastes as a function of paste concentration, temperature and shear rate.

The influence of ingredient moisture content and extrusion temperature on the resulting starch pasting characteristics have been investigated by Colonna et al. (1984) and Davidson et al. (1984) and illustrated in Figure 6. Native starch exhibits a rapid rise in viscosity with the onset of gelatinization. Extruded starches lack such a peak because they are gelatinized and show higher initial cold paste viscosity. This decreases as heating of the hydrated sample proceeds. Increased severity of the extrusion treatment occurring with low moisture and high extrusion temperature produces a starch with the lowest cold paste viscosity (Olkku and Vainionpää, 1980). This is reasonable as these conditions result in the greatest damage to the starch molecule.

The effects of extrusion temperature on the paste viscosity has been investigated by Launay and Lisch (1983) and Mason and Hoseney (1986). At constant feed moisture, maximum cold paste viscosity occurs at extrusion temperatures between 160-180°C, with lower values at either higher or lower temperatures. The effects of increased moisture at constant extruder barrel temperature was evaluated by El Dash et al. (1983), who showed a consistent increase in the

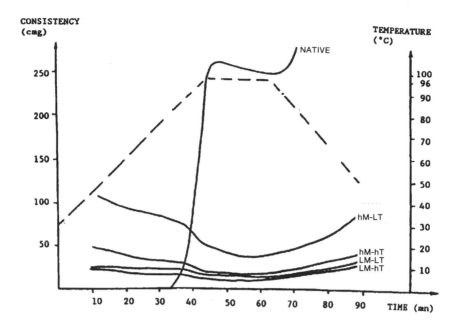

Fig. 6. Role of extrusion conditions on Brabender
ViscoamyloGrams of wheat starch: hM-LT = 34.7%$_{db}$MC,
125°C; hM-hT = 34.7%$_{db}$MC, 180°C; LM-LT = 24.4%$_{db}$MC,
130°C; LM=hT = 23.9%$_{db}$MC, 180°C. (Data from Colonna
et al., 1984. Used with permission.)

cold paste viscosity with increasing moisture. Meuser
et al. (1987) showed a reduction in viscosity and an
increase in solubility with increasing SME. The value
of SME appears to be a complex interaction between
extrusion temperature, moisture and shear.

Expansion

The creation of an expanded starch based matrix
by extrusion cooking is an important element in the
production of snacks, RTE cereals and pet foods. The
extent of expansion and the resulting structure depend
on the characteristics of the hot starch melt behind
the die and the die design. As illustrated in Figure
7, water within the dough nucleates when the pressure
on the dough drops as it passes through the die
restriction and the vapor formed creates bubbles and

the expanded structure. Die swell, or increase in radial dimension resulting from the elastic character- istic of molten starch based doughs, also contributes to expansion once the product leaves the die opening according to Guy and Horne (1988) and Harper (1986).

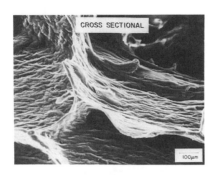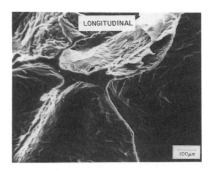

Fig. 7. Micrograph of extrusion expanded corn grits (Data from Harper, 1986. Used with permission.)

With the loss of moisture in the form of steam, cool- ing occurs to temperatures below the glass transition temperature (T_g) of starch causing solidification and retention of the expanded shape. It is not unusual for high moisture hot extrudates to expand extensively just beyond the die and then collapse before they cool sufficiently to retain their remaining expansion, usually resulting in a hard and dense product.

Product bulk density is directly related to the extent of expansion and is a very important parameter in the production of expanded food products. Bulk densities of most extruded foods range between 0.04- 0.38 g/cm^3 according to Colonna et al. (1989), with pore area between 230-480 cm^2/g, and nearly all pores accessible to the atmosphere. Without the addition of some coating on the surface of extruded pieces, they can be quite susceptible to oxidative deterioration. The range of pore sizes within the solidified extrudate is related to the uniformity of the free water distribution within the plasticized or melted starch in the dough. The cell wall in extruded starch foods has a density of 1.46-1.50 g/cm^3.

Product expansion can be visualized in Figure 8.

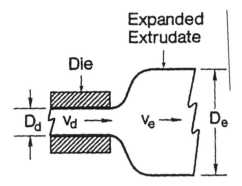

Fig. 8. Product expansion occurring when hot melted starch dough emerges from the extrusion die.

Die expansion has been characterized by Alvarez-Martinez *et al.* (1988) using three dimensionless parameters:

$$\text{Sectional Expansion Index:} \quad \text{SEI} = \left[\frac{D_e}{D_d}\right]^2$$

or the ratio of extrudate diameter (D_e) to die diameter (D_d);

$$\text{Longitudinal Expansion Index:} \quad \text{LEI} = \left[\frac{v_e}{v_d}\right]$$

or the ratio of extrudate velocity after expansion (v_e) to dough velocity in the die opening (v_d); and,

$$\text{Volumetric Expansion Index:} \quad \text{VEI} = \text{SEI} \cdot \text{LEI}$$

This complete description of expansion is necessary in order to understand how the extrusion process affects finished product characteristics. For example, an inverse relationship between SEI and LEI has been established by Launay and Lisch (1983) and Alvarez-Martinez *et al.* (1988). SEI increased at lower extrude moistures and higher extrusion temperatures while LEI decreased under the same conditions (Figure 9).

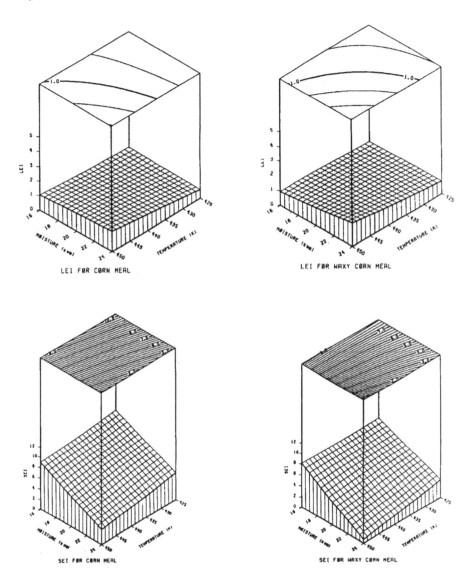

Fig. 9. The inverse relationship between SEI and LEI for native and waxy corn as a function of extrusion moisture and temperature. (Reprinted from Harper and Tribelhorn, 1991, by courtesy of Marcel Dekker, Inc.)

VEI is increased by reduced feed moisture more than increases in temperature and is the expansion parameter most directly correlated with bulk density of a finished product. Owusu-Anseh et al. (1984), Chinnaswamy and Hanna (1988a), Alvarez-Martinez et al. (1988), and Mercier and Feillet (1975) all determined that the expansion of cereal based extrudates was affected more significantly by feed moisture, then by temperature and, to a lesser extent, by screw speed. Maximum expansion was found to occur at temperatures near 170°C and moistures around 14% for corn-based products. However, this value varied with different types of starch and the protein content of the material being extruded. Individual piece size is therefore directly related to and controlled by changes in the extrusion conditions.

The ratio of amylose to amylopectin affects the expansion of extruded starch. Launey and Lisch (1983), Chinnaswamy and Hanna (1988b), and Mercier and Feillet (1975) showed that amylose required higher temperatures (225°C) for maximum expansion than did amylopectin (135°C). Harper and Tribelhorn (1991) found amylose tended to expand longitudinally while amylopectin tended to expand radially, giving a higher SEI. Higher amylopectin content in extruded starch results in an expanded product which has smaller pores and lower bulk density while higher amylose content products are typically less expanded, harder and more resistant to moisture uptake. All these characteristics are important to RTE cereal products.

Products which have received extensive shearing (indicated by high SME) associated with low moisture extrusion, have greater expansion, smaller pore sizes, and softer textures. Extrusion expanded snacks fit into this range of products. To products having the opposite characteristics, higher moisture extrusion is practiced. This requires that more of the heat energy

be added in the form of steam injection or heat transfer through the barrel wall, with a corresponding reduction in SME. The result is less expansion, larger individual pores, and increased resistance to moisture uptake. The addition of small lipid quantities (<3%) has only a small effect on expansion, while amounts over 5% result in rapid decreases in extrudate expansion as observed by de la Guérivière (1976). Paton and Pratt (1978) found that the gluten in wheat flour resulted in products having greater expansion than could be achieved by extruding wheat starch alone. The addition of protein usually increases product strength and provides a crisp texture.

Mechanical Properties

Mechanical properties of starch based extrudates have been measured. They serve as indirect measurements of such product attributes as hardness, crispiness, cohesiveness, *etcetera*. These include the same mechanical properties measured for elastic, plastic and brittle foams made from synthetic polymers. Mechanical measurements have been made on starch based extrudates with an impact tester (Van Zuilichem *et al.*, 1975) to determine break strength, the Instron Universal Testing Machine (Launay and Lisch, 1983) to determine compressive and tensile properties, and the Kramer Shear Press and the Warner-Bratzler shear press to measure shear strength at low rates of shear. The rates of shear possible with these latter two testers are considerably slower than those encountered in the mouth. Therefore, the results may not correlate well with sensory evaluations. All mechanical properties used are highly dependent upon the moisture content of the extrudate being tested, requiring that it be controlled and specified if meaningful comparisons are to be made between samples.

Van Zuilichem *et al.* (1975) found that corn meal extrudates had the maximum impact strength when the extrusion temperature is between 165-170°C. Impact strength increased with the moisture content of the extrudate. A direct relationship between extrudate bulk density and tensile, flexural and compressive strengths was observed by Hutchinson *et al.* (1987). The flexural strength was greater than the compressive or tensile strengths. Extrudate mechanical properties are also affected by the amount of starch degradation which occurs in the extruder due to shear and heat and the alignment of the molecules during their laminar flow through the extrusion screw and die. If the extrusion process renders the starch in a state where the starch molecules are depolymerized, the product loses mechanical strength and becomes soft. Typically, such extrusion conditions would be favored by low moisture and high shear, SME and temperature, and are those used for highly expanded snack foods.

The characteristics of the cell wall in the expanded starch extrudate influence its mechanical properties. Hutchinson *et al.* (1987) found the force to break the walls at the surface of the extrudate to be greater than that required for an internal pore wall. Initial collapse of the cellular structure at the surface during cooling may account for these differences. Differences in the botanical source of starch on strength of extrudates were measured by Hayter *et al.* (1986), who found corn extrudates to be mechanically stronger than wheat. Faubion and Hosenay (1982) showed that increases in gluten content in wheat flour extrudates strengthened them. The effects of bran, sucrose and magnesium carbonate on the properties of extrudates were studied by Moore *et al.* (1990). With all three ingredients, apparent density increased, with corresponding changes in the modules of deformity in bending. Bran caused an increase in the number of cells and decreased their average size.

SUMMARY

Extrusion represents an entirely different way of cooking and forming cereal starch based food products. The continuous process is characterized by low moisture, high temperature and shear applied under pressure for a short time duration. During extrusion, native starch is completely transformed into an amorphous mass of hot, melted and plasticized starch which is shaped by a die at the extruder discharge.

The characteristics of the extruded starch are greatly influenced by the specific extrusion conditions employed. More severe extrusion conditions indicated by high temperatures, shear and extensive mechanical energy results in starch depolymerization and molecular disruption (dextrinization), which affect expansion and reduce mechanical strength and paste viscosity of the final product.

LITERATURE CITED

Alvarez-Martinez, L., Kondury, K. P., and Harper, J. M. 1988. A general model for expansion of extruded products. J. Food Sci. 53:609.

Anderson, R. A., Conway, H. F., Pfeifer, V. F., and Griffin, L. E. J. 1969. Gelatinization of corn grits by roll- and extrusion cooking. Cereal Sci. Today 14:4.

Chang, K. L. B., and Halek, C. W. 1991. Shear and thermal history during co-rotating twin-screw extrusion. J. Food Sci. 56(2):518.

Chinnaswamy, R., and Hanna, M. A. 1988a. Optimum extrusion-cooking conditions for maximum expansion of corn starch. J. Food Sci. 53(3):834.

Chinnaswamy, R., and Hanna, M. A. 1988b. Relation ship between amylose content and extrusion-expansion properties of corn starches. Cereal Chem. 6-5(2):138.

Colonna, P., and Mercier, C. 1983. Macromolecular modifications of manioc starch components by extrusion cooking with and without lipids. Carbohydr. Polym. 3:87.

Colonna, P., Tayeb, J., and Mercier, C. 1989. Extrusion cooking of starch and starchy products. pp 247. In: Extrusion Cooking. C. Mercier, P. Linko and J. Harper, eds. American Association of Cereal Chemists, St. Paul.

Colonna, P., Doublier, J. L., Melcion, J. P., De Monredon, F., and Mercier, C. 1984. Physical and functional properties of wheat starch after extrusion-cooking and drum-drying. pp 96. In: Thermal Processing and Quality of Foods. P. Zeuthen, J.-C. Cheftel, C. Eriksson, M. Jul, H. Leniger, P. Linko, G. Varela, and G. Vos, eds.

Elsevier Applied Science Publ., London.

Davidson, V. J., Paton, D., Diosady, L. L., and
Larocque, G. J. 1984. Degradation of wheat starch
in a single screw extruder: Characteristics of
extruded starch polymers. J. Food Sci. 49:453.

De la Guérivière, J. F. 1976. Principles of the
extrusion-cooking process. Application to starchy
foods. Bull. Anc. El. Ec. Fr. Meun. 276:305.

Della Valle, G., Kozlowski, A., Colonna, P., and
Tayeb, J. 1989. Starch transformation estimated
by the energy balance on a twin screw extruder.
Lebensm. Wiss. Technol. 22(5)279.

Diosady, L. L. 1986. Review of recent studies in the
mechanism of starch extrusion. pp 143. In: Food
Engineering and Process Applications. Vol. 2.
Unit Operations. M. Le Maguer and P. Jelen, eds.
Elsevier Applied Science Publ., New York.

Diosady, L. L., Paton, D., Rosen, N., Rubin, L. J.,
and Athanassoulias, C. 1985. Degradation of wheat
starch in a single screw extruder: Mechanico-
kinetic breakdown of cooked starch. J. Food Sci.
50:1697.

Dziezak, J. D. 1989. Single- and twin-screw
extruders in food processing. Food Technol.
43(4):164.

El-Dash, A. A., Gonzales, R., and Ciol, M. 1983.
Response surface methodology in the control of
thermoplastic extrusion of starch. J. Food Ecg.
2:129.

Faubion, J. M., and Hoseney, R. C. 1982. High-
temperature short-time extrusion cooking of wheat
starch and flour. II. Effect of protein and lipid
on extrudate properties. Cereal Chem. 59:533.

Faubion, J. M., Hoseney, R. C. and Seib, P. A. 1982.
Functionality of grain components in extrusion.

Cereal Foods World 27(5):212.

Gomez, M. H., and Aguilera, J. M. 1983. Changes in the starch fraction during extrusion cooking of corn. J. Food Sci. 48:378.

Gomez, M. H., and Aguilera, J. M. 1984. A physico chemical model for extrusion of corn starch. J. Food Sci. 49:40.

Guy, R. C. E., and Horne, A. W. 1988. Extrusion and co-extrusion of cereals. pp 331. In: Food Structure - Its Creation and Evaluation. J. M. V. Blanshard and J. V. Mitchell, eds. Butterworths, London.

Harper, J. M. 1986. Extrusion texturization of foods. Food Technol. 40:70.

Harper, J. M. 1989a. Food extruders and their applications. pp 1. In: Extrusion Cooking. C. Mercier, P. Linko and J. Harper, eds. American Association of Cereal Chemists, St. Paul.

Harper, J. M. 1989b. Food extrusion. pp 271. In: Food Properties and Computer-aided Engineering of Food Processing Systems. P. Singh and A. Medina, eds. Kluwer Academic Publishers, Boston.

Harper, J. M. and Tribelhorn, R. E. 1991. Expansion of native cereal starch extrudates. pp 653. In: Food Extrusion Science/Technology. J. Kokini, ed. Marcel Dekker, Inc., New York.

Harper, J. M., Rhodes, T. P., and Wanniger, L. A. 1971. Viscosimetry model for cooked cereal doughs. AIChE Symp. Ser. 67(108):40.

Hayter, A. L., Smith, A. C., and Richmond, P. 1986. The physical properties of extruded food foams. J. Mater. Sci. 21:3729.

Huber, G. R., and Rokey, G. J. 1990. Extruded snacks. pp 107. In: Snack Food. R. Booth, ed.

Van Nostrand Reinhold, New York.

Hutchinson, R. J., Siodlak, G. D. E., and Smith, A. C. 1987. Influence of processing variables on the mechanical properties of extruded maize. J. Mater. Sci. 22:3956.

Jansen, G. R., O'Deen, L., Tribelhorn, R. E., and Harper, J. M. 1981. The caloric density of gruels made from extruded corn-soy blends. Food Nutr. Bull. 3:39.

Kim, J. C. 1984. Effect of some extrusion parameters on the solubility and viscograms of extruded wheat flours. pp 251. In: Thermal Processing and Quality of Foods. P. Zeuthen, J.-C. Cheftel, C. Eriksson, M. Jul, H. Leniger, P. Linko, G. Varela, and G. Vos, eds. Elsevier Applied Science Publ., London.

Launay, B., and Lisch, J. M. 1983. Twin screw extrusion cooking of starches: Behavior of starch pastes, expansion and mechanical properties of extrudates. J. Food Ecg. 2:259.

Linko, P., Linko, Y. Y., and Hakulin, S. 1984. Continuous extrusion processing of starchy materials for the production of sirups and ethanol. pp 122. In: Thermal Processing and Quality of Foods. P. Zeuthen, J.-C. Cheftel, C. Eriksson, M. Jul, H. Leniger, P. Linko, G. Varela, and G. Vos, eds. Elsevier Applied Science Publ., London.

Mason, W. R., and Hoseney, R. C. 1986. Factors affecting the viscosity of extrusion-cooked wheat starch. Cereal Chem. 63:436.

Mercier, C. 1977. Effect of extrusion cooking on potato starch using a twin screw French extruder. Staerke 29:48.

Mercier, C. 1980. Structure and digestibility alterations of cereal starches by twin-screw extrusion-cooking. pp 795. In: Food Process Engineering.

Vol. I. Food Processing Systems. P. Linko, Y. Mälkki, J. Olkku and J. Larinkari, eds. Applied Science Publishers, London.

Mercier, C., and Feillet, P. 1975. Modification of carbohydrate components by extrusion-cooking of cereal products. Cereal Chem. 52:283.

Mercier, C., Charbonniere, R., Gallant, D., and Guilbot, A. 1979. Structural modifications of various starches by extrusion cooking with a twin-screw French extruder. pp 153. In: Polysaccharides in Food. J. M. V. Blanshard and J. R. Mitchell, eds. Butterworths, London.

Mercier, C., Charbonniere, R., Grebaut, J., and de la Guérivière, J. F. 1980. Formation of amylose-lipid complexes by twin-screw extrusion cooking of manioc starch. Cereal Chem. 57:4.

Meuser, F., Van Lengerich, B., and Kohler, F. 1982. Einfluss der Extrusions Parameter auf funktionnelle Eigenschaften von Weisenstarke. Staerke 34:366.

Meuser, F., Pfaller, W., and Van Lengerich, B. 1987. Technological aspects regarding specific changes to the characteristic properties of extrudates by HTST extrusion cooking. pp 35. In: Extrusion Technology for the Food Industry. C. O'Connor, ed. Elsevier Applied Science Publ., London.

Miller, R. C. 1985. Extrusion cooking of pet foods. Cereal Foods World 30(5):323.

Miller, R. C. 1988. Continuous cooking of breakfast cereals. Cereal Foods World 33(3):284.

Miller, R. C. 1990. Unit operations and equipment. IV. Extrusion and extruders. pp 135. In: Breakfast Cereals and How They Are Made. R. B. Fast and E. F. Caldwell, eds. American Association of Cereal Chemists, Inc., St. Paul.

Moore, D., Saneis, A., Van Hecke, E., and Bouvier, J.

M. 1990. Effect of ingredients on physical/ structural properties of extrudates. J. Food Sci. 55(5):1383.

Olkku, J., and Vainionpää, J. 1980. Response surface analysis of HTST-extrusion texturized starch-protein-sugar paste. pp 821. In: Food Process Engineering, Vol. 1. P. Linko, Y. Mälkki, J. Olkku, and J. Larinkari, eds. Elsevier Applied Science Publ., London.

Owusu-Ansah, J., Van De Voort, F. R., and Stanley, D. W. 1984. Textural and microstructural changes in corn starch as a function of extrusion variables. Can. Inst. Food Sci. Technol. J. 17:65.

Paton, D., and Spratt, W. A. 1978. Component interactions in the extrusion cooking process. I. Processing of chlorinated and untreated soft wheat flour. Cereal Chem. 55:973.

Paton, D. and Spratt, W. A. 1984. Component interactions in the extrusion cooking process: Influence of process conditions on the functional viscosity of the wheat flour system. J. Food Sci. 49:1380.

Richmond, P., and Smith, A. C. 1985. The influence of processing on biopolymeric structures. Br. Polym. J. 17:246.

Schweizer, T. F., Reimann, S., Solms, J., Eliasson, A. C., and Asp, N. G. 1986. Influence of drum-drying and twin screw extrusion cooking on wheat carbohydrates. II. Effects of lipids on physical properties, degradation and complex formation of starch in wheat flour. J. Cereal Sci. 4:249.

Van Zuilichem, D. J., Lamers, G., and Stolp, W. 1975. Influence of process variables on quality of extruded maize grits. Proc. Eur. Symp. Engineering and Food Quality, 6th, Cambridge, UK.

Vergnes, B., and Villemaire, J. P. 1987. Rheological behaviour of low moisture molten maize starch.

Rheol. Acta 26:570.

Wen, L. F., Rodis, P., and Wasserman, B. P. 1990.
    Starch fragmentation and protein insolubilization
    during twin-screw extrusion of corn meal. Cereal
    Chem. 67(3):268.

STARCH HYDROLYZING ENZYMES

Ronald E. Hebeda and W. Martin Teague

Enzyme Bio-Systems Ltd.
Arlington Heights, IL 60005

INTRODUCTION

History

The use of carbohydrases for starch hydrolysis is certainly not new. In fact, as far back as the ninth century malt was used to convert arrowroot starch to a sweetener (Yoshizumi et al, 1986). In 1833, precipitated malt extract was observed to produce a sugar from starch. The enzyme responsible for the conversion was unknown at the time and the mysterious material responsible for the reaction was termed "diastase" from the French word meaning "separation" (Payen and Persoz, 1833). "Diastase" became the generic term for amylases. Trivial enzyme names were later derived by adding the suffix "ase" to the root of the word denoting the substrate or action of the particular enzyme.

Development and commercialization of alpha-amylases from fungal (Takamine, 1894) and bacterial (Boidin and Effront, 1917a, 1917b) sources occurred in the late 19th and early 20th centuries, respectively. By the 1930s, these enzymes were being used commercially in a variety of applications such as brewing, textiles, paper (Wallerstein, 1939) and corn syrup (Dale and Langlois, 1940). In the 1950s, the effectiveness of bacterial alpha-amylase as an anti-staling agent in baked goods was recognized and studied (Stone, 1952; Miller et al, 1953). A significant increase in amylase utilization occurred in the early 1960s when *Bacillus subtilis* bacterial alpha-amylase and *Aspergillus niger* glucoamylase were first used to replace acid catalysis in the production of dextrose from starch.

The process initially developed to utilize the *B. subtilis* alpha-amylase for starch liquefaction is diagrammed in Fig. 1. A starch slurry, at the proper pH and solids levels and containing the required amount of calcium and enzyme, was passed through a steam injection heater

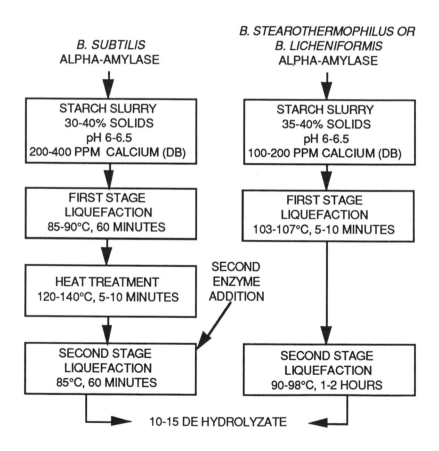

Fig. 1. Starch liquefaction processes.

at about 90°C and held at that temperature for about 60 minutes. Under these conditions, starch is simultaneously gelatinized and hydrolyzed to a low viscosity hydrolyzate. Since the alpha-amylase was limited to a maximum operational temperature of about 90°C, the liquefaction process required inclusion of a high temperature heat treatment step to completely solubilize residual starch present as a fatty acid/amylose complex. This treatment, at 120 - 140°C, inactivated the enzyme necessitating a second addition of alpha-amylase to continue hydrolysis to the required end-point. The resulting 10 - 15 DE hydrolyzate was then saccharified with glucoamylase at 60°C, pH 4 - 4.5, for 24 - 96 hours. Under these conditions, dextrose (glucose) units are liberated successively from the non-reducing end of starch polymers by the action of glucoamylase. The maximum dextrose level attainable under industrial conditions is limited to 95 - 96% due primarily to the formation of disaccharides by the reverse reaction of glucoamylase.

The thermostability of the *B. subtilis* alpha-amylase allowed efficient starch liquefaction at high temperature and the specificity of glucoamylase provided an increased yield of dextrose during subsequent saccharification. These two amylases were the primary carbohydrase enzymes in use 25 years ago.

Current Status

Today, alpha-amylase and glucoamylase are still the major carbohydrases used commercially. Together they represent about 79% of the total carbohydrate hydrolyzing enzyme sales in the U. S. (Wolnak, 1990); the remaining 21% includes pectinase (12%), beta-amylase (7%), cellulase (<2%) and invertase (<1%).

The primary commercial application for alpha-amylase and glucoamylase is the hydrolysis of corn, wheat and other starches to produce dextrose, fructose syrups and industrial ethanol. It is estimated that the total world market for alpha-amylases (both bacterial and fungal) and glucoamylase is about $60 MM (Teague and Brumm, 1991). About 80% of these sales occur in the U. S.

A small but growing application for carbohydrase enzymes is in the baking industry where alpha-amylases are used as anti-staling agents. The potential sales for amylases in white bread alone may be as much as $18 MM annually in the U. S. This figure is based on a per capita consumption of 28 pounds of white bread per year (Anonymous, 1991) and an estimated enzyme cost to the baker of $0.50/cwt flour. By including other types of bread and rolls, enzyme sales could even be higher. Consequently, the potential sales of starch hydrolyzing enzymes for U. S. corn wet-milling and baking applications could conceivably total $70 MM or more.

Developing new industrial enzymes requires considerable time and expenditure devoted to basic research, scale-up and regulatory approval. It has been estimated that the commercialization of a new enzyme requires five to ten years at a cost of $5 to $30 MM (Wolnak, 1990). Obviously, development of industrial enzymes for existing or new applications is not a trivial matter. Enzymes introduced to the marketplace today have, no doubt, been studied for years - probably for a decade or more.

These limitations, however, have not prevented the development of a number of new carbohydrases for the starch hydrolysis, baking and brewing industries. These include liquefying enzymes with increased tolerance to high temperature and low pH, enzymes that act in combination with glucoamylase to increase dextrose yield beyond the normal level, enzymes that have unique stability characteristics required for use in baking and an immobilized enzyme for use in brewing. In several instances, genetic engineering has been successfully employed to increase fermentation yields of these enzymes to economically feasible levels. Several of these developments will be discussed in this chapter.

NEW STARCH HYDROLYZING ENZYMES

Liquefaction Enzymes

The first improvement in starch liquefaction enzymes occurred with the development of a *Bacillus licheniformis* alpha-amylase that exhibits considerably greater thermo-stability than the *B. subtilis* amylase (Table I).

The *B. licheniformis* alpha-amylase was commercialized in the early 1970s using a two stage process

TABLE I
Maximum Operational Temperature for
Starch Liquefaction and Saccharification Enzymes

Enzyme	Enzyme Source	Temp., °C
Liquefaction Enzymes		
Alpha-Amylase	*B. subtilis*	90
Alpha-Amylase	*B. licheniformis*	110+
Alpha-Amylase	*B. stearothermophilus*	110+
Saccharification Enzymes		
Glucoamylase	*A. niger*	60
Isoamylase	*P. amyloderamosa*	55
Pullulanase	*K. planticola*[a]	55
Pullulanase	*B. acidopullulyticus*	60
Alpha/Transferase	*B. megaterium*	60

[a]Formerly identified as *K. pneumoniae, K. aerogenes* or *A. aerogenes.*

(Figure 1) that was developed to take advantage of the enzyme's thermostability (Slott and Madsen, 1975). The process involves a first stage reaction at 103 - 107°C for five to ten minutes followed by additional hydrolysis at 90 - 98°C for one to two hours. Since the enzyme maintains activity above 100°C in a temperature range where fatty acid/amylose complexes are dissociated and solubilized, efficient liquefaction is achieved without the need for the heat treatment step and second enzyme addition required with the *B. subtilis* enzyme. The simplified process gained widespread use throughout the starch hydrolysis industry.

A more recent development occurred in the mid 1980s when a second highly thermostable alpha-amylase was commercialized. The enzyme, derived from *Bacillus stearothermophilus*, is extremely stable at high temperatures. Based on literature data (Henderson and Teague, 1988; Rosendal et al, 1979), the enzyme exhibits half-lives that are 3 to 4 times greater than those observed with *B. licheniformis* amylase (Table II). At typical industrial reaction conditions (102 - 105°C, pH 6 - 6.5, 31 - 37% solids, 25 - 40 ppm calcium), the *B. stearothermophilus* amylase has a half-life of 2 - 3.5 hours. Based on this high degree of thermostability, it has been reported that enzyme dosage can be reduced by as much as 40% by merely increasing the initial liquefaction time from the standard 6 minutes to 10 minutes (Henderson and Teague, 1988).

The *B. stearothermophilus* enzyme also exhibits improved heat stability at low pH. For instance, in the second stage of liquefaction, the *B. stearothermophilus* alpha-amylase has been shown to retain 95-100% activity at the optimum pH of 5.8. For comparison, the *B. licheniformis* alpha-amylase retains 85% activity at its optimum pH of 6.5 -7.0 (Carroll and Swanson, 1990). As a result, the *B. stearothermophilus* alpha-amylase exhibits a lower minimum operational pH than the *B. licheniformis* amylase, i.e., 5.6 vs. 6.0 (Reeve, 1991).

It has been reported that the combined use of *B. stearothermophilus* and *B. licheniformis* alpha-amylases provides a degree of synergy during starch liquefaction. The mixture of enzymes allows low pH operation while

TABLE II
Stability of Alpha-Amylases Under Liquefaction Conditions

Temp., °C	pH	Added Calcium, ppm, as is	Solids, % w/w	Half-Life Minutes BS[a]	Half-Life Minutes BL[b]	Half-Life Ratio BS/BL
102	6	25	37.3	213	70	3.0
102	6	25	31.4	126	44	2.9
102	6	25	26.5	81	25	3.2
105	6.5	40	35	132	50	2.6
105	5.5	40	35	43	10	4.3
105	6.5	40	25	75	20	3.8
105	6.5	10	35	42	15	2.8

[a]*B. stearothermophilus alpha*-amylase (Henderson and Teague, 1988).
[b]*B. licheniformis alpha*-amylase (Rosendal et al, 1979).

minimizing the generation of insoluble polysaccharide sediment that could adversely affect refining. Efficient liquefaction can be achieved with the dual enzyme system at a pH below 6.0 while reducing overall dosage by 15 - 25% (Carroll and Swanson, 1990).

It is obvious that the trend in starch liquefying enzymes has been to develop amylases that exhibit both increased themostability and increased aciduricity. The next generation of improved liquefaction enzymes may be highly thermostable alpha-amylases that are even more effective at low pH, i.e., at the normal saccharification pH of about 4 - 4.5. In this way, both liquefaction and saccharification could be conducted without the need for an intermediate pH adjustment.

An example of this type of enzyme is a "hyperther-mostable" alpha-amylase produced by hyperthermophilic *Pyrococcus* archaebacteria isolated from thermal waters. This enzyme exhibits an optimum temperature activity in excess of 100°C (Brown et al, 1990) and is remarkably heat stable, retaining 70 or 80% activity in the presence of substrate even when heated at 110 or 100°C, respectively, for one hour (Antranikian et al, 1990). High temperature starch liquefaction is possible within a pH range of 4.2 - 4.8 without the need for the calcium that is normally added as a co-factor to maintain enzyme stability.

A second example of a low pH, thermostable alpha-amylase is a cyclodextrin glycosyltransferase (CGTase) de-rived from a genus of *Thermoanaerobacter*. This enzyme has been found to have an optimum temperature activity of 95°C at pH 5 and is able to liquefy starch efficiently at low pH (Starnes, 1990). Under simulated industrial liquefaction conditions, the enzyme is effective in thinning starch at pH 4.5 without the need for added calcium.

Reported advantages of these low pH, thermostable enzymes include the elimination of pH adjustment for sac-charification, decreased refining costs and reduced forma-tion of color and by-products such as maltulose.

Saccharification Enzymes

Although development of new liquefaction enzymes has continued over the last several decades, the same cannot be said for glucoamylase. Both the enzyme and saccharification process have remained virtually un-changed during this same period of time. What has occurred, however, is the development of enzymes that can be used in combination with glucoamylase to increase dextrose yield beyond the normal level.

Debranching enzymes that specifically hydrolyze alpha-1,6 linkages in starch have been used for this purpose for several years. During saccharification, glucoamylase not only hydrolyzes polysaccharides to dextrose but also catalyzes the reverse reaction by repolymerization of dextrose. Consequently, maximum dextrose yield is limited. By using a debranching enzyme in combination with glucoamylase, rate of alpha-1,6 hydrolysis is increased relative to reversion and a higher maximum dextrose yield is achieved. In this manner, isoamylase from *Pseudomonas amyloderamosa* and pullulanase from *Klebsiella planticola* (formerly identified as *K. pneumoniae, K. aerogenes* or *Aerobacter aerogenes*) have been found to have a positive effect on maximum dextrose yield. However, a maximum operational temperature of about 55°C (Table I) has prevented the use of these enzymes at the normal saccharification temperature of 60°C (Teague and Brumm, 1991).

More recently, a pullulanase derived from *Bacillus acidopullulyticus* has been found to be sufficiently stable for use at 60°C (Nielsen et al, 1985) and the enzyme has been commercialized. The use of this enzyme in saccharification provides an increase in dextrose yield of 0.5 to 1.5% (Jensen and Norman, 1984).

In the last few years, a completely new type of enzyme has been developed that provides the same advantages as pullulanase, but by a unique hydrolysis/transferase mechanism (David et al, 1987). This enzyme, produced from *Bacillus megaterium*, acts on highly branched oligosaccharides that are only slowly hydrolyzed by glucoamylase and accumulate during the saccharification process. As shown in Fig. 2, the *B. megaterium* amylase (BMA) has the ability to degrade the branched oligosaccharides to panosyl units which are then transferred to dextrose. The maltotetraose saccharide formed (6^3-alpha-glucosylmaltotriose) is readily hydrolyzed by glucoamylase to dextrose. The combined action of BMA and glucoamylase provides a number of possible advantages including increased dextrose yield, shorter reaction time, lower glucoamylase dosage and higher solids operation (Hebeda et al, 1988).

BMA also catalyzes a variety of hydrolytic and transfer reactions which may have potential use in other applications. For instance, BMA hydrolyzes alpha-1,4 glycosidic linkages in a wide range of substrates such as starch,

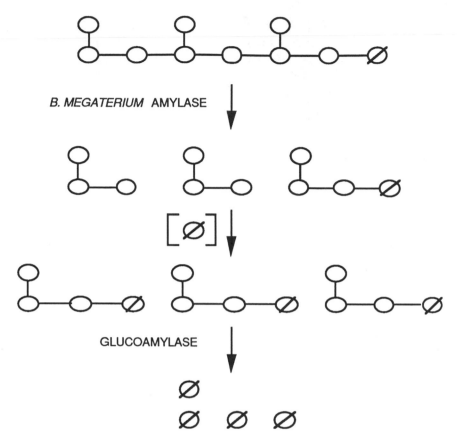

Fig. 2. Mechanism of BMA action during saccharification. Horizontal and vertical lines between dextrose units represent alpha-1,4 and alpha-1,6 linkages, respectively.

glycogen, pullulanase and cyclodextrins (Brumm et al, 1991a). The enzyme also catalyzes transfer reactions in which glucose or glucosides act as acceptor molecules (Brumm et al, 1991b).

Baking Enzymes

As discussed above, a high priority has been placed on the development of increasingly thermostable amylases for starch liquefaction. The opposite trend, however, has been followed in the baking industry where reduced enzyme stability is desired.

The action of alpha-amylases as anti-staling agents has been recognized for years. Amylases have the ability to disrupt the starch network formed during the production of baked goods, thereby inhibiting the rate of starch polymer association (retrogradation) and reducing the rate of product firming.

Traditional alpha-amylases from bacterial (*B. subtilis*) and fungal (*Aspergillus oryzae*) sources are used in baking for various purposes. However, neither is well suited as an anti-staling agent (Hebeda et al, 1990) because of an optimum temperature that is either too high (bacterial amylase) or too low (fungal amylase) (Table III).

The inherent high thermostability of bacterial alpha-amylase from *B. subtilis* can cause excessive hydrolysis during the baking process or even in the final product, resulting in an undesirable gummy or sticky texture. Conversely, the inadequate thermostability of *A. oryzae* amylase does not permit sufficient hydrolysis during starch

TABLE III
Optimum Temperature for Baking Enzymes

Enzyme Type	Enzyme Source	Temp., °C
Alpha-Amylase	*B. subtilis*	75+
Alpha-Amylase	*A. oryzae*	55
Alpha-Amylase	*A. niger*	70
Alpha/Transferase	*B. megaterium*[a]	70
Maltogenic	*B. stearothermophilus*[a]	70

[a]The gene for the enzyme has been cloned into and is produced by *B. subtilis.*

gelatinization to achieve an anti-staling affect.

Attempts have been made to increase the effectiveness of these conventional enzymes. For instance, the use of a debranching enzyme in combination with a bacterial amylase has been reported to be effective in hydrolyzing low molecular weight branched dextrins responsible for causing gumminess (Carroll et al, 1987). In another example, the stability of a fungal amylase was reportedly increased by mixing the enzyme with a sugar solution prior to use to achieve a stabilizing affect (Cole, 1982). Commercial products based on these two systems have been developed.

Recently, enzymes have been identified and commercialized that exhibit the proper inherent thermostability for baking. These amylases, derived from both fungal and bacterial sources, are often referred to as "intermediate temperature stability" (ITS) enzymes since they exhibit thermostability properties between those of traditional fungal and bacterial amylases (Hebeda et al, 1990). The ITS amylases exhibit maximum activity at about 70°C compared to 55°C for conventional *A. oryzae* fungal amylase and 75°C or higher for thermostable bacterial amylases (Table III).

The difference in stability is exemplified in Fig. 3 where temperature activity profiles of conventional amylases are compared to that of a typical ITS enzyme produced by *Aspergillus niger* (Hebeda et al, 1991). The *A. oryzae* alpha-amylase does not retain sufficient activity above the gelatinization temperature of starch and, therefore, does not provide significant anti-staling activity. The *B. subtilis* amylase maintains activity at high temperature and continues to act throughout the baking stage and in the final product. The ITS amylase, however, exhibits maximum activity above the starch gelatinization point, but is rapidly inactivated at higher temperatures.

The difference in enzyme stability was confirmed by determining the level of enzyme that survives oven temperatures in actual baking tests. Dough was prepared containing thermostable bacterial alpha-amylase (from

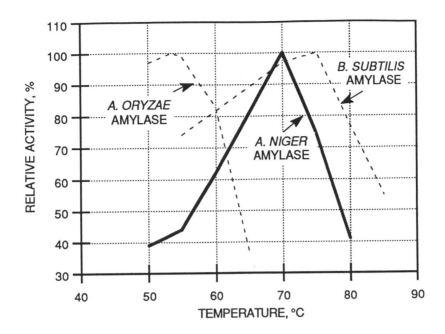

Fig. 3 Effect of temperature on enzyme activity.

either *B. subtilis* or *B. stearothermophilus*) or an ITS alpha-amylase (from *A. niger*). Bread crumb from final products was assayed for residual enzyme activity. It was determined that the ITS enzyme is completely inactivated in the oven, whereas, thermostable alpha-amylases from *B. subtilis* or *B. stearothermophilus* retain 100% activity under the same conditions. As a result, the ITS enzyme does not develop a gummy texture in the final product but does provide a significant increase in baked good shelf-life.

As an example, the firmness of white pan bread during several days storage is shown in Fig. 4 (Hebeda et al, 1991). Compared to the firmness of the control bread (no enzyme) at two days, the use of the *A. niger* ITS enzyme increases shelf-life by 1.5 to 4 days depending on dosage.

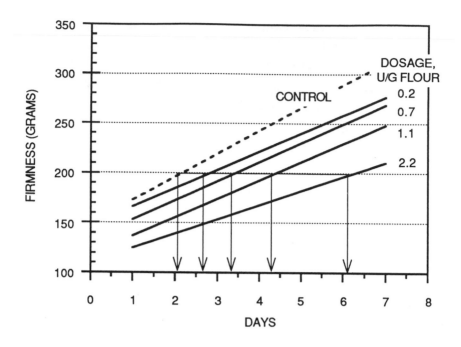

Fig. 4. Firming rate of white breads made with varied levels of an *A. niger* intermediate temperature stability alpha-amylase added to the sponge stage of a sponge/dough process. Dosage given in units per gram total flour.

Enzymes with temperature stability profiles similar to that of the *A. niger* ITS amylase have also been obtained from bacterial sources. These include an amylase from *B. megaterium* (Hebeda et al, 1991) and a maltogenic amylase (LaBell, 1991) presumedly from *B. stearothermophilus* (Outtrup, 1986). Both of these enzymes have been cloned into *B. subtilis* for commercial production.

In another interesting concept, a blend of an *A. niger* ITS enzyme and a conventional *B. stearothermophilus* thermostable alpha-amylase has been developed for baking purposes (Bowles, 1991). In this case, the very low level of thermostable enzyme minimizes the potential for gumminess in normal operation, but provides a synergy with the ITS enzyme for anti-staling.

Brewing Enzymes

Starch hydrolyzing enzymes are often added to the fermentation stage of the brewing process to increase the level of fermentable saccharides utilizable by yeast (Hebeda and Styrlund, 1989). In this way, a greater degree of fermentation is achieved and residual carbohydrate level is reduced resulting in a lower calorie, or "light" beer. Typically, a soluble glucoamylase is used to convert residual dextrins to dextrose that is then fermented by the yeast. The soluble enzyme must be inactivated by a pasteurization step to eliminate the possibility of continued action during beer storage.

In some operations, such as the production of draft light beer, the pasteurization step is eliminated and soluble enzyme, therefore, is not inactivated. A potential solution to this problem is the development of insolubilized enzymes that can be removed from the beer during the yeast filtration step.

Development of an immobilized glucoamylase for this purpose has been recently reported (Boyle, 1991). The product is produced by oxidation of the enzyme with periodic acid followed by precipitation with polyethylenimine. The insoluble enzyme is then washed to remove residual iodate and concentrated to the desired activity.

Studies have shown that the immobilized enzyme produces a light beer with the same carbohydrate profile and organoleptic properties as obtained with a soluble enzyme.

Safety tests have demonstrated the absence of toxic substances in the product. A food additive petition for the use of periodic acid and polyethylenimine as fixing agents for immobilization of glucoamylase was submitted to the Food and Drug Administration (FDA) in 1991.

Cloned Enzymes

Traditional mutation and screening techniques have been used for years to increase enzyme yields to a point where commercial production is economically feasible. These techniques, although often successful, involve a hit or miss approach. Through the recent advent of recombinant DNA (rDNA) technology, the ability to transfer specific genetic material from one organism to another has become common place. As a result, several starch hydrolyzing enzymes have been cloned into hosts that provide benefits in commercial production.

A typical cloning process in which a *B. stearothermophilus* alpha-amylase gene was transferred into a *B. subtilis* host is shown in Fig. 5 (Zeman and McCrea, 1985). The plasmid borne *B. stearothermophilus* gene was removed from the natural host and transferred into *Escherichia coli* using a vector. The exact location of the gene was determined, unnecessary *B. stearothermophilus* DNA removed, and a plasmid was used to clone the gene into an asporogenic strain of *B. subtilis*. The resultant DNA material contained needed information for expression and secretion of the protein. Subsequent studies involved characterization of the cloned enzyme to show identity to the natural enzyme and feeding studies to confirm safety of the product. These tests showed that the alpha-amylase from the rDNA strain was equivalent to the natural enzyme and established the safety of the product.

In 1984, this enzyme was the subject of the first petition filed with the FDA for affirmation of GRAS status for a genetically engineered food ingredient (Table IV). In 1986, the petition (CPC International, Inc. 1986) became the first accepted for filing by the FDA for a recombinant DNA derived food ingredient (Teague et al, 1990).

Several of the new enzymes discussed earlier have also been cloned and petitions for GRAS status accepted for filing by the FDA (Table IV). These include a *B. megaterium* amylase used in saccharification and baking and cloned into *B. subtilis* (Enzyme Bio-Systems Ltd., 1988), a *B. stearothermophilus* maltogenic enzyme used in baking

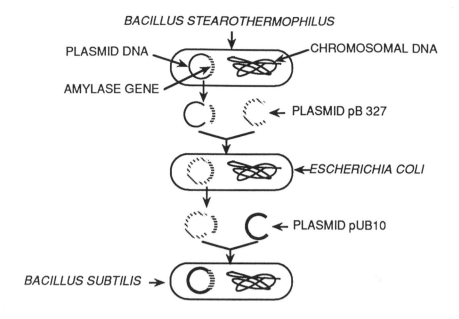

Fig. 5. Cloning of the *Bacillus stearothermophilus* alpha-amylase gene.

TABLE IV
GRAS Petitions Accepted for Filing by The FDA for rDNA
Derived Starch Hydrolysis Enzymes

	Organism	
Enzyme	Natural/Host	Year
Alpha-Amylase	*B. stear./B. subtilis*	1986
Alpha/Transferase	*B. megaterium/B. subtilis*	1988
Maltogenic	*B. stear./B. subtilis*	1990
Alpha-Amylase	*B. stear./B. licheniformis*	1991

and cloned into *B. subtilis* (Novo Laboratories, Inc., 1990) and a *B. stearothermophilus* alpha-amylase used for starch liquefaction and cloned into *B. licheniformis* (Novo Nordisk Bioindustrial, Inc., 1991).

It is a certainty that rDNA technology will continue to provide new and useful starch hydrolyzing enzymes for commercial applications in the future.

LITERATURE CITED

Antranikian, G., Koch, R. and Spreinat, A. 1990. Novel hyperthermostable alpha-amylase. International patent application PCT/DK90/00074.

Anonymous. 1991. Per capita bread consumption to increase 2% through '96. Milling & Baking News. 69:1.

Boidin, A. and Effront, J. 1917a. Process for treating amylaceous substances. U. S. patent 1,227,374.

Boidin, A. and Effront, J. 1917b. Process of manufacturing diastases and toxins by oxidizing ferments. U. S. patent 1,227,525.

Bowles, L. K. 1991. Enzyme composition for retarding staling of baked goods. U. S. patent 5,059,430.

Boyle, P. 1991. A new enzyme for brewing. Presented at the MBAA Conference, Guadalajara, Mexico.

Brown, S. H., Costantino, H. R. and Kelly, R. M. 1990. Characterization of amylolytic enzyme activities associated with the hyperthermophilic archaebacterium *Pyrococcus furiosus*. Appl. and Environ. Microbiol. 56:1985.

Brumm, P. J., Hebeda, R. E. and Teague, W. M. 1991a. Purification and characterization of the commercialized, cloned *Bacillus megaterium* alpha-amylase. Part 1: Purification and hydrolytic properties. Starch/Stärke 43:315.

Brumm, P. J., Hebeda, R. E. and Teague, W. M. 1991b. Purification and characterization of the commercialized, cloned *Bacillus megaterium* alpha-amylase. Part 2: Transferase properties. Starch/Stärke 43:319.

Carroll, J. O., Boyce, C. O. L., Wong, T. M. and
Starace, C. A. 1987. Bread antistaling method.
U. S. patent 4,654,216.

Carroll, J. O. and Swanson, T. R. 1990. Starch liquefaction
with alpha amylase mixtures. U. S. patent 4,933,279.

Cole, M. S. 1982. Antistaling baking composition.
U. S. patent 4,320,151.

CPC International, Inc. 1986. Filing of a petition for
affirmation of GRAS status (petition No. GRASP
4G0293). Fed. Reg. 51:10571.

Dale, J. K. and Langlois, D. P. 1940. Sirup and method of
making the same. U. S. patent 2,201,609.

David, M.-H., Günther, H. and Röper, H. 1987. Catalytic
properties of *Bacillus megaterium* amylase.
Starch/Stärke 39:436.

Enzyme Bio-Systems Ltd. 1988. Filing of a petition for
affirmation of GRAS status (petition No. GRASP
7G0328). Fed. Reg. 53:16191.

Hebeda, R. E., Styrlund, C. R. and Teague, W. M. 1988.
Benefits of *Bacillus megaterium* amylase in dextrose
production. Starch/Stärke 40:33.

Hebeda, R. E. and Styrlund, C. R. 1989. Concurrent
saccharification/fermentation of enzymatic corn syrups
in light beer production. Biotech. Letters 11:825.

Hebeda, R. E., Bowles, L. K. and Teague, W. M. 1990.
Developments in enzymes for retarding staling of baked
goods. Cereal Chem. 35:453.

Hebeda, R. E., Bowles, L. K. and Teague, W. M. 1991. Use
of intermediate temperature stability enzymes for retarding staling in baked goods. Cereal Chem. 36:619.

Henderson, W. E. and Teague, W. M. 1988. A kinetic model
of *Bacillus stearothermophilus* alpha-amylase under
process conditions. Starch/Stärke 40:412.

Jensen, B. F. and Norman, B. E. 1984. *Bacillus
acidopullulyticus* pullulanase: Application and
regulatory aspects for use in the food industry. Process
Biochem. Aug:129.

LaBell, F. 1991. Enzyme for bread extends freshness.
Food Proc. April:118.

Miller, B. S., Johnson, J. A. and Palmer, D. L. 1953. A comparison of cereal, fungal, and bacterial alpha-amylases as supplements for breadmaking. Food Tech. January:38.

Nielsen, G. C., Diers, I. V., Outtrup, H. and Norman, B. E. 1985. Debranching enzyme product, preparation and use thereof. U. S. patent 4,560,651.

Novo Laboratories, Inc. 1990. Filing of a petition for affirmation of GRAS status (petition No. GRASP 7G0326). Fed. Reg. 55:8772.

Novo Nordisk Bioindustrial, Inc. 1991. Filing of a petition for affirmation of GRAS status (petition No. GRASP 0G0363). Fed. Reg. 56:32435.

Outtrup, H. 1986. Maltogenic amylase enzyme, preparation and use thereof. U. S. patent 4,604,355.

Payen, A. and Persoz, J. F. 1833. Ann. Chim. (Phys.) 53:73.

Reeve, A. 1992. Starch hydrolysis:processes and equipment. pp. 79-120. In: Starch hydrolysis products: Worldwide technology, production and applications. F. W. Schenck and R. E. Hebeda, eds. VCH. New York, NY.

Rosendal, P., Nielsen, B. H. and Lange, N. K. 1979. Stability of bacterial alpha-amylase in the starch liquefaction process. Starch/Stärke 31:368.

Slott, S. and Madsen, G. B. 1975. Procedure for liquefying starch. U. S. patent 3,912,590.

Starnes, R. L. 1990. Industrial potential of cyclodextrin glycosyl transferases. Cereal Foods World. 35:1094.

Stone, I. 1952. Retarding the staling of bakery products. U. S. patent 2,615,810.

Takamine, J. 1894. Process of making diastatic enzyme. U. S. patent 525,823.

Teague, W. M., Metz, R. J. and Zeman, N. W. 1990. Safety evaluation of genetically engineered enzymes for food use. pp. 311-323. In: Biotechnology and food safety. D. D. Bills and S. Kung, eds. Butterworth-Heinemann, Boston.

Teague, W. M. and Brumm, P. J. 1992. Commercial enzymes for starch hydrolysis products. pp. 45-77. In: Starch hydrolysis products: Worldwide technology, production and applications. F. W. Schenck and R. E. Hebeda, eds. VCH. New York, NY.

Wallerstein, L. 1939. Ind. Eng. Chem. October:1218.

Wolnak, B. 1990. Forty years attacking barriers. pp. 3-25. In: Industrial use of enzymes. B. Wolnak and M. Scher, eds. Johnson Graphics. Decatur, MI.

Yoshizumi, S., Itoh, H. and Kokubu, T. 1986. Amami no keifu to sono kagaku (Genealogy and science of sweeteners, translated by S. Enokizono), pp. 24-25. Korin Publishing Co., Tokyo.

Zeman, N. W. and McCrea, J. M. 1985. Alpha-amylase production using a recombinant DNA organism. Cereal Foods World. 30:777.

CHARACTERIZATION OF STARCH NETWORKS
BY SMALL STRAIN DYNAMIC RHEOMETRY

Costas G. Biliaderis
Food Science Department, University of Manitoba
Winnipeg, Manitoba, Canada R3T 2N2

INTRODUCTION

There has been considerable research activity over the last decade in the area of fundamental mechanical properties of aqueous starch and other polysaccharide dispersions and gels with the objective to identify relationships between physical properties and structure of these materials. Knowledge of the thermomechanical behaviour of polysaccharides in pure and composite food systems is important in understanding and predicting texture, flow properties during processing as well as shelf-life and other quality attributes of carbohydrate-based products. In most cases, aqueous polysaccharide dispersions, even at low polymer concentrations, exhibit a rheological response which reflects both solid-like and liquid-like characteristics (viscoelastic materials). Measurements on viscoelasticity can thus provide insights into the structure formation processes involving polysaccharides with respect to both short- and long-range structural order. Such information can be useful for improved processing performance and end-product quality.

An important first step in the description of individual polysaccharide networks is the recognition that covalent cross-linking or formation of tertiary structures from fluctuating disordered chains in solution are required to establish a three dimensional network; the spaces not filled by the polymer chains are occupied by the solvent. The ability of the chains to interchange at any point in the network, and the lifetime of the cross-links depend on the nature of the cross-link points or "junction zones". Covalently cross-linked polymers have permanent structures which remain coherent in the presence of unlimited solvent. In contrast, the junction zones in non-covalent polysaccharide networks could vary in their stability depending on the conformational state of the polymer chains and the extent of associations between ordered chain segments which can yield higher levels of structural organization (aggregates) (Clark and Ross-Murphy, 1987; Ross-Murphy, 1987; Oakenfull, 1987). Factors which affect interchain associations between polysaccharide molecules are the structural regularity of chains, nature and extent of branching or substitution, solvent quality, and presence of specific co-solutes (e.g. ions, sugars, other biopolymers) (Rees, 1969; Morris and Norton, 1983; Morris, 1985, 1986; Ross-Murphy, 1984). In such physically cross-linked systems, the density and lifetime of the junction zones will govern the mechanical properties of the network and its responses to applied stress or strain (deformation). In this context, one has to realize that measurements of rheological properties are constrained to certain limited deformation levels and time scales; the stress-strain relationships depend not only on temperature but on time and polysaccharide concentration. Consequently, any assessment of the mechanical properties of hydrated gel networks (e.g. starch pastes and gels) must be carried out under carefully chosen conditions of stress or strain and temperature-time regimes. This is usually accomplished by employing oscillatory (dynamic) tests which are non-destructive to the structure of the specimen tested. With respect to the influences of experimental temperature and time on the mechanical properties of polymers, some simplification arises from the well known time-temperature superposition principle (Ferry, 1980); i.e. data obtained over accessible time scales at various temperatures can be used to prepare a master curve at a fixed temperature over a very broad time scale. This time-temperature superposition principle can be expressed analytically in terms of the WLF (Williams-Landel-Ferry) equation which is generally applicable to amorphous polymers (Ferry, 1980).

Recent reviews have discussed the structural and mechanical properties of biopolymer gels as well as the application of small-deformation rheological testing to characterize biopolymer networks (Clark and Ross-Murphy, 1987; Oakenfull, 1987; Clark, 1987; Kokini and Plutchok, 1987; Bell, 1989; Hamann et al., 1990). The present chapter will be limited to recent studies on rheological characterization of starch networks by dynamic rheometry; these systems represent physical gels formed by molecular ordering and subsequent chain aggregation/crystallization. In this respect, emphasis will be placed on the advantages of using small strain mechanical testing to unravel structural changes and interactions of starch macromolecules as affected by composition, polymer structure and temperature-time history. Complementary analytical data are also presented to enhance our understanding of the molecular processes involved in structure formation.

SMALL AMPLITUDE SINUSOIDAL SHEAR MEASUREMENTS

Although the use of small strain rheometry in the study of viscoelastic properties of starch is not new (Myers et al., 1962; Myers and Knauss, 1965), dynamic mechanical measurements on starch and other polysaccharide systems are now widely practised due to the development of reasonably priced commercial instruments. Modern rheometers, capable of dynamic rheological testing, can distinguish between the viscous and elastic components of a viscoelastic specimen, as well as quantify stress, deformation, and rate of deformation in fundamental units. These instruments also allow a description of the frequency and strain dependence of the viscoelastic parameters over a frequency range of two to three decades. Moreover, precise control of temperature makes feasible to perform experiments over a range of static or dynamic temperature-time conditions. Structure formation or breakdown in macromolecular systems, as a result of conformational changes, state transitions or interactions with other constituents, can be thus probed by monitoring continuously the rheological responses of the sample under small strain.

When a material is deformed sinusoidally at a frequency ω (rad·s^{-1}) the shear strain can be expressed as

$$\gamma(t) = \gamma_0 \sin(\omega t) \quad (1)$$

where γ_0 is the strain amplitude (maximum deformation), ω is the angular frequency and t is the time (Fig. 1). The strain rate, or shear rate, will be the first derivative of strain with respect to time,

$$\frac{d\gamma}{dt} = \dot{\gamma} = \omega\gamma_0\cos(\omega t) \quad (2)$$

The shear stress function, σ, will also be sinusoidal and have some maximum value, σ_0 (Fig. 1). For an ideal elastic material (Hookean solid), the stress will be exactly in phase with the imposed strain, while for a purely viscous liquid (Newtonian fluid), σ will be exactly 90° out of phase with γ. If the material is viscoelastic, such as a polysaccharide solution or gel, the phase angle (δ) between the σ and γ functions varies within 0° and 90°. The shear stress can be then written as

$$\sigma(t) = \sigma_0 \sin(\omega t + \delta) \quad \text{or}$$
$$\sigma(t) = \sigma_0 [\sin(\omega t)\cos\delta + \sin\delta\cos(\omega t)] \quad (3)$$

For linear viscoelastic materials, where σ_0 is proportional to γ_0, equation (3) may be split into an elastic and a viscous component according to:

$$\sigma(t) = \gamma_0 [\sigma_0/\gamma_0 \sin(\omega t)\cos\delta + \sigma_0/\gamma_0 \sin\delta\cos(\omega t)] \quad (4)$$

The first part of equation (4) constitutes the part of stress in phase with the strain (elastic component), while the second term constitutes the part of stress out of phase with the strain (viscous component).

90

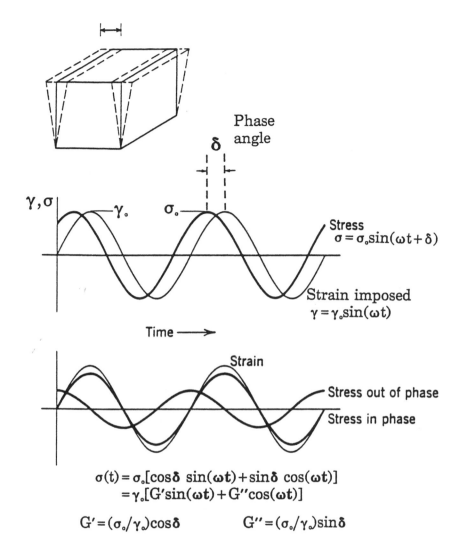

Fig. 1. Typical time profiles of sinusoidally imposed deformation, resultant stress and its in-phase and out-of phase stress components characterizing a small strain dynamic experiment. Adapted from Ferry (1980), with permission from John Wiley & Sons.

The ratio of in-phase stress to strain is the storage modulus (G'), and the 90° out-of-phase stress to strain is the loss modulus (G"):

$$G' = (\sigma_o/\gamma_o) \cos\delta \qquad (5)$$
$$G" = (\sigma_o/\gamma_o) \sin\delta \qquad (6)$$

The stress function, as described by equation (4), can be thus written as

$$\sigma(t) = \gamma_o \left[G'\sin(\omega t) + G"\cos(\omega t) \right] \qquad (7)$$

The storage modulus is a measure of the energy stored in the material on sinusoidal deformation and recovered per cycle. On a molecular basis, the magnitude of G' is dependent upon what rearrangements can take place within the period of oscillation (Ferry, 1980), and is taken as an indicator of the solid or elastic character of the material. The loss modulus is a measure of the energy dissipated or lost (as heat) per cycle of deformation; this behaviour is usually exhibited by non-permanent cross-linked networks which result from chain entanglement and undergo high levels of molecular rearrangements on deformation. Strong interactions between chains (i.e. having long relaxation times) contribute to G', whereas weak, rapidly relaxing bonds contribute only to G". Bonds with relaxation times in the time scale of the measurements will contribute to both G' and G".
    It should be kept in mind that linear viscoelasticity assumes that the stress is linearly proportional to strain and this condition must be verified experimentally. The linear viscoelastic dynamic moduli are functions of frequency and as such are sensitive probes of the structure of biopolymer solutions or gels (Morris, 1984; Clark and Ross-Murphy, 1987). For a dilute polysaccharide solution, the elastic stress relaxes and viscous stress dominates (G">G'), particularly at low frequencies (Fig. 2a). Both moduli increase with increasing frequency, but G' increases more quickly; i.e. with increasing frequency (short-time scale), intramolecular motions become more important, compared to molecular rearrangements, and thus G' approaches G"

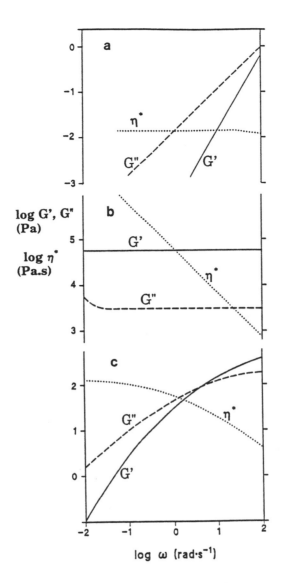

Fig. 2　Frequency dependence of G', G'', and η^* for: (a) a dilute polysaccharide solution (5% w/w dextran); (b) a strong gel (2% w/v agar); (c) a concentrated polysaccharide solution (5% w/v λ-carrageenan). Adapted from Morris (1989).

at high frequencies. For a gel system (Fig. 2b), in which the junction zones are stable on a relatively long time scale, the stress cannot relax and both moduli show very little dependence on frequency; G' is also very much greater than G" throughout the usually accessible frequency range (10^{-2} to 10^2 rad·s^{-1}). In the case of an entanglement network system (e.g. concentrated polysaccharide solution), G" is greater than G' at low frequencies; i.e. at long-time scales molecular rearrangements are feasible (resulting in flow) to accommodate the applied strain (Fig. 2c). At high frequencies, however, such transient networks behave like a gel with G'>G"; i.e. interchain entanglements do not have sufficient time to come apart within the period of one oscillation cycle.

A viscoelastic system is also characterized by two dynamic viscosity coefficients, η' and $\eta"$, again related to the in phase and out of phase stresses:

$$\eta' = G"/\omega, \qquad \eta" = G'/\omega \qquad (8)$$

The complex viscosity, η^*, is defined as a ratio of total stress to the frequency of oscillation

$$\eta^* = G^*/\omega = (G'^2 + G"^2)^{1/2}/\omega \qquad (9)$$

For many disordered polysaccharide solutions the frequency dependence of η^* and the shear rate dependence of η are superimposable (Morris et al., 1981); this empirical relationship was first reported by Cox and Merz (1958). However, for systems with long-range ordering and structure (e.g. polymer gels or highly filled polymer melts), η^* is higher than η.

Another parameter which is useful in characterizing the physical state of a viscoelastic material is the loss tangent or tanδ. This dimensionless parameter is the ratio of the energy lost to the energy stored for each cycle of deformation; tanδ = G"/G'. The tanδ is a more sensitive parameter than G' and G" in probing changes in the viscoelastic character of a polymer network.

VISCOELASTIC BEHAVIOUR OF STARCH SYSTEMS

Starch is a mixture of two structurally distinct polysaccharides, amylose (mainly linear polymers of α-D-glucose linked through 1→4 linkages) and amylopectin (branched polymers containing short DP chains which are interlinked via α-(1→6) linkages at certain anhydroglucose moieties). For most granular starches, amylose accounts for some 20-25% of the polymeric starch components, although starches of unusually low (e.g. waxy, about 1%) or high (e.g. amylomaize, 50-60%) amylose content are also found. The polydispersed starch macromolecules are densely packed as partially crystalline arrays of left-handed co-axial double helices in the granules which range in size between 1 and 100μ (Banks and Greenwood, 1975; French, 1984). Evidence of structural order in granular starch is provided by optical birefringence, X-ray diffraction, electron microscopy, small angle neutron and laser light scattering, differential scanning calorimetry, and spectroscopic methods such as ^{13}C-NMR (Blanshard, 1987). Starch owes much of its functionality (thickening agent, flavour carrier, and binder in food systems) to these two polysaccharide components as well as to their physical organization in the granule.

When aqueous starch suspensions are heated above the gelatinization temperature, irreversible swelling of the granules occurs along with a concomitant loss of structural order. As the granules continue to expand, amylose leaches out into the aqueous intergranular phase; these processes bring about a substantial increase in viscosity. If the starch concentration is high enough (usually > 6.0% w/w), the mixture of swollen granules and exuded amylose behaves as a viscoelastic paste. Upon cooling, the polymer mixture sets as a gel in which further thickening and rigidity development occur during storage as a result of chain aggregation (retrogradation). From the above description, starch gels can be regarded as a polymer composite in which swollen granules (particles) are embedded in and reinforce a continuous matrix of entangled amylose molecules (Ring, 1985). Consequently, the mechanical properties of starch gels are influenced by the rheological characteristics of the amylose gel matrix, the volume fraction and rigidity of gelatinized granules as well as the interactions between the starch components (Eliasson, 1986a). Granule size, amylose/amylopectin ratio, physical organization of the granule, minor constituents (granular lipids and phosphorus), presence of solutes (salts, sugars), pH, starch concentration, and shear-temperature-time regimes, are all important contributing factors to the viscoelasticity of starch dispersions and account for the variation in rheological responses frequently reported for a particular material. A comprehensive discussion on the parameters affecting the rheological behavior of starch can be found in recent reviews (Launay et al. 1986; Doublier, 1990). Because of the complex nature of starch gels, it is convenient to examine the rheological properties of amylose and amylopectin as well as the mechanisms underlying the gelation processes of these polymers separately in order to elucidate their contribution to starch rheology.

Amylose Gelation

The evolution of shear modulus (G') as a function of time (25°C, 5-10% strain, 5 rad$\cdot$s^{-1}) for amylose solutions in 0.2M KCl is shown in Figure 3a (Doublier and Choplin, 1989). The G'-time profiles are typical for amylose gelation, where an initial rapid rise in G' is followed by a much slower increase at longer times (pseudoplateau region). The initial rise in modulus is faster and a plateau value is attained at shorter times with increasing amylose concentration (Clark et al., 1989; Doublier and Choplin, 1989). The kinetics of G' development are also dependent on the chain length of amylose. For nearly monodisperse amyloses of DP (degree of polymerization) between 250 and 1100, Clark et al. (1989) observed that plateau G' values are attained faster with shorter chain length amyloses. Moreover, amyloses of very high DP (2550, 2800) exhibited very slow development of modulus under identical gelation conditions. For such long amylose chains, it is possible that initial formation of

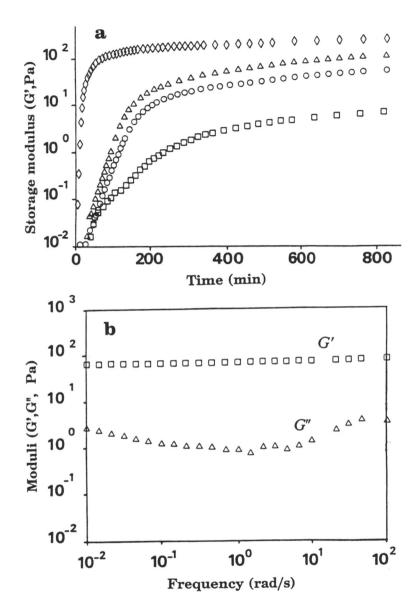

Fig. 3. Top, evolution of shear storage modulus for amylose solutions of varying polymer concentration in 0.2 M KCl at 25°C (□, 1.03%; ○,1.33%; △,1.48%; ◇, 1.78%). Bottom, frequency dependence of dynamic moduli (G',G") for 1.33% amylose gel (0.2 M KCl; gel curing condition 25°C-15 h). Adapted from Doublier and Choplin (1989).

relatively few cross-links significantly retards chain mobility, thereby slowing down any subsequent cross-linking and increases in modulus. These data are consistent with the view that amylose gel networks are non-equilibrium, kinetically constrained structures; the very slow increases in G' over longer times may reflect contributions from diffusion-controlled chain aggregation processes. The non-equilibrium nature of the system is evident from the data of Ellis and Ring (1985); rapidly quenched amylose gels (3.4% w/w, DP 3080, cure time 24 h) to 25°C had a limiting modulus value of 3500 Pa, whereas a slowly cooled gel (over 4 h) attained a modulus of 5300 Pa. The mechanical properties of amylose gels are thus dependent on polymer concentration, molecular size and gel curing conditions (time-temperature cooling regimes), presence of electrolytes, etc. Once formed, amylose gels are thermally stable networks up to the temperature range of 140-160°C (Biliaderis et al., 1985); indeed, the shear modulus of amylose gels could not be reversed by heating at 100°C (Miles et al., 1985b).

Once the modulus of amylose gels has reached a certain plateau value, the G' and G'' are found essentially independent of frequency, as shown in Figure 3b (Doublier and Choplin, 1989). Moreover, the G' value is at least one order of magnitude greater than G''; i.e. a typical mechanical spectrum of a "true" gel. The slight dip in G'' at ~1 rad·s^{-1} has been previously reported for other biopolymer gels including gelatin (Clark et al., 1983). According to Clark et al. (1989), 2% amylose gels (DP: 250-2550) exhibit linear viscoelasticity under small deformations in the range of 1-5% strain. Doublier and Choplin (1989) have reported even greater range of strains (5-10%) for the linear viscoelastic domain of amylose gels (DP: 2400, amylose concentration > 1.03%).

The plateau values of G' for amylose gels are strongly dependent on polymer concentration. Figure 4 summarizes some of the published data in this respect. Early rheological investigations on pea amylose (DP 3080) by Ellis and Ring (1985) identified a minimum amylose concentration for gelation (C_o) of

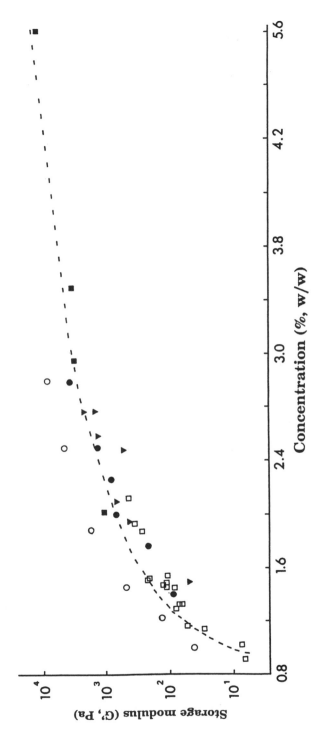

Fig. 4. Experimental plateau values of G' vs. concentration for amylose gels: ◄ DP ~ 3080 (Ellis and Ring, 1985); ○ DP ~ 1100, ● DP ~ 660 (Clark et al., 1989); □ DP ~ 2400 (Doublier and Choplin, 1989); ■ DP ~ 1150 (Biliaderis and Zawistowski, 1990). Reproduced with permission.

~1.0%; C_0 is defined as the critical concentration below which no macroscopic gel is formed. The experimental values presented in Figure 4 (although represent amylose samples of different origin and molecular weights, and different methods of preparation) illustrate the marked concentration dependence of modulus at the lowest range of concentrations; when G' is extrapolated on the horizontal axis it yields a concentration of 0.9%. Doublier and Choplin (1989) suggested that this value can be taken as an estimate of the lowest concentration of amylose where gelation is possible (C_0). Previous studies by Clark et al. (1989) with amyloses of varying chain length (DP: 250-2550) did indicate similar C_0 values (0.8-1.1%), thus suggesting very little dependence of C_0 on the molecular size of amylose. These data clearly demonstrated that amylose gelation can occur at concentrations below that of coil overlap (C^*), a parameter determined from viscosity measurements. Thus, molecular entanglements do not necessarily determine the critical gelling concentration of amylose as originally proposed by Miles et al. (1985a) and often found in the case of synthetic polymers for which overlap of macromolecules is a necessary prerequisite for gel formation (Boyer et al., 1985). The non-equivalence of C_0 and C^* has been also reported for globular proteins by Clark and Ross-Murphy (1987), who noted that gelation at suboverlap concentrations is possible when the cross-links have finite energy.

With respect to the mechanism of network self-organization from amylose solutions, a number of studies have provided recently direct kinetic evidence for the occurrence of several distinct processes (Miles et al., 1985a; Gidley and Bulpin, 1989; Clark et al., 1989; Gidley, 1989; Doublier and Choplin, 1989; Morris, 1990). Kinetic experiments at amylose concentrations between 0.3% (below this concentration precipitation of amylose occurs without gelation) and 2.0% (above this concentration gelation proceeds very fast, making difficult to distinguish the sequential stages of chain organization events) are usually conducted to resolve all steps leading to polymer self-assembly from the homogeneous sol. In this respect, many probes of structural order have been applied (turbidity measurements, dilatometry, light scattering, X-ray diffraction, optical rotation, NMR spectroscopy, and small strain rheometry) which respond differently to various levels of molecular organization.

For a polydisperse pea amylose preparation (DP 3080), the turbidity development (evidence for onset of chain aggregation) was found to slightly precede the G' growth (Miles *et al.*, 1985a). Similarly, Doublier and Choplin (1989) observed a concurrent rise in modulus and turbidity for amylose solutions (DP 2400) in 0.5M KCl. These results support the hypothesis that amylose gelation is initiated by phase separation in the homogeneous sol; i.e. demixing of the sol occurs yielding a polymer-rich phase interspersed within a solvent-rich phase. Depending on concentration, molecular size, solvent quality and quenching temperature, such demixing may be a relatively slow process. In the polymer-rich microphase chain ordering, involving at least some segments of amylose molecules (presumably in the form of double helical structures), gives rise to a tenuous network. Using both small angle-scattering (SAXS) and wide-angle diffraction X-ray techniques to probe the time-dependent molecular associations during gelation of amylose solutions, I'Anson *et al.* (1988) have suggested that nucleation followed by growth are the dominant processes for chain ordering in the polymer-rich phase. These events also mark the onset of the viscoelastic response of the network; further chain associations (growth) result in sharp increases of G'. The final period is characterized by a slowing rate of G' development. This step most likely reflects the development of crystallites (long-range order) in the amylose network. Crystallinity development (as monitored by the intensity of the 100 diffraction peak) is a substantially slower process than gelation and continues well after the G' has reached pseudoplateau values (Miles *et al.*, 1984, 1985a). Based on the turbidimetric and rheological data of Clark *et al.* (1989) on monodisperse amyloses, Morris (1990) has proposed that growth and coarsening (helix bundling) of amylose gel networks may develop in different ways depending on the molecular size and the conditions used during gelation. For high molecular weight amyloses, poor chain matching and associations over relatively short segment length occur, favouring the formation of a fine network that coarsens thereafter. In contrast, shorter chains may initially foster lateral associations of chains due to a better matching upon double helix formation; this process is followed by interlinking of the coarse aggregates to form a network.

The relationship between G' and amylose concentration shown in Figure 4 has been the subject of many investigations (Ring, 1985; Ellis and Ring, 1985; Doublier and Choplin, 1989; Clark et al., 1989; Biliaderis and Zawistowski, 1990). Using a Saunder-Ward gelometer and a pulse shearometer (200 Hz), Ring and co-workers (Ring, 1985; Ellis and Ring, 1985) have first reported a C^7 dependence of modulus on concentration within the range of 1.5-7.0% (amyloses from different sources with DP between 3080 and 7650). Our experimental findings (Biliaderis and Zawistowski, 1990) with potato amylose (DP 1150, concentration 1.9-8.8%, 0.2 Hz, and 2% strain) suggested a $C^{3.1}$ dependence. Using a mechanical spectrometer (10 rad·s^{-1}, 2% strain) Clark et al. (1989) reported exponent values of 5.9 (DP 1100), 4.7 (DP 600) and 4.4 (DP 300) for monodisperse amyloses within 1.0-3.0% concentration. The apparent inconsistencies in the observed exponent values among these studies may reflect the differences in polymer concentration range, molecular size and polydispersity of the amylose as well as conditions employed during gelation. As pointed out by Clark and Ross-Murphy (1987), such high exponent values represent a clear deviation from the theoretical relationship of $G' \propto C^{2.0}$ predicted for biopolymer gels. Such deviation, however, is not uncommon when the experimental G' values are obtained for concentrations where C/C_0 is less than 10; indeed, this seems to be the case for most published data on amylose gels. These exponents should converge eventually to values close to 2.0 as the polymer concentration increases.

Dynamic rheological data are also available for the branched polymeric component of starch, amylopectin. In contrast to amylose, amylopectin gelation requires much higher polymer concentrations, well above the coil overlap concentration of this biopolymer ($C^* \sim 0.9\%$). Ring et al. (1987) observed gelation for amylopectin solutions (at 1°C) at concentrations as low as 10% (w/w), the gel taking, however, several weeks to approach a limiting value. With increasing polymer concentration gelation proceeded at faster rates. A linear relationship was also shown between shear modulus of amylopectin gels and concentration in the range 10-25%; amylopectin gels behaved as Hookean solids at strains less than 0.1 (Ring et al., 1987).

Immediately after gelatinization (100°C), the G' and G" of aqueous dispersions of waxy starch (40% w/w) exhibit strong dependence on frequency (Fig. 5, inset a). The overall kinetics of the gelation process, as monitored by small strain oscillatory rheometry (0.2 Hz, 2% strain), generally follow a sigmoidal G'-time curve. Upon attainment of a limiting modulus value, the gel modulus (G') is raised by at least one order of magnitude (compared to amylopectin melt) and both G' and G" become independent of frequency (Fig. 5, inset b).

To explore the gelation mechanism, similar experiments to those described for amylose were carried out by Ring et al. (1987) using several techniques to monitor structure development in amylopectin networks. While the turbidity of 20% amylopectin solutions (at 1°C) approached a limiting value after 4-5 days, structure formation as probed by dilatometry, differential scanning calorimetry (DSC) and dynamic rheometry (200 Hz) was a much slower process; i.e. volume changes (ΔV), enthalpy of gel melting (ΔH), and storage modulus (G') approached limiting values after 30-40 days. Furthermore, the ΔV, ΔH, and the later part of G' development followed parallel time-scales. X-ray diffraction data of aging amylopectin gels also indicated similar kinetic profile of B-crystal formation to that of ΔH

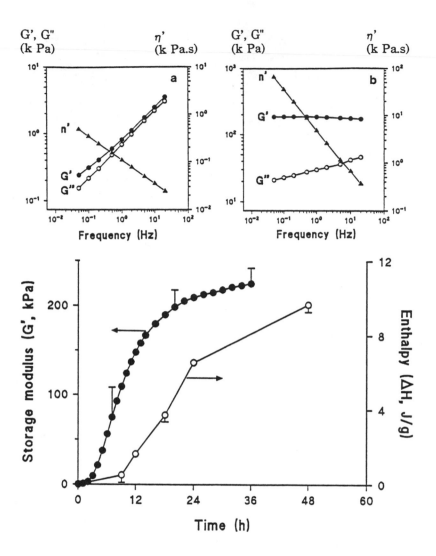

Fig. 5. Storage modulus (G') and melting enthalpy (ΔH) of aging waxy maize (amylopectin) starch gels at 40% (w/w). Measurements of G' at 0.2 Hz, strain < 2.0%, 8°C, and ΔH following storage at 6°C. Insets: (a) frequency dependence of G', G", and η' at the start; (b) frequency dependence following storage at 8°C for 36 h (author's unpublished data).

development. These results suggested that the increase of modulus in an amylopectin network is closely related to ordering and crystallization of the short DP (15-20) chains of the molecule (Ring *et al.*, 1987). The concurrent growth of G' and ΔH were also observed for 40% (w/w) amylopectin gels (Biliaderis and Zawistowski, 1990). However, while G' reaches almost to a limiting value after 36 h (6°C), ΔH continues to increase over a longer time-scale (Fig. 5). In would appear, therefore, that once the gel network is established, subsequent chain ordering and crystallization add very little to the rigidity of the gel. Since diffusion of macromolecules is hindered considerably in such high viscosity medium, these late structural changes in the aging network are likely of a localized nature (within the short DP chain bundles). The gelation of amylopectin is thermoreversible since the gel structure and crystallinity are abolished by heating at 100°C (Miles *et al.*, 1985b).

The results of studies on the temperature dependence of rigidity development of 40% (w/w) amylopectin and 5% (w/w) amylose gels are shown in Figure 6. The most important feature of these data is the strong temperature dependence of the rate of G' development in amylopectin gels compared to amylose. This implies that gelation of amylopectin follows nucleation kinetics in a manner typical of polymer crystallization in the presence of a diluent (Mandelkern, 1964). These observations add further support to the notion that while amylose gelation mainly involves rapid formation of double-helical junction zones, a network-based structure of amylopectin is due to reordering and crystallization of the short chains in this polymer. Crystallite melting and the loss of gel structure, with the accompanying changes in mechanical properties, are thus concomitant processes for the branched polymeric component of starch.

The botanical source and fine structure of amylopectin seem to have an influence on its gelling behaviour. Cereal amylopectins (from wheat, barley, and maize starches) were found to exhibit slower rates of modulus development than tuber (potato, canna) and

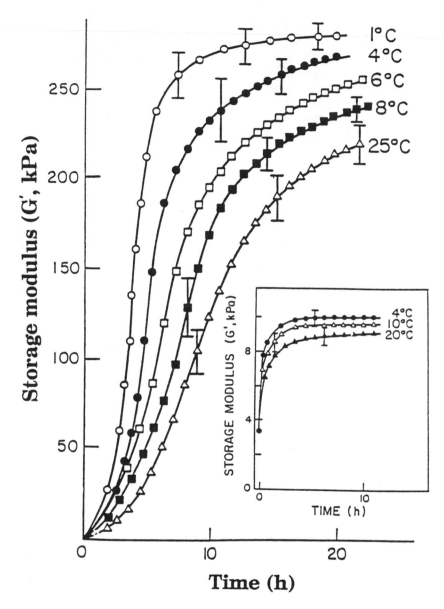

Fig. 6. The temperature dependence of storage modulus
(G') development in 40% (w/w) waxy maize amylopectin
gels. Inset shows the respective kinetics for 5%
(w/w) potato amylose (DP 1150) gels; strain < 2.0%,
frequency 0.2 Hz. Adapted from Biliaderis and
Zawistowski (1990).

pea starch amylopectins, presumably due to the shorter chain length of the former (Kalichevsky *et al.*, 1990). However, there was no simple relationship between the gel shear modulus and the structure of the amylopectin; for 25% gels in 0.05M NaCl, the limiting modulus values ranged between 8-36 kPa.

Amylopectin gelation is also influenced by the presence of small molecular weight sugars (Fig. 7). Understanding these effects is important for improving the texture and shelf life of high sugar content starch products (cakes, cookies, etc). Among the sugars tested (starch : sugar : water at 1:0.5:1.5), ribose, maltotriose, and sucrose suppressed the growth rate of G' and staling endotherm (ΔH of gel melting), while fructose and glucose appeared to accelerate these processes. It is also obvious from the data of Figure 7 that the relative ranking of sugars in promoting or retarding G' and ΔH development varies between the two techniques; e.g. sucrose vs. maltotriose and glucose vs. fructose. In this respect, one must note the different time scales over which the DSC and rheological measurements are made. Moreover, it is possible that the two techniques differ in their responses to the various levels of structural order present in the aging networks. Confirming earlier studies by Maxwell and Zobel (1978), I'Anson *et al.* (1990) reported reduced firmness (large deformation testing) and crystallinity of aging starch gels in the presence of sugars (starch : sugars : water, 1:1:1); the relative effectiveness of sugars in retarding crystallization and rigidity followed the order ribose > sucrose > glucose. The mechanisms by which sugars modify the amylopectin and starch gel networks remain obscure. The suggestion that polyols may affect the phase separation of starch polymers in an aqueous medium (I'Anson *et al.*, 1990) or the postulate that sugars exert an anti-plasticizing effect in the amorphous gel matrix (elevate the glass transition temperature and thereby retard crystallization; Slade and Levine, 1987) certainly require further experimentation to test their validity.

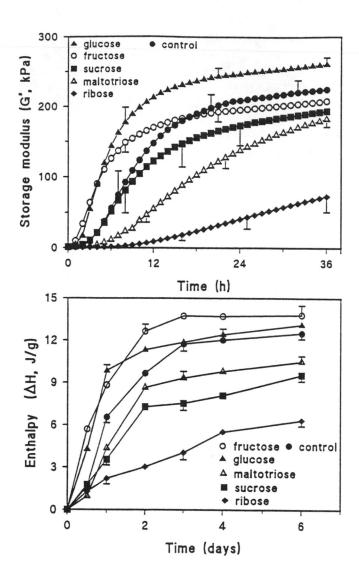

Fig. 7. Effect of added sugars on storage modulus (G′) and melting enthalpy (ΔH) of aging waxy maize amylopectin gels. Storage temperature for gelation 8°C and crystallization 6°C; frequency 0.2 Hz, strain < 2.0%. Sugars were incorporated at a ratio of 1:0.5:1.5 for starch:sugar:water mixtures. (author's unpublished data).

Viscoelastic Properties of Starch Gels

Rheological Responses upon Heating and Cooling

Dynamic viscoelastic measurements have provided an excellent means for studying the rheological changes during gelatinization and cooling of aqueous starch dispersions (Eliasson, 1986a, b; Autio, 1990; Hansen *et al.*, 1990; Svegmark and Hermansson, 1990, 1991a). Eliasson (1986a) studied the heat-induced changes in 10% wheat, maize, waxy barley, and potato starch dispersions using heating conditions similar to viscoamylography. For all starches, G' and G" sharply increased at gelatinization (Fig. 8). The rise in moduli was attributed to granule swelling; along with this process there was also release of small amounts of amylose for the non-waxy starches. Further heating (stage corresponding to the later part of the DSC gelatinization endotherm) caused a decrease in G' and G" which was interpreted as softening of granules due to crystallite melting. While potato and waxy barley showed a continuous decline in moduli during this later phase, the profiles of wheat and maize starches were characterized by a second peak around 90-95°C. Furthermore, hot wheat starch pastes had a more viscous character than maize; within 75-95°C, G" was greater than G' for wheat, but the reverse for maize. Interestingly, the amount of leached amylose was higher for wheat than for maize. Eliasson (1986a) suggested that the different rheological responses between wheat and maize starches might arise from differences in granule size and shape of gelatinized granules (filler particles) in the gel matrix as well as interactions between the continuous and dispersed phases. Autio (1990) has reported on the dynamic rheological properties and microstructure of heated oat and barley starch dispersions (8-20%). The heating temperature greatly affected the viscoelastic behaviour of the cooled gels for both starches; gelation at 95°C yielded lower storage modulus and higher tanδ than heating at 90°C. Microscopic examination of barley and oat starch gels (8%, 90°C) showed the typical morphology of a composite gel matrix (swollen granules embedded in an amylose-based network). In contrast to barley, oat starch granules

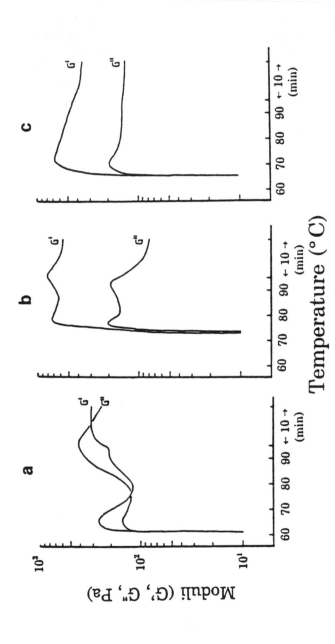

Fig. 8. Changes in storage (G') and loss (G'') moduli of 10% starch dispersions upon heating (25–95°C at 1.5°C·min⁻¹ and then kept at 95°C for 10 min); frequency 10 rad·s⁻¹, strain 10%. (a) wheat starch, (b) maize starch, and (c) potato starch. Adapted from Eliasson (1986a).

when heated to 95°C seemed to undergo disintegration with substantial amylopectin solubilization. This is in agreement with earlier observations by Doublier *et al.* (1987a) who reported co-leaching of amylose and amylopectin from oat starch granules during pasting.

The effect of heating and shear on the mechanical properties of potato, maize, and wheat starch dispersions were recently investigated by Svegmark and Hermansson (1990, 1991a) using small strain rheometry. Starch pastes were prepared in a Brabender viscoamylograph and transferred at various stages of the heating-pasting cycle to the rheometer for rheological measurements; pastes taken at the stage of "peak viscosity" of the amylogram are defined as low shear (LS), while those at the stage of "breakdown viscosity" (samples stirred at 75 rpm for 30 min at 90°C in the amylograph) are defined as high shear (HS). For potato starch dispersions (5%), heating at 90°C for 30 min alone lowered the G^* to about 55% of its maximum value at the start of the heating cycle (Svegmark and Hermansson, 1990). The impact on lowering the modulus was greater when heating was combined with shear; G^* dropped to 5% of its maximum value. These findings were subsequently supported by studies on wheat and maize starches (Svegmark and Hermansson, 1991a). In this last study, samples of equivalent maximum Brabender peak viscosity (1200 B.U.) were examined (i.e. 4% potato starch, 10% maize starch, 11% wheat starch), as shown in Figure 9. Shear brought about a substantial reduction in G^* for all starch dispersions, however, the most pronounced being for potato (note the logarithmic scale of modulus). In this respect, cooled (25°C) potato starch pastes (4-10% concentration) changed from a weak gel (LS, δ : 6-13°) to a more viscous state (HS, δ : 25-27°) by shear. Much smaller differences in the phase angles (δ) between LS and HS maize and wheat starch pastes were observed (Svegmark and Hermansson, 1991a). The above observations are in agreement with earlier reports of the impact of shear on paste and gel properties of wheat and maize starches (Wong and Lelievre, 1981; Doublier *et al.*, 1987b). Overall, these data clearly demonstrated how important is to control processing conditions (temperature-shear) when

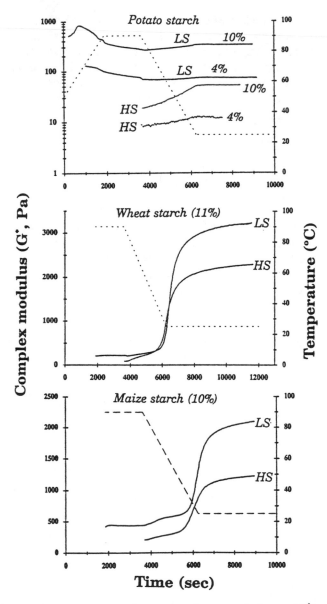

Fig. 9. Changes in the complex modulus (G*) of low shear (LS) and high (HS) starch pastes (frequency 1.0 Hz, maximum strain 1.5 x 10⁻³). The samples were subjected to a temperature-time cycle in the rheometer (heating and cooling rates of 1.5°C·min⁻¹) as specified in each diagram by the broken lines. Adapted from Svegmark and Hermansson (1991a).

assessing the gel network properties of a particular starch specimen.

The data of Figure 9 also reveal the transition in the viscoelastic properties of cereal starch pastes when cooled (rapid rise in G^*) which has been interpreted as reflecting the gelation process of amylose (Svegmark and Hermansson, 1991a). In view of the gelation mechanisms of the linear and branched starch molecules, as reviewed above, and the well accepted composite nature of starch gels (Ring, 1985), attempts have been made recently to examine the morphology of starch gel networks in terms of distribution (phase separation) of amylose and amylopectin components (Langton and Hermansson, 1989; Autio, 1990; Svegmark and Hermansson, 1991b). Thermodynamic incompatibility between gelling biopolymers (Tolstoguzov et al., 1974; Grinberg and Tolstoguzov, 1972; Clark et al., 1982, 1983; Kalichevsky et al., 1986), including amylose and amylopectin (Kalichevsky and Ring, 1987), is a fairly common phenomenon which leads to phase separation and establishment of a mixed gel network; i.e. when interactions between two polymers are not favored there is a tendency for each to exclude the other from its domains. When phase separation occurs, the effective concentration of both polymers in their respective microphases is raised and this may have a profound influence on the mechanical properties of the composite system. For example, it is possible to observe viscosity enhancement or gelation at concentrations where neither one of the two polymers is capable of bringing about such effects alone. In terms of this mixed type gel model and the feasibility for phase separation of amylose and amylopectin upon heating of concentrated aqueous starch dispersions, there is direct microscopic evidence for morphological differences between potato and cereal starch gels. In low shear wheat starch pastes (Langton and Hermansson, 1989; Autio, 1990), amylose seems to diffuse outside the granule and forms upon cooling a continuous network where the gelatinized amylopectin-enriched granules are included. In contrast, heating of potato starch dispersions results in pastes where most of the amylose is retained in the peripheral layer of the swollen granules; i.e. amylose gelation is restricted due to insufficient separation of the two polymeric starch constituents (Svegmark and Hermansson, 1991b). Even when gels were prepared under high shear, wheat starch amylopectin did not appear to solubilize (still present within the granular fragments), compared to potato starch pastes where the continuous phase was a mixture of solubilized amylose and amylopectin (Langton and Hermansson, 1989; Svegmark and Hermansson, 1991b). These morphological features were discussed in relation to the viscoelastic behaviour of wheat and potato starch gels by Svegmark and Hermansson (1991a, b). According to these authors, cereal starches form stronger gels than potato starch due to a more efficient separation of amylose and amylopectin during heating which promotes gelation of the linear starch fraction. Leloup et al. (1991) have recently shown that in mixed amylose-amylopectin gels the supermolecular organization and mechanical properties of the gel matrix varies with the amylose to amylopectin ratio (r). Mixed gels exhibited behaviour similar to pure amylopectin for r < 0.43 or to pure amylose for higher r values. The sharp transition in gel properties at r = 0.43 was assumed to reflect an inversion in the polymeric composition of the continuous and dispersed microphases (amylopectin being the continuous phase at r < 0.43 and amylose above this value). For such mixed gel starch systems there is a need for more fundamental work on the structural aspects and physical properties of the composite networks.

Linear Viscoelasticity and Frequency Dependence
        In exploring the viscoelastic properties of starch pastes and gels, dynamic measurements must be conducted under small deformation to eliminate structure breakdown or interference with the formation of network structure. The linear viscoelastic range of starch gels has been found to depend on the botanical origin of starch, concentration range, shear history and measuring temperature (Evans and Haisman, 1979; Svegmark and Hermansson, 1990, 1991a), as shown in Table I. At equivalent concentration, potato

TABLE I
Linear Viscoelastic Ranges of Gelatinized
Starch Pastes

Material	Concentration (%)	Linear Strain Region
Maize[a]	3.72-7.44	0.015
Maize[b]	10.0 (LS)	0.050[c]
Maize[b]	10.0 (HS)	0.170[c]
Wheat[b]	11.0 (LS)	0.020[c]
Wheat[b]	10.0 (HS)	0.020[c]
Wheat flour[a]	4.53-10.15	0.008
Tapioca[a]	3.54	0.006
Potato[a]	0.47-7.28	0.004
Potato[b]	4.0 (LS)	0.002[d]
Potato[b]	4.0 (HS)	0.080[c]
Potato[b]	10.0 (LS)	0.020[c]
Potato[b]	10.0 (LS)	0.005[e]

[a] Evans and Haisman, 1979; starch dispersions were
minimally sheared, measurements at 60°C.
[b] Svegmark and Hermansson, 1991a; LS and HS, low- and
high-shear dispersions.
[c] Measurements at 90°C; [d] measurements at 73°C;
[e] measurements at 67°C.

starch dispersions are the most sensitive to the
amount of strain applied, showing the narrowest linear
strain region. The viscoelastic behaviour of potato
starch is also highly dependent on pH (2-4%, w/w
aqueous gelatinized dispersions exhibit maximum G^* and
η' at pH 8.5) and ionic strength of the medium
(Muhrbeck and Eliasson, 1987); for the later,
electrolytes suppress the electrostatic repulsions
between phosphate groups present in potato amylopectin
and thereby reduce the G^* of the gel. On the other

hand, the rheological properties of cassava starch
were not affected by pH or electrolytes, presumably
because of the low phosphate content of this starch
(Muhrbeck and Eliasson, 1987).

The frequency dependence of shear moduli, G' and
G", may also give valuable information on the nature
of the network structure of a starch paste (Wong and
Lelievre, 1981; Lindahl and Eliasson, 1986; Muhrbeck
and Eliasson, 1987; Svegmark and Hermansson, 1990;
Biliaderis and Tonogai, 1991). In two studies on the
viscoelastic behaviour of thin potato starch
dispersions (concentration 4-8%), a strong dependence
of G' and G^* on frequency was observed (Muhrbeck and
Eliasson, 1987; Svegmark and Hermansson, 1990); also,
the longer the time the pastes were heated and
sheared, the stronger the frequency dependence.
However, with increasing starch concentration, well
above that corresponding to the close-packing of
swollen gelatinized granules (2.8% for maize, 0.25%
for potato, 1.7% for cassava, and 3.8% for wheat
flour; Evans and Haisman, 1979), there is less
dependence of the moduli on frequency for every starch
specimen. The mechanical spectra of 20 and 35% (w/w)
starch gels (Fig. 10) indeed show very little
frequency dependence of G' and G" over the range 0.1-
20.0 Hz.

Amylose Content, Concentration, and Time Dependence of
Gel Modulus

For cereal starches, where amylose leaches out
of the granule upon gelatinization and contributes to
the viscoelasticity of the continuous phase, it is
interesting to explore the relationships between
amylose content and rheological properties of starches
from a common botanical origin. In this respect, rice
starches offer a unique model system to test such
relationships. Since there are no major differences
in shape, size, and distribution of granules among
low-, intermediate- and high-amylose rice starches,
one would expect similar contributions from the
dispersed phase to the rheological behaviour of the
respective gels. If the rigidity of the gelatinized
granules varies with the amylose content, this would
be also reflected in the dynamic moduli of the

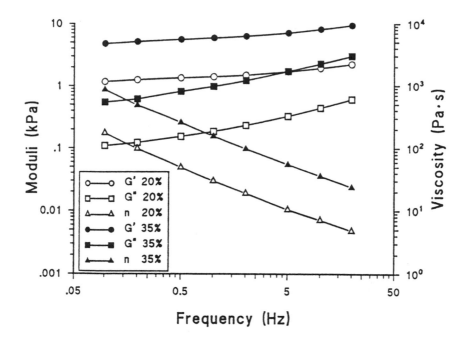

Fig. 10. Frequency dependence of dynamic moduli (G′, G″) and viscosity (η') of freshly prepared wheat starch gels (20 and 30%, w/w) at 25°C. Reprinted, with permission, from Biliaderis and Tonogai (1991). Copyright 1991 American Chemical Society.

composite gels. Figure 11 shows the dependence of G′ and tanδ on amylose content for 25% (w/w) gels of starches from 43 different rice cultivars. An inverse linear relationship between tanδ and amylose content was found; r = -0.73, p < 0.01. Also, there was a general trend for increased G′ values with increasing amylose content. However, the observed variation in modulus at the intermediate- to high-amylose content range clearly indicates that other factors contribute to the mechanical properties of these gels. In a related study, Wong and Lelievre (1981) concluded that differences in the dynamic mechanical properties of wheat starch pastes (< 7% w/w) from several cultivars may be due to variations in the swelling capacity and number fraction of large granules.

The relationship between gel modulus and

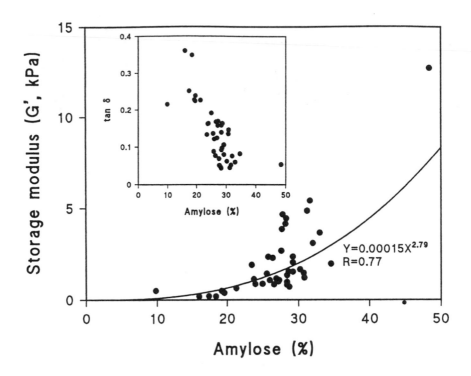

Fig. 11. Experimental G' values (at 1.0 Hz, 25°C) of 25% (w/w) rice starch gels as a function of amylose content of starch. Inset shows the relationship between tanδ and amylose content for the same group of samples. (Biliaderis and Juliano, unpublished data).

concentration is often examined for physically cross-linked gelling biopolymer systems. In this context, since G' varies with time as cross-linking proceeds, one has to assume that gel curing has gone long enough to ensure very little additional increases in G' with time. Particularly for starch pastes or gels, the modulus is highly dependent on time due to retrogradation of starch molecules. This necessitates the rheological testing to be carried out immediately after gelation to control such time dependent changes in rigidity. Similarly, measurements must be taken at a fixed frequency and temperature. Clark and Ross-Murphy (1987) provided a theoretical analysis for the power law dependence of G' on polymer concentration according to which a limiting $C^{2.0}$ relationship is predicted at high C/C_0 ratios. For starch gels, using a Ward and Saunder's U-tube, Ring (1985) found linear relationships between shear modulus and concentration of potato, maize, and pea starch gels in the concentration range 6-30%; at any given concentration, the rigidity modulus of the gels was in the order of pea starch > maize starch > potato starch. For corn and potato starch pastes (2-8%), linear relationships were also reported by Evans and Haisman (1979) using small strain oscillatory measurements (0.2 Hz); similar relationships were shown for the G". These authors, however, observed a slightly stronger concentration dependence for wheat flour. Using an oscillatory coaxial cylinder rheometer, Wong and Lelievre (1981) found an exponential dependence of G' on concentration for wheat starch pastes (2-6%, frequency 0.05-1.00 rad·s^{-1}); similar trends were also shown for η'. These authors derived an equation to describe the experimental data of G' based on the empirical equation:

$$G' = A \ (C-C_s)^B$$

where A and B are constants and C_s is the close packing concentration of starch; estimates of B varied within 1.44-1.48. Figure 12 shows the concentration dependence for two different starches over a broad concentration range (8-40%); i.e. at much greater C/C_0 than in the previous studies. Reasonably linear plots were obtained when data were plotted on a log-log scale (r > 0.94, p < 0.01). From the slopes of the linear regression lines, it was found that G' varied as $C^{2.3}$ and $C^{2.4}$ for garbanzo bean and rice starches, respectively. These exponents are closer to the theoretical relationship G' $\propto$ $C^{2.0}$ than those previously reported (Ring, 1985; Evans and Haisman, 1979; Wong and Lelievre, 1981). However, it is unlikely that composite networks, such as starch gels, can be rationalized in terms of a $C^{2.0}$ concentration dependence of modulus. Their viscoelastic behaviour is different than that of pure amylose and amylopectin networks. In addition to cross-link density in the continuous phase, the rigidity, volume fraction and

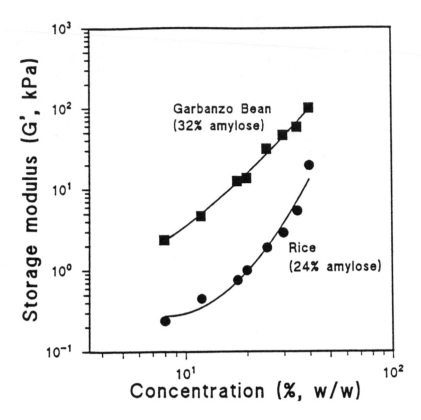

Fig. 12. Experimental G' values of gels (1.0 Hz, strain < 1.4%, 25°C) as a function of starch concentration. Adapted from Biliaderis and Tonogai (1991).

spatial distribution of swollen granules are also important. In concentrated starch gels there is a close contact between granules. Thus, with increasing concentration, greater restrictions are imposed to granule swelling and phase separation between starch molecules. Granules of limited swelling would raise the modulus due to their increased rigidity (Ring and Stainsby, 1982). On the other hand, a lower concentration of leached amylose would yield a weaker continuous phase. It is perhaps the relative contribution of these opposing effects that ultimately determines the magnitude of the gel modulus.

Dynamic measurements are particularly useful in monitoring structural changes in starch networks with time. Depending on the concentration and storage temperature, the overall kinetics of G' growth of aging gels may vary. In general, the G'-time profiles reveal a biphasic gelation process: an initial rapid rise in modulus, followed by a phase of slower G' development. The first phase corresponds to amylose gelation, while the later has been attributed to amylopectin crystallization (Miles *et al.*, 1985b; Biliaderis and Zawistowski, 1990). Supporting evidence for the origin of the long-term increases in shear modulus was provided by heating 10 and 20% pea starch gels aged for 7 days (26°C) at 95°C (Miles *et al.*, 1985b); heating reduced the stiffness and crystallinity of the gels to their original values. Since the long-term increases in modulus and crystallinity were thermo-reversible, recrystallization of the outer short chains of amylopectin was implicated for these changes. Miles *et al.* (1985b) suggested that recrystallization of amylopectin within the gelatinized granules enhances their rigidity, thereby reinforcing the composite network. Storage modulus data of aging wheat starch pastes (at concentrations < 7.0%) have been treated by means of the Avrami equation (Wong and Lelievre, 1982). However, as pointed by Doublier (1990), the Avrami analysis, although applicable to crystallization of synthetic polymers, is at best an empirical approach in describing kinetically starch gelation and recrystallization data; its theoretical development is based on the assumption of a single macromolecular species present in the crystallization medium, which is not the case for starch. A comparison in the initial rate of modulus development (20°C) for 30% gelatinized starch dispersions from various botanical sources revealed the order: pea > maize > wheat > potato; this ranking was related to the amount of amylose solubilized during gelatinization (Orford *et al.*, 1987). However, the rate of the long-term modulus increase followed the order: pea > potato > maize > wheat; the higher rates for pea and potato may reflect the lack of endogenous lipids from these starches which are known to retard amylopectin recrystallization.

Effects of Additives and other Constituents

Lipids have long been known to affect gelatinization and texture of starch-based products. Monoacyl lipids, in particular, form inclusion complexes with starch (Krog, 1971; Evans, 1986) and it is likely that texture modification is brought about via such interactions. The viscoelastic behaviour of heated starch dispersions (10% w/w, wheat, maize, potato, and waxy barley) in the presence of emulsifiers was first investigated by Eliasson (1986b). Glycerol monostearate and sodium stearoyl lactylate delayed the onset of G' and G" rise on gelatinization; this effect was less pronounced with the waxy barley starch. All lipids also reduced granule swelling and amylose leaching, presumably due to complexation with starch. The resultant viscoelastic responses for most starch pastes were characterized by increased tanδ, implying less organized network structures. In a subsequent study, (Eliasson et al., 1988), a cationic surfactant (cetyltrimethylammonium bromide, CTAB) was found to increase the G' of native and modified maize starch pastes (7.5%, 0.01-1.0 Hz, 25°C); instead, saturated monoglycerides increased the G' for normal maize starch, decreased the G' for cross-linked waxy maize and had no effect on an acetylated high-amylose maize starch derivative. Interactions between lipids [sodium dodecyl sulfate (SDS), glycerol monostearate, CTAB, and lysophosphatidylcholine (LPC); at 2-4% on starch basis] were also studied in thermoset gel networks at high starch concentrations (20-35%, w/w) by dynamic shear measurements (0.1-20.0 Hz, 25°C) and DSC (Biliaderis and Tonogai, 1991). Although rice and wheat starch gels exhibited higher storage modulus (G') values when lipids were included, smaller changes in the viscoelastic properties were observed for the legume starches; the higher amylose content of pea and garbanzo bean starches seemed to dominate the rheological responses of their composite gels. Among the lipids tested, LPC exerted the greatest effect in increasing the G' and lowering the tanδ of cereal starch gels. Kinetic experiments on the development of storage modulus and recrystallization (staling endotherm) in aging rice and pea starch gels (35% w/w) indicated that lipids retard both processes (Fig. 13).

The gelation process of corn starch dispersions (4%) in the presence of galactomannans (0.1-0.5%) was recently investigated by Alloncle and Doublier (1990) using a controlled stress rheometer (1.0 Hz, 25°C). The addition of galactomannans accelerated the gelation kinetics as well as resulted in higher G' for the final gel networks compared to the starch alone. Moreover, the mixed gels were characterized by greater tanδ values, implying a less elastic network. The mechanical spectra of the mixed gels were also modified, showing a slightly greater dependence of G' on frequency. The rheological data of this study were rationalized in terms of thermodynamic incompatibility between amylose and galactomannans in the continuous phase of the gel matrix. Gudmundsson et al. (1991) have studied the effect of adding soluble rye arabinoxylan (at 2.0% on starch basis) on the viscoelastic properties of 40% (w/w) potato starch gels (0.05-10.0 Hz, 25°C). Although fresh gels with added arabinoxylan had lower G' than pure potato starch gels, the reverse was observed for 7 days aged gels. In a related study, we have examined the influence of two purified wheat and rye arabinoxylans (2.0% on α-D-glucan basis) on retrogradation kinetics (probed by dynamic rheometry, DSC, X-ray diffraction) of concentrated amylose (20%, w/w), wheat starch and amylopectin (40%, w/w) gels (Biliaderis and Izydorczyk, 1992). For the amylose- and wheat starch-arabinoxylan mixtures, the G' was higher than for amylose and starch alone (Fig. 14). In contrast, both arabinoxylan preparations markedly retarded the gelation rate of amylopectin. These observations suggested that arabinoxylans interfere with inter- and intramolecular processes responsible for the development of a three-dimensional amylopectin gel network.

Since starch gels are composites, one approach to manipulate the viscoelasticity of the system would be to interrupt the interchain associations between amylose molecules. In this context, we have examined the effect of incorporating starch hydrolyzates of

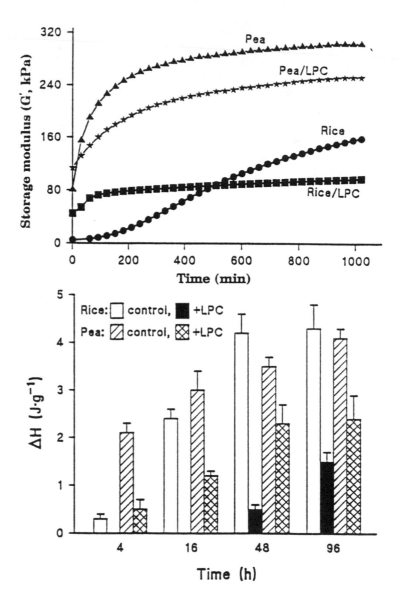

Fig. 13. Effect of added L-α-lysophosphatidylcholine (4% w/w on starch basis) on storage modulus (G', at 0.2 Hz, 8°C) and magnitude of staling endotherm (ΔH, storage temperature 6°C) for rice and pea starch gels (35%, w/w). Adapted from Biliaderis and Tonogai (1991).

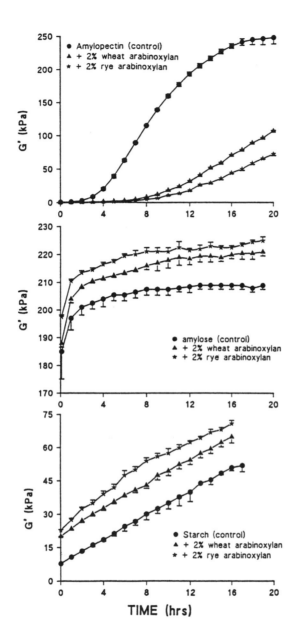

Fig. 14. Effect of added wheat and rye arabinoxylans (2% on α-D-glucan basis) on G' development in 40% (w/w) amylopectin, 20% (w/w) amylose and 40% (w/w) wheat starch gels (0.2 Hz, 8°C). Reprinted from Biliaderis and Izydorczyk (1992) by permission of Oxford University Press.

varying DE (dextrose equivalent) on the gel network properties of wheat starch (Fig. 15) (Biliaderis and Zawistowski, 1990). Starch hydrolyzates added at 20% (w/w, on starch basis) greatly reduced the rigidity of the composite gels. There have been no clear trends, however, regarding the magnitude of this effect in relation to DE. Chain polydispersity and degree of branching of maltodextrin products may be important determinants of the rheological properties of starch-maltodextrin systems.

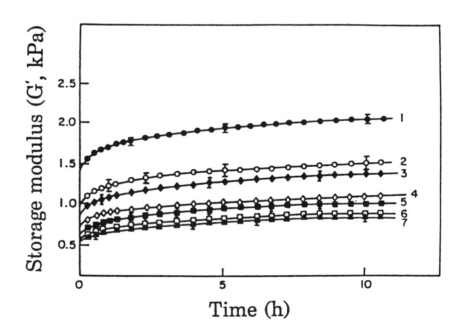

Fig. 15. Storage modulus of 18% (w/w) wheat starch gels vs time in the presence of starch hydrolyzates (20%, w/w on starch basis): 1 control; 2 wheat starch/Maltrin M200 (DE 20); 3 wheat starch/Maltrin M040 (DE 4.7); 4 wheat starch/Fro-Dex 42 (DE 42); 5 wheat starch/Maltrin M150 (DE 15); 6 wheat starch/Lo-Dex 5 (DE 5); 7 wheat starch/Star-Dri 1 (DE 1). Measurements at 0.2 Hz, 2% strain, 6°C. Reprinted, with permission, from Biliaderis and Zawistowski (1990).

Low DE maltodextrins have the ability of forming thermally-reversible gels with fat-mimetic properties. The sol-gel transition and gel properties have been characterized by thermal, mechanical, X-ray diffraction, and NMR-relaxation methods (Braudo *et al.*, 1979; Reuther *et al.*, 1983; Bulpin *et al.*, 1984; German *et al.*, 1989). As revealed by wide-angle X-ray scattering, gelation of aqueous maltodextrin dispersions proceeds via chain aggregation and crystallization (Reuther *et al.*, 1984); the crystalline domains consist of double helices (B-polymorph). The presence of sufficiently long chain segments is also important as they can act as interconnecting elements between crystallites. For low DE potato maltodextrins, Bulpin *et al.* (1984) have identified and isolated both high (mainly branched) and low (linear) molecular weight fractions. Gel curing experiments, using dynamic rheometry, indicated that the low molecular weight fraction, although itself non-gelling, is intimately involved in the maltodextrin gel network. The gelation behaviour of mixed low DE potato maltodextrin-gelatin system was recently explored by Kasapis *et al.* (1992). Evidence was presented (rheometry, DSC, and light microscopy) that mixed gels of these materials show phase inversion from a gelatin-continuous network at low maltodextrin concentration to a maltodextrin-continuous network at higher concentrations. Furthermore, the observed moduli (G') of the composite gels were related to the individual moduli of the constituent phases using polymer blending laws. These theories have been first applied to biopolymer networks by Clark and co-workers (Clark *et al.*, 1983; Clark, 1987).

Finally, the viscoelastic properties of starch have been examined in the presence of wheat proteins and dough ingredients under varying processing conditions in an attempt to understand the rheological behaviour of complex starch-based food systems (Lindahl and Eliasson, 1986; Dreese *et al.*, 1988a, b; Kokini *et al.*, 1989). Such studies are expected to provide a sound framework for texture modification and quality improvement of cereal-derived products.

ACKNOWLEDGMENTS

The author wishes to express his appreciation to the Natural Sciences and Engineering Research Council of Canada for supporting his work through operating and equipment research grants. Thanks are also due to Marta Izydorczyk for her assistance in the typesetting of this manuscript.

LITERATURE CITED

Alloncle, M., and Doublier, J.-L. 1990. Rheology of starch-galactomannan gels. Page 111 in: Gums and Stabilisers for the Food Industry, vol. 5. G.O. Phillips, D.J. Wedlock, and P.A. Williams, eds. Elsevier Applied Science, London.

Autio, K. 1990. Rheological and microstructural changes of oat and barley starches during heating and cooling. Food Structure 9:297.

Banks, W., and Greenwood, C.T. 1975. Starch and its Components. Edinburgh University Press, Edinburgh.

Bell, A.E. 1989. Gel structure and food biopolymers. Page 251 in: Water and Food Quality. T.M. Hardman, ed. Elsevier Applied Science, London.

Biliaderis, C.G., and Izydorczyk, M.S. 1992. Observations on retrogradation of starch polymers in the presence of wheat and rye arabinoxylans. In: Gums and Stabilisers for the Food Industry, G.O. Phillips, D.J. Wedlock, and P.A. Williams, eds. IRL Press, Oxford. (in press).

Biliaderis, C.G., Page, C.M., Slade, L., and Sirett, R.R. 1985. Thermal behaviour of amylose-lipid complexes. Carbohydr. Polym. 5:367.

Biliaderis, C.G., and Tonogai, J.R. 1991. Influence of lipids on the thermal and mechanical properties of concentrated starch gels. J. Agr. Food Chem. 39:833.

Biliaderis, C.G., and Zawistowski, J. 1990. Viscoelastic behaviour of aging starch gels: effects of concentration, temperature, and starch hydrolyzates on network properties. Cereal Chem. 67:240.

Blanshard, J.M.V. 1987. Starch granule structure and function: a physicochemical approach. Page 16 in: Starch: Properties and potential. T. Galliard, ed. John Wiley & Sons, New York.

Boyer, R.F., Baer, E., and Hiltner, A. 1985. Concerning gelation effects in atactic polystyrene solutions. Macromolecules 18:427.

Braudo, E.E., Belavsteva, E.M., Titova, E.F., Plashchina, I.G., Krylov, V.B., Tolstoguzov, V.P., Schierbaum, .R., Richter, M., and Berth, G. 1979. Structure and properties of maltodextrin hydrogels. Starch 31:188.

Bulpin, P.V., Cutler, A.N., and Dea, I.C.M. 1984. Thermally-reversible gels from low DE maltodextrins. Page 475 in: Gums and Stabilisers for the Food Industry, vol. 2. G.O. Phillips, D.J. Wedlock, and P.A. Williams, eds. Pergamon Press, Oxford.

Clark, A.H. 1987. The application of network theory to food systems. Page 13 in: Food Structure and Behaviour. J.M.V. Blanshard and P. Lillford, eds. Academic Press, London.

Clark, A.H., Gidley, M.J., Richardson, R.K., and Ross-Murphy, S.B. 1989. Rheological studies of aqueous amylose gels: the effect of chain length and concentration on gel modulus. Macromolecules 22:346.

Clark, A.H., Richardson, R.K., Ross-Murphy, S.B., and Stubbs, J.M. 1983. Structural and mechanical properties of agar/gelatin co-gels. Small deformation studies. Macromolecules 16:1367.

Clark, A.H., Richardson, R.K., Robinson, G., Ross-Murphy, S.B., and Weaver, A.C. 1982. Structure and mechanical properties of agar/BSA co-gels. Prog. Fd. Nutr. Sci. 6:149.

Clark, A.H. and Ross-Murphy, S.B. 1987. Structural and mechanical properties of biopolymer gels. Adv. Polym. Sci. 83:60.

Cox, W.P., and Merz, E.H. 1958. Correlation of dynamic and steady-flow viscosities. J. Polym. Sci. 28:619.

Doublier, J.-L. 1990. Rheological properties of cereal carbohydrates. Page 111 in: Dough Rheology and Baked Product Texture. H. Faridi and J.M. Faubion, eds. Van Nostrand Reinhold, New York.

Doublier, J.-L., and Choplin, L. 1989. A rheological description of amylose gelation. Carbohydr. Res. 193:215.

Doublier, J.-L., Llamas, G., and Le Meur, M. 1987b. A rheological investigation of cereal starch pastes and gels. Effect of pasting procedures. Carbohydr. Polym. 7:251.

Doublier, J.-L., Paton, D., and Llamas, G. 1987a. A rheological investigation of oat starch pastes. Cereal Chem. 64:21.

Dreese, P.C., Faubion, J.M., and Hoseney, R.C. 1988a. Dynamic rheological properties of flour, gluten, and gluten-starch doughs. I. Temperature-dependent changes during heating. Cereal Chem 65:348.

Dreese, P.C., Faubion, J.M., and Hoseney, R.C. 1988b. Dynamic rheological properties of flour, gluten, and gluten-starch doughs. II. Effect of various processing and ingredient changes. Cereal Chem 65:354.

Eliasson, A.-C. 1986a. Viscoelastic behaviour during the gelatinization of starch. I. Comparison of wheat, maize, potato and waxy-barley starches. J. Text. Stud. 17:253.

Eliasson, A.-C. 1986b. Viscoelastic behaviour during the gelatinization of starch. II. Effects of emulsifiers. J. Text. Stud. 17:357.

Eliasson, A. -C., Finstad, H., and Ljunger, G. 1988. A study of starch -lipid interactions for some native and modified maize starches. Starch 40:95.

Ellis, H.S., and Ring, S.G. 1985. A study of some factors influencing amylose gelation. Carbohydr. Polym. 5:201.

Evans, I.D. 1986. An investigation of starch/ surfactant interactions using viscometry and differential scanning calorimetry. Starch 38:227.

Evans, I.D., and Haisman, D.R. 1979. Rheology of gelatinized starch suspensions. J. Text. Stud. 10:347.

Ferry, J.D. 1980. Viscoelastic Properties of

Polymers, 3rd ed. John Wiley & Sons, New York.

French, D. 1984. Organization of starch granules. Page 267 in: Starch: Chemistry and Technology. R.L. Whistler, J.N. BeMiller, and E.F. Paschall, eds. Academic Press, New York.

German, M.L., Blumenfeld, A.L., Yuryev, V.P., and Tolstoguzov, V.B. 1989. An NMR study of structure formation in maltodextrin systems. Carbohydr. Polym. 11:139.

Gidley, M.J. 1989. Molecular mechanisms underlying amylose aggregation and gelation. Macromolecules 22:351.

Gidley, M.J., and Bulpin, P.V. 1989. Aggregation of amylose in aqueous systems: the effect of chain length on phase behaviour and aggregation kinetics. Biopolymers 22:341.

Grinberg, V. Ya, and Tolstoguzov, V.B. 1972. Thermodynamic compatibilty of gelatin with some D-glucans in aqueous media. Carbohydr. Res. 25:313.

Gudmundson, M., Eliasson, A.-C., Bengtsson, S., and Aman, P. 1991. The effects of water soluble arabinoxylan on gelatinization and retrogradation of starch. Starch 43:5.

Hamann, D.D., Purkayastha, S., and Lanier, T.C. 1990. Applications of thermal scanning rheology to the study of food gels. Page 306 in: Thermal Analysis of Foods. V.R. Harwalkar and C.-Y. Ma, eds. Elsevier Applied Science, London.

Hansen, L.M., Hoseney, R.C., and Faubion, J.M. 1990. Oscillatory probe rheometry as a tool for determining the rheological properties of starch-water systems. J. Text. Stud. 21:213.

I'Anson, K.J., Miles, M.J., Morris, V.J., Besford, L.S., Jarvis, D.A., and Marsh, R.A. 1990. The effects of added sugars on the retrogradation of wheat starch gels. J. Cereal Sci. 11:243.

I'Anson, K.J., Miles, M.J., Morris, V.J., and Ring, S.G. 1988. A study of amylose gelation using a synchrotron X-ray source. Carbohydr. Polym. 8:45.

Kalichevsky, M.T., Orford, P.D., and Ring, S.G. 1986. The incompatibilty of concentrated aqueous solutions of dextran and amylose and its effect on amylose gelation. Carohydr. Polym. 6:145.

Kalichevsky, M.T., Orford, P.D., and Ring, S.G. 1990. The retrogradation and gelation of amylopectins from various botanical sources. Carbohydr. Res. 198:49.

Kalichevsky, M.T., and Ring, S.G. 1987. Incompatibility of amylose and amylopectin in aqueous solution. Carbohydr. Res. 162:323.

Kasapis, S., Morris, E.R., and Norton, I.T. 1992. Physical properties of maltodextrin/gelatin systems. In: Gums and Stabilisers for the Food Industry. G.O. Phillips, D.J. Wedlock, and P.A. Williams, eds. IRL Press, Oxford. (in press).

Kokini, J.L., Baumann, G., Breslaner, K., Chedid, L.L., Herh, P.K., Lai, P.K., and Madeka, H. 1989. A kinetic model for starch gelatinization and effect of starch/protein interactions on rheological properties of 98% amylopectin and amylose rich starches. Page 109 in: Engineering and Food, vol 1. W.E.L. Spiess and H. Schubert, eds. Elsevier Applied Science, New York.

Kokini, J. L., and Plutchok, G.J. 1987. Viscoelastic properties of semisolid foods and their biopolymeric components. Food Technol. 41 (3):89.

Krog, N. 1971. Amylose complexing effect of food grade emulsifiers. Starch 23:206.

Langton, M., and Hermansson, A.-M. 1989. Microstructural changes in wheat starch dispersions during heating and cooling. Food Microstructure 8:29.

Launay, B., Doublier, J.-L., and Cuvelier, G. 1986. Flow properties of aqueous solutions and dispersions of polysaccharides. Page 1 in: Functional Properties of Food Macromolecules. J.R. Mitchell and D.A. Ledward, eds. Elsevier Applied Science, London.

Leloup, V.M., Colonna, P., and Buleon, A. 1991. Influence of amylose-amylopectin ratio on gel properties. J. Cereal Sci. 13:1.

Lindahl, L., and Eliasson, A.-C. 1986. Effects of wheat proteins on the viscoelastic properties of starch gels. J. Sci. Food Agric. 37:1125.

Mandelkern, L. 1964. Crystallization of Polymers. McGraw-Hill, New York.

Maxwell, J.L., and Zobel, H.F. 1978. Model studies on cake staling. Cereal Foods World 23:124.

Miles, M.J., Morris, V.J., and Ring, S.G. 1984. Some recent observations on the retrogradation of amylose. Carbohydr. Polym. 4:73.

Miles, M.J., Morris, V.J., and Ring, S.G. 1985a. Gelation of amylose. Carbohydr. Res. 135:257.

Miles, M.J., Morris, V.J., Orford, P.D., and Ring, S.G. 1985b. The roles of amylose and amylopectin in the gelation and retrogradation of starch. Carbohydr. Res. 135:271.

Morris, E.R. 1984. Rheology of hydrocolloids. Page 57 in: Gums and Stabilisers for the Food Industry, vol 2. G. Phillips, D.J. Wedlock, and P.A. Williams, eds. Pergamon Press, New York.

Morris, E.R. 1989. Polysaccharide solution properties: origin, rheological characterization and implications for food systems. Page 132 in: Frontiers in Carbohydrate Research 1: Food Applications. R.P. Millane, J.N. BeMiller and R. Chandrasekaran, eds. Elsevier Applied Science.

Morris, E.R., Cutler, A.N., Ross-Murphy, S.B., and Price, J. 1981. Concentration and shear rate dependence of viscosity in random coil polysaccharide solutions. Carbohydr. Polym. 1:5.

Morris, E.R., and Norton, I.T. 1983. Polysaccharide aggregation in solutions and gels. Page 549 in: Aggregation Processes in Solution. E. Wyn-Jones and J. Gormally, eds. Elsevier/North Holland, Amsterdam.

Morris, V.J. 1985. Food gels - roles played by polysaccharides. Chem. Ind. (London) 1.59.

Morris, V.J. 1986. Gelation of polysaccharides. Page 121 in: Functional Properties of Food Macromolecules. J.R. Mitchell and D.A. Ledward, eds. Elsevier Applied Science, London.

Morris, V.J. 1990. Starch gelation and retrogradation. Trends in Food Sci. and Technol. 1:2.

Muhrbeck, P., and Eliasson, A.-C. 1987. Influence of pH and ionic strength on the viscoelastic properties of starch gels - A comparison of potato and cassava starches. Carbohydr. Polym. 7:291.

Myers, R.R., Knauss, C.J., and Hoffman, R.D. 1962. Dynamic rheology of modified starches. J. Applied Polym. Sci. 6:659.

Myers, R.R., and Knauss, C.J. 1965. Mechanical properties of starch pastes. Page 393 in: Starch: Chemistry and Technology, vol 1. R.L. Whistler and E.F. Paschall, eds. Academic Press, New York.

Oakenfull, D. 1987. Gelling agents. CRC Crit. Rev. Food Sci. Nutr. 26:1.

Orford, P.D., Ring, S., Carroll, V., Miles, M.J., and Morris, V.J. 1987. The effect of concentration and botanical source on the gelation and retrogradation of starch. J. Sci. Food Agric. 39:169.

Rees, D.A. 1969. Structure conformation and mechanism in the formation of polysaccharide gels and networks. Adv. Carbohydr. Chem. Biochem. 24:267.

Reuther, F., Damaschun, G., Gernat, Ch., Schierbaum, F., Kettlitz, B., Radosta, S., and Nothnagel, A. 1984. Molecular gelation mechanism of maltodextrins investigated by wide-angle X-ray scattering. Colloid & Polymer Science 262:643.

Reuther, F., Gernat, Ch., Damaschun, G., and Schierbaum, F. 1983. Crystal structure of thermally reversible maltodextrin gels. Studia Biophysica 97:143.

Ring, S.G. 1985. Some studies on starch gelation. Starch 37:80.

Ring, S.G., Colonna, P., I'Anson, K.J., Kalichevsky, M.T., Miles, M.J., Morris, V.J., and Orford, P.D. 1987. The gelation and crystallization of amylopectin. Carbohydr. Res. 162:277.

Ring, S.G., and Stainsby, G. 1982. Filler reinforcement of gels. Prog. Fd. Nutr. Sci. 6:323.

Ross-Murphy, S.B. 1984. Rheological methods. Page 138 in: Biophysical Methods in Food Research. H.W.-S. Chan, ed. Blackwell, London.

Ross-Murphy, S.B. 1987. Physical gelation of

biopolymers. Food Hydrocolloids 1:485.

Slade, L., and Levine, H. 1987. Recent advances in starch retrogradation. Page 387 in: Industrial Polysaccharides. S.S. Stilva, V. Crescenzi, and I.C.M. Dea, eds. Gordon and Breach Science, New York.

Svegmark, K, and Hermansson, A.-M. 1990. Shear induced changes in the viscoelastic behaviour of heat treated potato starch dispersions. Carbohydr. Polym. 14:29.

Svegmark, K, and Hermansson, A.-M. 1991a. Changes induced by shear and gel formation in the viscoelastic behaviour of potato, wheat, and maize starch dispersions. Carbohydr. Polym. 15:151.

Svegmark, K, and Hermansson, A.-M. 1991b. Distribution of amylose and amylopectin in potato starch pastes: effects of heating and shearing. Food Microstructure 10:117.

Tolstoguzov, V.B., Belkina, V.P., Gulov, V.Ja., Titova, E.M., and Belavsteva, E.M. 1974. State of phase, structure and mechanical properties of the gelatinous system water-gelatin-dextran. Starch 26:130.

Wong, R.B.K. and Lelievre, J. 1981. Viscoelastic behaviour of wheat starch pastes. Rheol. Acta. 20:299.

Wong, R.B.K. and Lelievre, J. 1982. Effect of storage on dynamic rheological properties of wheat starch pastes. Starch 34:231.

THERMAL ANALYSIS OF CARBOHYDRATES AS ILLUSTRATED BY AQUEOUS STARCH SYSTEMS

John Lelièvre
Department of Food Science
Acadia University
Wolfville, Nova Scotia, Canada   B0P 1X0

INTRODUCTION

Thermal analysis covers five main topics (Wunderlich, 1990) :-
a) thermometry,
b) differential thermal analysis (DTA) which includes differential scanning calorimetry (DSC),
c) calorimetry,
d) thermomechanical analysis (TMA) which includes dilatometry,
e) thermogravimetry.

These techniques have been used in a very large number of studies of carbohydrate systems ranging from monosaccharides to polymers (Biliaderis, 1990). The data so obtained have been analysed according to theories of thermodynamics, irreversible thermodynamics and kinetics. In view of the breadth of the subject, a complete discussion of all the applications of thermal analysis to carbohydrates is clearly beyond the scope of a single article. Hence the present paper seeks only to consider a part of the total research area. Aspects of the thermal analysis of starch-water mixtures have been selected for review since these systems are of both scientific and industrial interest.

The early pioneers who studied starch were among the first to characterize a macromolecule. To some extent, these researchers laid a part of the foundation on which modern polymer science is built. However since that time knowledge of synthetic polymers has increased enormously. Now the trend is to use the principles established for well-defined synthetic macromolecules, in an attempt to understand the behavior of the complex polysaccharides found in starch.

Thermal analysis techniques have proved to be powerful means of probing the transitions that occur when synthetic polymers are heated (Jenkins, 1972; Wunderlich, 1980). For this reason the methods have been widely used to investigate the gelatinization of starch. The first such study, which unfortunately was never published in full, employed DSC and considered whether gelatinization is a glass, or melting, transition (Zobel et al., 1965). Some twenty-five years later, this question is still debated in the literature.

The current review gives an outline of the crystalline structures formed by synthetic polymers, and their study by thermal analysis, since this provides a basis on which to consider starch polysaccharides. Problems in the interpretation of starch thermal analysis data are assessed. Emphasis is placed on the evidence for a glass transition since this is a key issue in deciding the mechanism of gelatinization. No attempt is made to consider every single aspect of the thermal analysis of starch as detailed reviews of the entire subject are available elsewhere (Lund, 1984; Biliaderis, 1990; Biliaderis, 1991).

CRYSTALLINE AND SEMICRYSTALLINE STRUCTURES FORMED BY SYNTHETIC POLYMERS

Most polymers do not usually form perfectly-aligned, completely crystalline, structures. Instead a semicrystalline arrangement, which contains both crystalline and amorphous material, is produced (Jenkins, 1972). The fact that a structure is partly crystalline implies that it is metastable, i.e. not in equilibrium. Thermodynamics requires that in a single-component polymer system, the crystalline and amorphous states can only be in equilibrium at the melting point (Wunderlich, 1990).

A semicrystalline polymer structure arises because it is difficult for an entire macromolecule to become incorporated into a crystal lattice. The requisite full extension of the molecular chains is hindered (Wunderlich, 1973; Wunderlich, 1976). Hence crystals of limited size are formed. The small crystal size means that the surface free energy is significant and represents an appreciable source of instability. As a result of this instability, the semicrystalline polymer melts at a lower temperature than that of an equilibrium crystal where surface effects are negligible (Wunderlich, 1990).

Linear macromolecules, because of their regular structures, would be expected to be the most likely to crystallize completely. However, even these polymers do not usually form perfectly aligned arrangements. In cases where the molecular structure is irregular, the crystallinity may be reduced further. For example, in branched polymers the branch points are not normally accommodated in the crystalline arrays formed by linear sections of the molecule. The junctions between the crystallizable and non-crystallizable units are positioned between the crystalline and amorphous phases. This further limits crystal size.

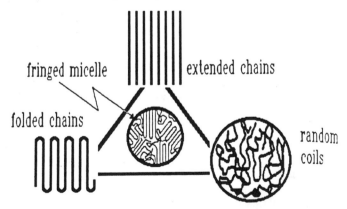

Fig. 1. Polymer macroconformations (from Wunderlich, 1990, with permission).

Figure 1 uses the procedure of Wunderlich (1990) to summarize the possible chain conformations found in crystals of macromolecules. In the equilibrium crystal with extended-chain molecules the thickness of the crystal in the chain direction, and the molecular length, match. A regular shape also occurs in the folded-chain macroconformation. The non-crystalline material may be entangled and has a random coil conformation. The fringed micelle structure combines parts of these three limits, the proportions of which depend on the crystallization conditions. The fringed micelle is a metastable structure since on further crystallization it is converted into a more ordered form with larger, and hence more stable, crystalline regions.

A refinement of the fringed micelle structure has been suggested (Wissler and Crist, 1980; Jin et al., 1984). This proposes that within the amorphous fraction, two types of amorphous domains can be recognized when in the rubbery state. These are a bulk, relatively mobile, amorphous fraction and a more restrained component immediately adjacent to the crystalline zones.

THE STRUCTURE OF STARCH GELS

It has long been recognized that the formation and retrogradation of starch gels involves the crystallization of the polymer they contain (Katz and Van Itallie, 1930). When the concentration of starch is sufficiently high, the swollen granule gel fragments formed during gelatinization pack together and overlap (Evans and Haisman, 1979; Wong and Lelièvre, 1981). Crystallization leads to the formation of bridges between the fragments and ultimately a continuous three-dimensional network. If the concentration of polymer is insufficient, the swollen particles do not overlap. Crystallization occurs within, but not between, granules (Callaghan et al., 1983). Hence a continuous network is not established and a paste is produced.

THE STRUCTURE OF STARCH GRANULES

The fringed micelle is often selected as a model for the starch granule (Blanshard, 1987). The ordered zones are considered to be formed by helical polysaccharide arrangements which contain water as an integral part of their structure (French, 1972). These ordered zones function as physical junction points which link the amorphous material together. A single starch chain can have ordered helical segments located within one or more of the crystalline microphases and random coil segments in one or more amorphous regions. It has been suggested that two types of amorphous polymer, with different mobilities in the rubbery state, are present in granules just as occur in linear homopolymers (Biliaderis et al., 1986). Another view is that a range of different amorphous regions, with a broad spectrum of mobilities, is present (Slade and Levine, 1987).

The precise way in which the starch polymers contribute to the crystalline and amorphous regions is uncertain. As far as amylopectin is concerned, some evidence suggests that the branched parts of the molecule lie in the amorphous domains while the linear parts form crystalline arrangements (French, 1984). Since amylose is essentially linear it is not subject to the same constraints. The flat shape of the amylopectin molecule encourages a more ordered structure to develop (Callaghan and Lelièvre, 1986; Lelièvre et al., 1986).

GLASS TRANSITIONS IN SYNTHETIC POLYMERS

Glass transitions take place in amorphous polymer as reviewed in depth by Wunderlich (1990). In pure amorphous macromolecules the transition entails a change, over a temperature range of about 5 to 20°C, from an immobile solid to a rubbery liquid. In semicrystalline materials, the presence of microcrystals or strain tends to increase the temperatures and range over which the softening occurs (Wunderlich, 1990). To describe the glass transition the temperature, T_g, at which the polymer is midway between the rubbery and glassy states is specified. However a range of values is possible for a given polymer since the glass to rubber transformation depends on kinetic factors and is not an equilibrium process. T_g is a function of the previous thermal history of the sample. For example, the rate at which the polymer is cooled influences its value (Wunderlich, 1990).

The glass transition defines the region in which micro-Brownian motion ceases or commences (Wunderlich, 1990). The change resembles a 'second-order' thermodynamic transition and always takes place at a lower temperature than melting. For many polymers, the ratio of the melting temperature, T_m, to the glass transition (°K) is about 1.5 to 2.0 (Wunderlich, 1990). The annealing and crystallization of polymer only occurs in the region between the glass and melting temperatures (Wunderlich, 1976).

Glass softening is much more subtle than melting and consequently more difficult to detect. DSC traces normally signal the presence of a glass transition by a shift in the heat capacity in the absence of an enthalpy change. The heat capacity increase is generally of the order of 11J/K.mole. Dilatometry and TMA measurements indicate the occurrence of a glass transition, by a discontinuity in the change of volume and mechanical properties with temperature, respectively (Wunderlich, 1990).

Plasticization of amorphous polymer depresses the temperature at which the glass-to-rubber transformation occurs. The plasticizer increases the segmental motion of the random coil chains. Low molar mass diluents which act as plasticizers have a high free volume, so that at the molecular level diluent-polymer interactions lead to an increased free volume and increased chain motion. Thus the ability of a diluent to depress T_g decreases as the molar mass increases. Water is a very effective plasticizing agent and small quantities can cause T_g to decrease markedly (Moy and Karasz, 1980; Carfagna et al., 1982).

GLASS TRANSITIONS IN STARCH GELS

It is generally agreed that the features of the glass transition in starch gels are typical of those exhibited by synthetic polymers. Water is mixed intimately with the amorphous material in these gels and functions as a plasticizer. Accordingly T_g decreases as the moisture content increases until the structure is fully plasticized. Figure 2 shows how T_g is predicted to depend on the weight fraction of water according to a theory developed for copolymers (Van Den Berg, 1986). The figure suggests that relatively small quantities of water are needed to cause a marked decrease in T_g at low moisture contents. Thus T_g decreases by about 6°C for each percentage water increase for the first 10% moisture. According to the figure, T_g is

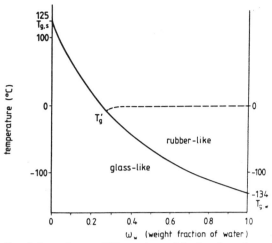

Fig. 2. The predicted dependence of T_g on the weight fraction of water (from Van Den Berg, 1986, with permission).

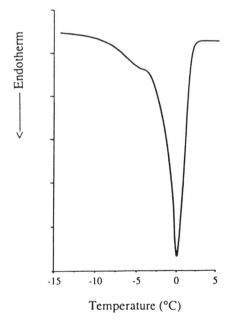

Fig. 3. The glass transition in a frozen starch gel (from Liu and Lelièvre, 1991a, with permission)

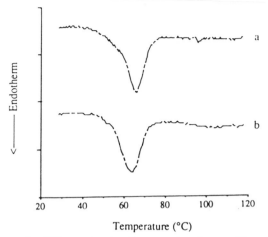

Fig. 4. DSC traces for 50% starch gels. Curve *a*, rice; *b*, corn (from Liu and Lelièvre, 1991a, with permission).

below the freezing point of pure water in many cases. This has been confirmed in DSC studies of frozen starch gels (Slade and Levine, 1988; Liu and Lelièvre, 1991a). A shoulder, that precedes and is superimposed on the ice melt, signals the heat capacity change accompanying the amorphous softening, see Figure 3.

A heat capacity change also accompanies the melting of the crystalline regions in starch gels that takes place at higher temperatures (Liu and Lelièvre, 1991a), see Figure 4. Clearly this heat capacity change is not due to a glass transition since annealing of the gels occurs well below the melting point (Slade and Levine, 1988).

GLASS TRANSITIONS IN STARCH GRANULES

There is considerable controversy as to the location of the glass transition in starch-water mixtures. To some extent, this simply reflects the fact that in all polymer-diluent systems where crystalline domains, in effect, crosslink amorphous polymer, the glass-rubber transformation occurs over a wide temperature range and is relatively difficult to detect (Wunderlich, 1990).

The Evidence From Different Experimental Techniques

Various experimental techniques have been employed in an attempt to determine T_g in starch granule systems.
a) Dilatometry studies.

Although TMA instruments have been used to make the measurements the expansion of the material, rather than the change in the mechanical properties, has been determined (Biliaderis et al., 1986; Maurice et al., 1985). There appears to be

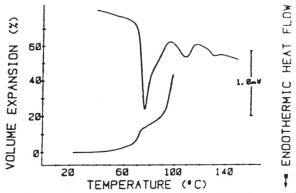

Fig. 5. DSC and dilatometry traces of rice starch (from Biliaderis et al., 1986, with permission. Copyright 1986 American Chemical Society)

no published work on the thermal expansion of relatively dilute starch suspensions, presumably because of experimental difficulties. In concentrated mixtures a two-stage process is observed, see Figure 5. The temperature range of the initial stage of expansion approximately corresponds to that of the leading edge of the endotherm while the final, more rapid, stage appears to take place towards the completion of melting as signalled by the levelling of the DSC trace. An exact comparison between the traces shown in the figure is not possible since different heating rates were used and this influences the pattern of the changes occurring when starch is heated (Shiotsubo and Takahashi, 1984).

The first investigators to use dilatometry suggested that the two stages are due to a glass transition followed by a melt (Maurice et al., 1985). A more elaborate explanation was proposed later in which four different regions of volume change were noted (Biliaderis et al., 1986). In stage 1, which is from ambient temperature to the suggested glass transition point, the sample is considered to undergo reversible thermal expansion. In stage 2, the granules are thought to swell irreversibly due to relaxation of the amorphous polymer fraction which is not bound or loosely connected to the crystallites. In the third intermediate plateau region, it was proposed that partial melting without dimensional change takes place. The final steep change in the volume expansion curve was attributed to melting. A different explanation, namely that the shape of the TMA curves is due to non-uniform melting, has recently been proposed (Liu et al., 1991). In any event, the non-uniform expansion shown in Figure 5 appears to be dominated by swelling of granules rather than by expansion of the amorphous regions of the particles (Slade and Levine, 1987).
b) X-ray studies.

X-ray measurements do not provide a direct means of assessing the changes that happen in amorphous starch polymer during a glass transition. However qualitative X-ray results have been reported to indicate that the initial portion of DSC endotherms occurs without a significant loss in crystallinity (Maurice et al., 1985). Consequently the shift in the DSC trace has been attributed to a major heat capacity

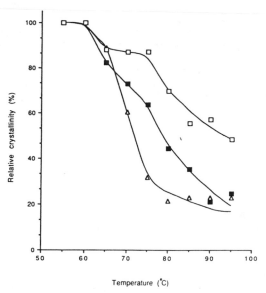

Temperature (°C)

Fig. 6. The crystallinity versus temperature for 30% ▲, 50% ■, and 60% □ rice starch suspensions (from Liu et al., 1991, with permission).

change and plasticization of amorphous polymer (Maurice et al., 1985). The qualitative X-ray experiments suggest that, after holding starch suspensions (45% water) at either 70°C or 80°C for 20 minutes, similar patterns are obtained and peak heights are only slightly reduced compared to controls. At higher temperatures, peak heights were reduced considerably. The magnitude of the corresponding crystallinity loss was not determined and this cannot be assessed from X-ray traces by simple visual inspection.

Quantitative X-ray measurements can be made using the crystallinity index procedure (Wakelin et al., 1959). This method compares the crystallinity of an unknown sample with that of crystalline and amorphous reference samples. Figure 6 shows how the crystallinity of starch suspensions changes with temperature. The data were obtained using intact native granules, and a 2% starch suspension that was heated to 95°C for 60 minutes and then freeze dried, as crystalline and amorphous standards respectively (Liu et al., 1991). The figure shows that in relatively dilute suspensions, crystallinity is lost monotonically while in more concentrated systems a two-step process is occurring. In the latter case there appears to be an initial decline in crystallinity, followed by an inflection leading to an intermediate plateau section, followed by a region of rapid crystallinity loss. This quantitative picture is consistent with the previous qualitative data. As discussed in more detail in the section on starch melting in this review, it is possible to demonstrate that the onset of crystallinity loss corresponds to the onset of the DSC gelatinization endotherm (Liu et al., 1991).

c) Heat capacity measurements.

Some of the first evidence consistent with a heat capacity change was reported by Biliaderis *et al.* (1986). These investigators studied eight starch samples and found that one gave a minor baseline shift at the start of the major endotherm associated with the loss of crystalline order. The other seven did not. The single DSC trace certainly suggests two events are taking place at the leading edge of the endotherm. However these may be two stages of melting rather than a heat capacity change followed by a melt (Liu and Lelièvre, 1991a). Figure 7 shows how the rate of change of the percentage of birefringent granules with temperature varies as a function of the temperature. The plot was constructed using birefringence data for wheat starch reported previously (Liu et al., 1991). The figure demonstrates that with some populations of starch particles, a number of less stable granules may gelatinize prior to the bulk of the sample. If this number is sufficiently large, two separate melting events would be evident at the leading edge of the endotherm.

Other early evidence for a heat capacity change associated with gelatinization was obtained in isothermal calorimetry measurements (Shiotsubo and Takahashi, 1986). A very small difference in the heat capacity of gelatinized and native starch was reported. A high degree of uncertainty was associated with the results, $\Delta C_p = -0.06 \pm 0.13J/\,g.K$.

In a key paper Zeleznak and Hoseney (1987) reported a baseline shift in DSC traces of samples containing less than about 20% moisture, see Figure 8. The figure shows the position of the heat capacity change is extremely sensitive to the moisture level. In samples containing more than about 20% moisture, the transition occurs below room temperature. Comparison of the granular and pregelatinized starches demonstrates that in the partly crystalline samples the temperature of the baseline shift is higher. This would be expected from synthetic polymer experiments

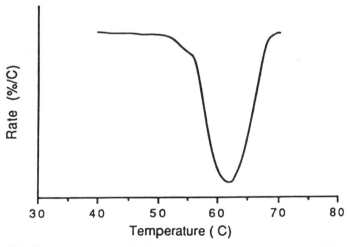

Fig. 7. d% birefringent granules/dt [temperature] versus temperature (Liu and Lelièvre, unpublished data).

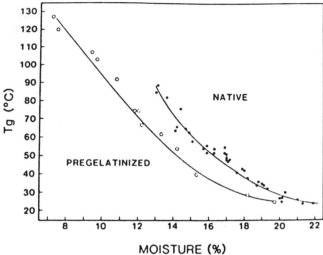

MOISTURE (%)

Fig. 8. Glass transition temperature of native and pregelatinized starches versus moisture content (from Zeleznak and Hoseney, 1987, with permission).

(Wunderlich, 1990). Zeleznak and Hoseney also showed that the heat capacity change is reversible in that starch polymer may be heated above, and then cooled below, the point of the baseline shift and on reheating the shift happens again. Recent work has confirmed these findings (Liu and Lelièvre, 1991a).

The method of Slade and Levine (1988) can be adapted to study the relationship between the course of the heat capacity change and of the DSC endotherm (Liu and Lelièvre, 1991a). The procedure entails heating the sample to a given temperature, cooling to the starting point and then rescanning to about 140°C. The difference between the two scans is then computed. The maximum temperature reached in the partial initial scan is first kept below that where the endotherm begins. In subsequent runs this temperature is gradually raised until the end point of the DSC trace is reached. A plot of the relative magnitude of baseline shift prior to the rescan trace versus temperature, see Figure 9, clearly demonstrates that the heat capacity change starts at the same point as the endotherm and increases in size until the phase transition is complete (Liu and Lelièvre, 1991a).

An earlier report suggested that the heat capacity change is located exclusively at the leading edge of the first DSC peak since after heating to any given temperature in this region, rescanning gives a baseline difference between the first and second scans (Slade and Levine, 1988). In addition, rescanning after heating to the peak maximum yields a comparatively symmetrical endotherm with a relatively flat baseline. This was considered to represent the melting transition of microcrystalline regions of fully water-plasticized starch (Slade and Levine, 1988). However, the data in the report shows that when the initial scanning temperature is increased further the baseline difference at the start of the two scans continues to increase.

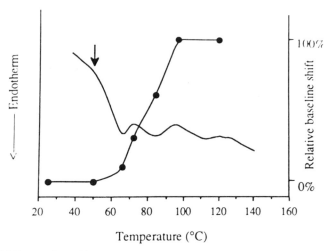

Fig. 9. DSC trace for a 50% corn starch-water mixture and corresponding baseline shift (from Liu and Lelièvre, 1991a, with permission).

The trend shown in Figure 9 would be expected if the heat capacity of the crystalline and amorphous starch polymer is not the same (Liu and Lelièvre, 1991a). The baseline difference would not be expected to increase on the grounds that there is more amorphous polymer due to gelatinization prior to the rescan, even if the glass transition of amorphous polymer in the ungelatinized granules were located at the leading edge of the first endothermic peak. This is because the amorphous material in gelatinized starch has a glass transition temperature below the melting point of ice (Slade and Levine, 1988; Liu and Lelièvre, 1991a). Furthermore, even if it is assumed that for every granule T_g does approximately equal T_m, given that starch particles have a range of stabilities and that some do not commence to melt until the trailing edge of the DSC endotherm is almost complete (Burt and Russell, 1983), the heat capacity would be expected to change as the entire endotherm is traversed.
d) Annealing experiments.

Since annealing can only take place when polymer is in the mobile rubbery state, its temperature range helps to define the glass transition point. It is well established that annealing occurs most rapidly just below the temperature at which gelatinization commences, however the phenomenon has been reported at temperatures down to 25°C (Slade and Levine, 1987). DSC traces of annealed samples show the peak temperature is increased and the peak width is decreased compared to controls (Yost and Hoseney, 1986; Knutson, 1990).

It has recently been demonstrated that annealed granules exhibit a heat capacity change when they are gelatinized (Liu and Lelièvre, 1991a). A sample of rice starch (50w%) was held at 50°C for one week and then heated by DSC. The temperature of the peak maximum increased from 76°C to 86°C and a baseline shift accompanied the endothermic trace.

Integration and Discussion of the Experimental Results

DSC demonstrates that a reversible baseline shift characteristic of a glass transition occurs in limited water systems (Zeleznak and Hoseney, 1987). The transition takes place below ambient temperature once the moisture content exceeds about 30%. There is some theoretical support for a glass transition in this temperature zone (Orford et al., 1989). In addition, comparison with synthetic polymers (Wunderlich, 1990) suggests that the temperature at which the glass transition occurs in starch would be less than the melting point, i.e. the gelatinization temperature. Granules can undergo annealing below the gelatinization point since the temperature is above T_g and below T_m.

It appears that the heat capacity change accompanying the major endothermic peak is not due to a glass transition, since the softening of amorphous polymer has already occurred (Zeleznak and Hoseney, 1986; Liu and Lelièvre, 1991a). The fact that this heat capacity change is given by samples that have undergone annealing, and are therefore above T_g, confirms this viewpoint (Liu and Lelièvre, 1991a). There is no quantitative X-ray evidence for a baseline shift in the absence of melting, suggesting a glass transition, in this region. In addition, the quantitative X-ray evidence suggests TMA curves reflect a two-stage melting process rather than a glass followed by a melt (Liu et al., 1991). The molecular events responsible for the heat capacity change accompanying gelatinization have yet to be established. One suggestion is that the change is attributable to the displacement of hydrogen bonds between starch crystallites and water molecules (Shiotsubo and Takahaski, 1986).

A central feature of the model proposing that the heat capacity change associated with gelatinization is due to a glass transition, is that an imbalance of water inside [10% (w/w)] and outside [100% (w/w)] the granule is considered to occur (Slade and Levine, 1988). Accordingly the water content inside the granule is not thought to reach the level required to fully plasticize the amorphous polymer below the gelatinization temperature. However, as discussed previously (Liu and Lelièvre, 1991a), there is extensive evidence demonstrating that the water content does reach 27% (w/w), at which point T_g < ambient temperature (Slade and Levine, 1988), in the time required to prepare DSC samples for measurement. Firstly, the gel phase of native starch granules is accessible to molecules with a molecular weight of less than about 1000 daltons (Brown and French, 1977). Secondly, NMR studies demonstrate that the surfaces of starch granules do not significantly limit water diffusion (Basler and Lechert, 1974) and it can be calculated that water enters into granules in a matter of seconds (Blanshard, 1979). Thirdly, tracer dilution measurements show the water content of intact granules to be about 45-50% on a dry basis (Dengate et al., 1978; Evans and Haisman, 1982). BeMiller and Pratt (1981) found a range of water contents (28-33%) for granular starch in aqueous suspension at room temperature. Finally, experiments using deuterium show that the hydroxyl hydrogens in the amorphous regions of native starch particles exchange readily (Taylor et al., 1961). In contrast, there appears to be no evidence that the water content of starch granules remains at about 10% (w/w) when they are suspended in water. However the possibility that a superheated glass is present cannot be ruled out solely by the presence of sufficient moisture to depress T_g, since the water could be present without plasticization occurring (Sears and Darby, 1982).

The melting of polymers has been explained in detail by Wunderlich (1990). The process is a first-order thermodynamic transition in which a crystalline-amorphous phase change occurs. When monitored by DSC, a change in the enthalpy is evident.

Melting is more complex in polymers than in low molar mass materials. Consider the case of an equilibrium crystal, i.e. one where the polymers are perfectly aligned and surface effects are negligible. Figure 10 illustrates the traces obtained with different heating rates. At low rates a sharp peak, similar to that obtained for pure organic reference materials, is found (Wunderlich, 1990). Melting occurs at, or very close to, the equilibrium melting point. However as the heating rate is increased, the trace broadens and the apparent melting point may increase by as much as 30°C. The beginning of melting is practically independent of the heating rate, this is typical of superheating (Wunderlich, 1990). Superheating takes place when heat is supplied to the crystal faster than it can melt. The interior of the crystal heats above the melting temperature, and finally melts when the crystal-melt interface progresses sufficiently (Wunderlich, 1990).

Superheating can also occur in imperfect crystals. These structures can reorganize and perfect themselves under slow heating regimes but not significantly when the temperature increases rapidly (Wunderlich, 1990). Figure 11 shows an example of how the melting temperature of a metastable crystal increases due to annealing . With such irreversible melting, slow heating gives more DSC peaks than fast, see Figure 12. The figure shows that with the slow heating profile, the initial

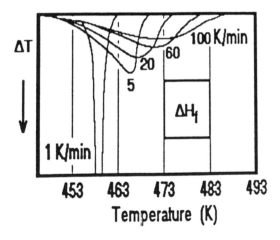

Fig. 10. DTA curves of extended chain polymer crystals heated at different rates (from Wunderlich, 1990, with permission).

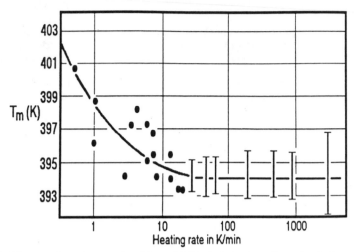

Fig. 11. Heating dependence of melting of folded-chain, lamella polyethylene (from Wunderlich, 1990, with permission).

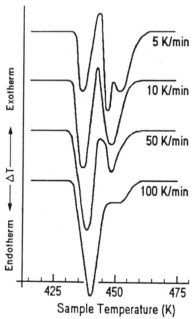

Fig. 12. Irreversible melting of a metastable crystal (from Wunderlich, 1990, with permission).

endotherm due to melting is interrupted as recrystallization occurs and an exothermic peak is manifest. A second endotherm signals the melting of the crystallized polymer. As the heating rate is raised, the exotherm and second melting peak gradually disappear until the limiting case is reached where the second melt is only visible as a weak shoulder.

In the presence of diluent, the melting point of polymer is depressed. The relationship between the equilibrium melting temperature and the volume fraction of diluent is given by the Flory-Huggins melting expression (Flory, 1953).

$$\frac{1}{T_m} - \frac{1}{T_m^o} = \frac{R}{\Delta H_u} \frac{V_u}{V_1} [\upsilon_1 - \chi_1 \upsilon_1^2] \qquad (1)$$

where R is the gas constant, ΔH_u is the enthalpy of fusion per repeat unit, V_u/V_1 is the ratio of molar volumes of the repeat unit and the diluent and χ_1 is the polymer-diluent interaction parameter.

When non-equilibrium melting takes place, as the volume fraction of diluent is increased, the melting behaviour follows a similar trend to the equilibrium case (Flory, 1953).

MELTING TRANSITIONS IN STARCH GELS

There is general agreement in the literature that when completely amorphous starch is allowed to recrystallize, the resultant gel melts on heating and gives a single endothermic peak (Slade and Levine, 1988). The position of the peak depends on the previous thermal history of the sample (Blanshard, 1987). It has been suggested that since the water is uniformly distributed throughout the gel, the melting takes place at the equilibrium T_m (Slade and Levine, 1988). Presumably this suggestion requires that the polymer crystals be in the equilibrium state and that superheating be avoided.

MELTING TRANSITIONS IN STARCH GRANULES

Although it is widely accepted that the endothermic transition that occurs during gelatinization is due, at least in part, to a melt there is considerable controversy as to the exact nature of the molecular events taking place (Biliaderis, 1991). Part of the controversy stems from the fact that gelatinization is a relatively complex event and DSC traces vary with factors such as the moisture content, heating rate and sample size.

The Evidence from Various DSC Experiments

a) The effect of moisture content.

DSC studies show that at relatively high water levels where the volume fraction of granule polymers is less than about 0.5, a single endotherm is apparent at about 60°C . The precise position of the peak depends on the starch variety being investigated (Biliaderis, 1991). When the volume fraction of starch polymer is increased, two endothermic transitions become evident. The first occurs at the same

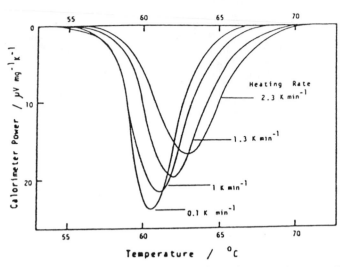

Fig. 13. Effect of heating rate on DTA peaks for a dilute starch suspension (Shiotsubo and Takahashi, 1984, with permission).

temperature as before. The temperature of the second increases as the water content decreases. If the volume fraction of polymer is raised still further, the lower temperature endotherm disappears while the temperature of the second peak continues to rise (Donovan, 1979). In addition to the above endotherms, transitions due to lipid-amylose complexes are found at about 120°C (Biliaderis et al., 1986; Biliaderis and Seneviratne, 1990). The exact temperature depends on the moisture content. The latter transitions will not be discussed further here as they have been the subject of a substantial review by Biliaderis (1991).
b) Heating rate and sample size.

Figure 13 shows that as the heating rate is increased the peaks move to a higher temperature and broaden (Shiotsubo and Takahashi, 1986). When the heating rate is raised to 20°C/min the peak temperature is increased by about 10°C (Shiotsubo and Takahashi, 1986). The beginning of the peak is practically independent of the heating rate. This behaviour is typical of superheating. The data in Figure 13 were obtained using an approximately 8% potato starch sample in which water would be expected to be in excess during the entire gelatinization process.

When more concentrated suspensions are heated at various rates a different set of changes occurs in the DSC profiles, see Figure 14. Slow heating gives more peaks than fast. In many polymer systems this is evidence for irreversible melting and certainly the order-disorder transition in starch is a non-equilibrium process under normal DSC conditions (Nakazawa et al., 1984; Slade and Levine, 1988). However the traces in Figure 14 do not necessarily mean that at slow heating rates recrystallization, or perhaps short-range ordering, is taking place after initial partial melting and that when heating is fast (30°C/min.) there is insufficient time for these processes to

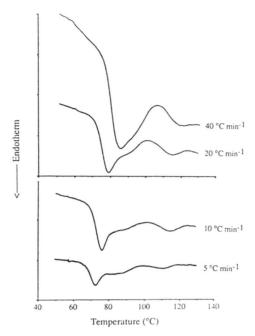

Fig. 14. Effect of heating rate on DSC traces for a concentrated starch suspension (from Liu and Lelièvre, 1991b, with permission).

occur. Experiments in which the sample size is varied show that large samples give single peaks with slow heating rates, while small samples give dual peaks with fast heating rates (Liu and Lelièvre, 1991b). This suggests that changes in the resolution of the DSC instrument with heating rate may be contributing to the shape of the traces given in Figure 14. Another factor may be that fast heating rates do not permit water redistribution in the samples (Liu and Lelièvre, 1991b). In any event, X-ray measurements fail to show evidence of recrystallization during the heating regimes normally employed in DSC instruments (Zobel et al., 1988; Liu et al., 1991). Indeed, amylopectin would not be expected to crystallize rapidly. However, short-range ordering which would not be detected by X-ray measurements, could occur in DSC experiments (Biliaderis, personal communication, 1992).
c) Annealing.

If the heating rate is sufficiently slow, partial melting followed by crystallization occurs. The temperature of the endotherm is then increased (Nakazawa et al., 1984). This phenomenon is easiest to demonstrate in experiments where annealing and crystal perfection are allowed to take place when the starch suspension is held at a given temperature (Nakazawa et al., 1984).

Discussion of the Experimental Results

A number of explanations of the molecular processes responsible for starch DSC traces have been proposed (Blanshard, 1987; Biliaderis, 1991). The suggestion that recrystallization of polymer accounts for biphasic endotherms has already been considered herein. Other investigators suggest that the lower temperature peak is a consequence of chain mobilization in the amorphous regions of the granule (Maurice et al., 1985; Biliaderis et al., 1986; Slade and Levine, 1988). According to the model the glass transition, which is responsible for the change in heat capacity, is located at the leading edge of the first DSC peak. Completion of the glass transition permits the crystalline domains to undergo a non-equilibrium melting process giving the second endothermic trace (Slade and Levine, 1988). However, as discussed previously in this article, considerable evidence indicates that the heat capacity change occurs during both endothermic peaks and is not associated with a glass transition (Zeleznak and Hoseney, 1987; Liu and Lelièvre, 1991a; Liu et al., 1991).

Donovan (1979) suggested that the initial DSC peak in biphasic endotherms results from the the stripping of polymer chains from the surfaces of crystallites due to stress, while the second peak represents melting at low diluent volume fractions. According to this model, in excess water swelling takes place and the resultant stress means that only the stripping mechanism operates. However the melting point of polymers under stress tends to increase, rather than decrease (Mandelkern, 1964), which is inconsistent with this model (Lelièvre, 1985). Since Donovan considered different mechanisms to be responsible for each section of the biphasic endotherm, the peaks were labelled G and M respectively. However Zobel has questioned this nomenclature and has emphasised that X-ray data do not indicate that different molecular processes are responsible for each peak (Zobel et al., 1988).

Another explanation of gelatinization is based on the fact that the crystalline zones in different granules have different stabilities (Evans and Haisman, 1982; Liu and Lelièvre, 1992). Water migrates from one location to another within the sample as the various granules gelatinize. Thus the peaks correspond to melting transitions of crystalline material with different stabilities, at different diluent levels. According to the theory if the water content of the suspension is sufficiently high, i.e. above about 65%, each granule absorbs moisture without restriction and a single endothermic peak is observed. If the water content is less than this level, there is competition between the granules for water. In this case the least stable granules melt first, absorb water and so deplete the remaining particles of diluent. The latter particles melt at higher temperatures partly because they are more stable and partly because the effective volume fraction of diluent is reduced. The associated endotherm initially occurs in the form of a trailing shoulder on the first peak. However as the water content of the sample is reduced, fewer granules are able to gelatinize in an unrestricted water environment and so the first endotherm decreases in size while the second increases and shifts to a higher temperature. If the volume fraction of diluent is reduced sufficiently, only the higher temperature endotherm is apparent (Evans and Haisman, 1982; Liu and Lelièvre, 1992).

The melting explanation of gelatinization is supported by X-ray crystallinity measurements (Zobel et al., 1988; Liu et al., 1991). However, direct comparison of data from the two techniques is not possible since the small size of DSC samples

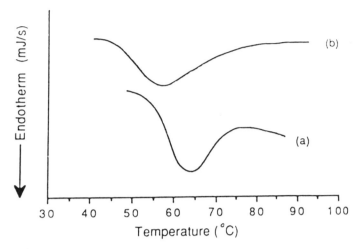

Fig. 15. DSC traces (a) for a 30% starch suspension and (b) the corresponding trace calculated from X-ray data (from Liu et al., 1991, with permission).

precludes their use for X-ray measurement (Maurice et al., 1985). The DSC enthalpy change equals the sum of the endothermic melt, the exothermic heat of hydration of the disordered polymer and the contribution from lipid-amylose complex formation (Yost et al., 1986). Any change in heat capacity will also be evident in the DSC trace. If it is assumed that a given decrease in ordered polymer gives rise to a constant enthalpy change, then X-ray crystallinity versus temperature relationships may be converted into a form that is comparable with the calorimeter output (Liu et al., 1991). Since DSC measures dH/dt, where H is the heat, the first derivative of the X-ray crystallinity curves is taken (Liu et al., 1991).

Relatively dilute starch suspensions exhibit a single peak in the X-ray derivative trace, see the data in Figure 15 for a mixture containing 30% polymer (Liu et al., 1991). Given the different thermal treatments received by the samples, the range over which the order-disorder transition occurs in the X-ray experiment should be about 8°C lower than found in the corresponding DSC run (Shiotsubo and Takahashi, 1984). Thus there is broad agreement between the derivative X-ray trace and the DSC output for the 30% suspension. The X-ray data for more concentrated suspensions clearly give a biphasic effect, see the example for a starch-water mixture containing 60% polymer in Figure 16. The transition again shifts to lower temperatures than found by DSC. In the low-moisture mixtures, rapid heating causes non-equilibrium water gradients with the result that a greater fraction of granules gelatinize at higher temperatures (Blanshard, 1987). Thus as shown in the figure, more particles gelatinize at lower temperatures in the X-ray experiments due to the higher heating rates (Liu et al., 1991).

DSC experiments on rice-wheat starch blends also provide evidence that water migrates from one region to another when mixtures containing relatively low moisture

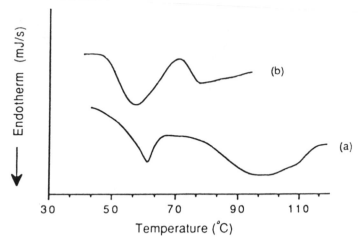

Fig. 16. DSC trace (a) for a 60% starch suspension and (b) the corresponding trace calculated from X-ray crystallinity data (from Liu et al., 1991, with permission).

levels are gelatinized (Liu and Lelièvre, 1992). These varieties were studied since they gelatinize at different temperatures and can be distinguished by microscopy. In excess water, where a single endotherm occurs, the DSC output is simply the sum of the traces for each of the components in the mixture. The wheat starch endotherm takes place about 10°C below that for the rice starch.

In more concentrated suspensions, for example those containing 50% starch, the situation is more complicated, see Figure 17. The temperature of the first peak in the endotherm is always about the same [63°C]. Microscopy suggests this peak is solely attributable to gelatinization of wheat granules. The area of the peak begins to increase and then decreases as the weight fraction of the wheat starch in the blend reduces. The temperature of the second peak, which is mainly due to the gelatinization of rice starch, increases as the weight fraction of rice in the blend decreases. The second peak becomes smaller as the wheat:rice ratio becomes larger.

The DSC trace for the pure wheat starch given in Figure 17 shows a trailing shoulder since water is restricted. This trailing shoulder is almost absent in the 85:15 wheat-rice blend. The rice component has a significantly higher melting temperature and does not compete effectively with the gelatinizing wheat particles for water. To some extent, the rice granules act as an inert filler in the suspension when the wheat starch gelatinizes. Hence the behaviour of the wheat granules is equivalent to that in a suspension containing about 45% [i.e. 0.85/1+0.85] starch, and about 55% water. With the water being, in effect, less restricted during the gelatinization of the wheat starch, the first peak is prominent and the second trailing peak is almost absent. The area of this former peak is greater than that for the corresponding endotherm in the 100% wheat sample even though the mass of wheat granules is less. Presumably this is because in the 100% wheat sample, the limited

156

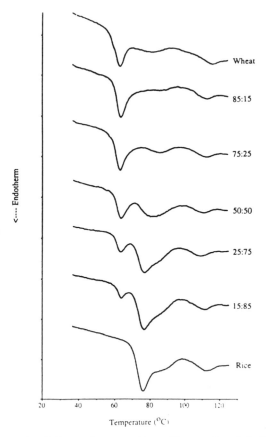

Fig. 17. DSC traces for 50% starch suspensions. The numbers labelling the traces give the ratios of the weight of wheat to rice in the blends (from Liu and Lelièvre, 1992, with permission).

water conditions cause a smaller mass of granules to gelatinize at the lower temperature. When the rice starch in the 85:15 blend melts most of the water has been absorbed by the gelatinized wheat granules so the transition shifts to a temperature which is about 25°C higher than for the corresponding pure rice case (Liu and Lelièvre, 1992).

As the proportion of wheat starch in the blend is reduced further the wheat endotherm remains at the same temperature since the water is, in effect, still unrestricted. The peak area reduces in size as fewer granules now undergo the melt. The corresponding endotherm for the rice firstly increases in size as a greater mass of particles gelatinizes, and secondly shifts to lower temperatures as the effective diluent volume fraction increases. However the first rice peak in the traces of the blends, corresponding to initial melting in excess water, is never particularly prominent as the wheat granules have already absorbed part of the diluent (Liu and Lelièvre, 1992).

Gelatinization and the Flory-Huggins Expression

Gelatinization is an order-disorder melting transition. Hence the Flory-Huggins expression might be expected to be an appropriate, though highly idealised, model of the process if polymer is annealed sufficiently and equilibrium melting conditions apply. However it may not be possible to obtain the requisite equilibrium crystals with a highly branched polymer such as amylopectin since the branch points limit crystal growth (Wunderlich, 1990). The Flory-Huggins approach would therefore need to be modified to take account of the effect of the surface free energy, as has been done for copolymers (Wunderlich, 1990). With rapid heating regimes, such as usually occur under industrial conditions, melting is a non-equilibrium process (Wunderlich, 1990). In this case application of the Flory-Huggins expression will predict trends in melting behavior that may be useful in practice but the values of the parameters obtained do not agree with the equilibrium situation and have limited theoretical basis.

OVERALL CONCLUSION

Partly crystalline polymers show two characteristic transitions on heating. As the temperature is raised from a low value a glass transition occurs first. The amorphous polymer goes from a glass to a rubber in a quasi second-order transition in which kinetic factors play a role. When the temperature rises considerably further, a melt takes place which entails a first order transition from partly-crystalline solid to amorphous liquid. Both of these transformations happen in starch systems where water, and other solvents, have a marked influence of the temperature at which the changes happen.

Much remains to be learnt about gelatinization and current views on the mechanism of the process may well change. In this connection, work by Hoseney et al. (1986) suggests that pure amylopectin crystals can give biphasic endotherms. This would not be expected unless some kind of gel structure is present causing water to migrate from one location to another. Another possibility is that amylopectin forms two kinds of ordered domains, namely crystallites and 'double helices' associated with short-range order (Hoseney et al., 1986; Russell, 1987). In concentrated granular systems these may melt at different temperatures (Hoseney et al., 1986; Russell, 1987). More research is needed to evaluate these interesting findings.

ACKNOWLEDGEMENT

Financial support from the Natural Sciences and Engineering Research Council of Canada is gratefully acknowledged.

Basler, W., and Lechert, H. 1974. Wide-line NMR of water in starch gels at 295K. Starke 26:39.

BeMiller, J. N., and Pratt, G. W. 1981. Sorption of water, sodium sulfate, and water-soluble alcohols by starch granules in aqueous suspension. Cereal Chem. 58:517.

Biliaderis, C. G. 1990. Thermal analysis of food carbohydrates. Page 168 in: Thermal Analysis of Food. V. R. Hatwalker and C.-Y. Ma ed. Elsevier Applied Science Publishers, Barking, UK.

Biliaderis, C. G. 1991. The structure and interaction of starch with food constituents. Can. J. Physiol Pharmcol. 69:60.

Biliaderis, C. G., Page, C. M., Maurice, T. J., and Juliano, B. O. 1986. Thermal characterization of rice starches: A polymeric approach to phase transitions of granular starch. J. Agric. Food Chem. 34:6.

Biliaderis, C.G. and Seneviratne, H. D. 1990. On the supermolecular structure and metastability of glycerol monostearate-amylose complex. Carbohydr. Polym. 13:185.

Blanshard, J. M. V. 1979. Physiochemical aspects of starch gelatinization. Page 139 in: Polysaccharides in Food: J. M. V. Blanshard and J. R. Mitchell ed. Butterworths, London.

Blanshard, J. M. V. 1987. Starch granule structure and function, a physicochemical approach. Page 16 in: Starch: properties and potential T. Galliard ed. John Wiley and Sons, New York.

Brown, S. A., and French, D. 1977. Specific adsorption of starch oligosaccharides in the gel phase of starch granules. Carbohydr. Res. 59:203.

Burt, D. J., and Russell, P. L. 1983. Gelatinization of low water content wheat starch- water mixtures. Starke 35:354.

Callaghan, P. T., Jolley, K. W., Lelièvre, J., and Wong, R. B. K. 1983. Nuclear magnetic resonance studies of wheat starch pastes. J. Colloid and Interface Sci. 92:332.

Callaghan, P. T., and Lelièvre, J. 1986. The influence of polymer size and shape on self-diffusion of polysaccharides and solvents. Analytica Chemica Acta 189:145.

Carfagna, C., Apicella, A., and Nicolais, L. 1982. The effect of the prepolymer composition of amino-hardened epoxy resins on the water sorption behavior and plasticization. J. Appl. Polym. Sci. 27:105.

Dengate, H. N., Baruch, D. W., and Meredith, P. 1978. The density of wheat starch granules: A tracer dilution procedure for determining the density of an immiscible dispersed phase. Starke 30:80.

Donovan, J. W. 1979. Phase transitions of starch-water systems. Biopolymers 18:263.

Evans, I. M., and Haisman, D. R. 1979. Rheology of gelatinized starch suspensions. J. Texture Studies 10:347.

Evans, I. D., and Haisman, D. R. 1982. The effect of solutes on the gelatinization temperature range of potato starch. Starke 34:224.

Flory, P. J. 1953. Principles of Polymer Chemistry. New York, Cornell University Press.

French, D. 1972. Fine structure of starch and its relationship to the organization of the granules. J. Jpn. Soc. Starch Sci. 19:8.

French, D. 1984. Organization of starch granules. Page 83 in: Starch Chemistry and Technology R. L. Whistler, J. N. BeMiller and E. F. Paschall ed. Academic Press, Orlando.

Hoseney, R. C., Zeleznak, K. J., and Yost, D. A. 1986. A note on the gelatinization of starch. Starke 38:407.

Jenkins, A. D. 1972. Polymer Science. Amsterdam, North-Holland Publishing Company.

Jin, X., Ellis, T. S., and Karasz, F. E. 1984. The effect of crystallinity and crosslinking on the depression of the glass transition temperature in nylon 6 by water. J. Polym. Sci.: Polym. Phys. 22:1701.

Katz, J. R., and Van Itallie, T. B. 1930. All varieties of starch have similar retrogradation spectra. J. Phys. Chem. A 150:90.

Knutson, C. A. 1990. Annealing of maize starches at elevated temperatures. Cereal Chem. 67:376.

Lelièvre, J. 1985. Gelatinization of crosslinked potato starch. Starke 37:267.

Lelièvre, J., Lewis, J., and Marsden, K. 1986. The size and shape of amylopectin: A study using ultracentrifugation. Carbohydr. Res. 153:195.

Liu, H., and Lelièvre, J. 1991a. A differential scanning calorimetry study of glass and melting transitions in starch suspensions and gels. Carbohydr. Res. 219:23.

Liu, H., and Lelièvre, J. 1991b. Effects of heating rate and sample size on differential scanning calorimetry traces of starch gelatinized at intermediate water levels. Starch 43:225.

Liu, H., and Lelièvre, J. 1992. A differential scanning calorimetry study of melting transitions in aqueous suspensions containing blends of wheat and rice starch. Carbohydr. Polym. 17:145.

Liu, H., Lelièvre, J., and Ayoung-Chee, W. 1991. A study of starch gelatinization using differential scanning calorimetry, X-ray, and birefringence measurements. Carbohydr. Res. 210:79.

Lund, D. B. 1984. Influence of time, temperature, moisture, ingredients and processing conditions on starch gelatinization. Crit. Rev. Food Sci. Nutr. 20:249.

Mandelkern, L. 1964. Crystallization of polymers. New York, McGraw-Hill Book Company.

Maurice, T. J., Slade, L., Sirett, R. R., and Page, C. M. 1985. Polysaccharide-water interactions-thermal behaviour of rice starch. Page 211 in: Properties of Water in Foods D. Simatos and J. L. Multon ed. Martinus Nijhoff Publishers: Dordrecht.

Moy, P., and Karasz, F. E. 1980. Water in polymers. ACS Symp. Series 127, Washington DC, American Chemical Society.

Nakazawa, F., Nogochi, S., Takahashi, J., and Takada, J. 1984. Thermal equilibrium state of starch-water mixture studied by differential scanning calorimetry. Agric. Biol. Chem 48:2647.

Orford, P. D., Parker, R., and Ring, S. G. 1989. Effect of water as a diluent on the glass transition behaviour of malto-oligosaccharides, amylose and amylopectin. Int. J. Biol. Macromol. 11:91.

Russell, P. L. 1987. Gelatinization of starches of different amylose/amylopectin content. A study by differential scanning calorimetry. J. Cereal. Sci. 6:133.

Sears, J. K., and Darby, J. R. 1982. The Technology of Plasticizers. New York, John Wiley and Sons.

Shiotsubo, T., and Takahashi, K. 1984. Differential thermal analysis of potato starch gelatinization. Agric. Biol. Chem. 48:9.

Shiotsubo, T., and Takahashi, K. 1986. Changes in enthalpy and heat capacity associated with the gelatinization of potato starch, as evaluated from isothermal calorimetry. Carbohydr. Res. 158:1.

Slade, L., and Levine, H. 1987. Recent advances in starch retrogradation. Page 387 in: Industrial Polysaccharides. S. S. Stivala, V. Crescenzi and I. C. M. Dea ed. Gordon and Breach Science Publishers, New York.

Slade, L., and Levine, H. 1988. Non-equilibrium melting of native granular starch. Part I. Temperature location of the glass transition associated with the gelatinization of A-type cereal starches. Carbohydr. Polym. 8:183.

Taylor, N. W., Zobel, H. G., White, M., and Senti, F. R. 1961. Deuterium exchange in starches and amylose. J. Phys. Chem. 65:1816.

Van Den Berg, C. 1986. Water activity. Page 11 in: Concentration and drying of foods. D. McCarthy ed. Elsevier Applied Science Publishers, London.

Wakelin, J. H., Virgin, H. S., and Crystal, E. 1959. Development and comparison of two X-ray methods for determining the crystallinity of cotton cellulose. J. Appl. Phys. 30:1654.

Wissler, G. E., and Crist, B. 1980. Glass transition in semicrystalline polycarbonate. J. Polym. Sci.: Polym. Phys. Ed (USA) 18:1257.

Wong, R. B. K., and Lelièvre, J. 1981. Viscoelastic behaviour of wheat starch pastes. Rheol. Acta 20:299.

Wunderlich, B. 1973. Crystal structure, morphology, defects. In: Macromolecular Physics ed. Academic Press, New York.

Wunderlich, B. 1976. Crystal nucleation, growth, annealing. In: Macromolecular Physics ed. Academic Press, New York.

Wunderlich, B. 1980. Crystal melting. In: Macromolecular Physics ed. Academic Press, New York.

Wunderlich, B. 1990. Thermal Analysis. Boston, Academic Press.

Yost, D. A., Burkhard, R. K., and Hoseney, R. C. 1986. Heat-burst calorimetry of heated starch. Starke 38:366.

Yost, D. A., and Hoseney, R. C. 1986. Annealing and glass transition of starch. Starke 38:289.

Zeleznak, K. J., and Hoseney, R. C. 1987. The glass transition in starch. Cereal Chem. 64:121.

Zobel, H. F., Senti, F. R., and Brown, D. S. 1965. Studies on starch gelatinization by differential thermal analysis. 50th Annual Meeting of American Association of Cereal Chemists.

Zobel, H. F., Young, S. N., and Rocca, L. A. 1988. Starch gelatinization: An X-ray diffraction study. Cereal Chem. 65:443.

NUCLEAR MAGNETIC RESONANCE ANALYSIS
OF CEREAL CARBOHYDRATES

Michael J. Gidley
Unilever Research Laboratory
Colworth House
Sharnbrook, Bedford, MK44 1LQ, UK

INTRODUCTION

Nuclear magnetic resonance (NMR) is finding
increasing application in the study of cereal
carbohydrates. This review seeks to summarise some
recent progress in this field concentrating on starch
and the characteristic non-starch polysaccharides of
cereals, arabinoxylans and mixed linkage β-glucans.
Many advances have been made possible by two separate
developments in NMR technology, namely ever higher
stable magnetic fields and greatly increased
electronic and computational facilities. The former
leads to greater spectral resolution and sensitivity,
and the latter has resulted in an explosion of NMR
techniques (particularly solution state) which are
now being exploited by workers in many fields
(Sanders and Hunter, 1987). A further development
over the last 10-15 years has been the availability
of commercial systems for obtaining high resolution
NMR spectra from solid materials; cereal grains and
starches have been studied using these techniques.
       The following sections describe (i) the use of
solution state NMR for determining molecular
structure and solution conformation, (ii) solid state
NMR analyses of starches and related systems
including conformational analysis, (iii) NMR

relaxation methods for studying cereal carbohydrates and (iv) application of all these techniques to the study of the important functional properties of gelatinization, retrogradation and gelation.

HIGH RESOLUTION SOLUTION STATE NMR

In principle, NMR spectra contain separate signals for all magnetically (in practise, chemically) inequivalent atomic sites within the system. Only those nuclei with a single unpaired nuclear spin give such simple spectra, but fortunately both ^{1}H and ^{13}C (and ^{31}P) fulfil this criterion. ^{1}H is a much more sensitive (ca. 10^4 times) NMR nucleus than ^{13}C, but the latter has a wider spectral range (expressed in parts per million of the applied magentic field). The choice of technique varies with application, with complementary information often being obtained by using both nuclei. The two major uses of solution state NMR in cereal carbohydrates are to establish (and quantify) structural features and to investigate solution conformations; each will be addressed in the following sections.

Molecular Structure

Starch
Determination and quantification of primary structure features by NMR requires accurate spectral assignments. Early work in this field (e.g. Usui et al, 1973, 1974) led to complete assignments for ^{13}C (Heyraud et al, 1979) and ^{1}H (Morris and Hall, 1982) spectra of linear starch oligomers. The ^{1}H-NMR spectrum of a model compound for amylopectin branch points, isopanose (α-Glc(1$\rightarrow$4)(α-Glc(1$\rightarrow$6))Glc) has also been assigned (Bock and Pedersen, 1984). These low molecular weight compounds give narrow spectral signals in comparison with their parent polymers which are sufficiently broadened as to give no more information than that from a simple combination of features due to interior chain, reducing and non-

reducing termini and branch points. Nevertheless
such information, if quantifiable, gives valuable
information on native and degraded starches. Using
only anomeric proton signals it is possible to
quantify ratios of both branch points and reducing
termini to main chain residues in starch-based
samples (Gidley, 1985). Thus the degree of branching
of amylopectins (and potentially amylose), and the DE
value together with the degree of branching for
starch degradation products can be readily obtained
using this non-destructive technique. As some
starches (notably potato) contain covalently-bound
phosphate groups, they are amenable to ^{31}P NMR
analysis. First results (Muhrbeck and Tellier, 1991)
indicate that, for a number of potato varieties, C-3
phosphorylation is nearly constant but that C-6
phosphorylation shows significant varietal
differences.

Mixed linkage glucans
        The general structure of cereal (1→3), (1→4)
glucans involves cellotriose or cellotetraose units
interspersed by β-(1→3) linkages. High field ^{13}C NMR
spectroscopy has been shown to resolve sufficient
structural features to allow any differences in
polymer composition to be assessed (Dais and Perlin,
1982). In particular all glycosidic linkage sites
could be discriminated and quantitative estimates for
each obtained. The limited resolution obtained from
^{1}H spectra did not yield any further information.
More recently (Bock et al, 1991), ^{1}H and ^{13}C spectra of
oligosaccharides obtained by incubation of barley β-
glucan with an endo-[1→3(4)]-β-D-glucanase have been
completely assigned. Structures of two tetrasaccharides (β-D-Glcp-(1→4)-β-D-Glcp-(1→3)-β-D-Glcp-(1→4)-
D-Glcp, and β-D-Glcp-(1→3)-β-D-Glcp-(1→4)-β-D-Glcp-
(1→4)-D-Glcp) and one trisaccharide (β-D-Glcp-(1→3)-
β-D-Glcp-(1→4)-D-Glcp) were completely determined by
NMR methods, primarily 2-dimensional techniques which
allow through-bond ^{1}H connectivities to be established
in a sequential manner (COSY-type). To complement
this information, through-space connections were

established _via_ nuclear Overhauser effects (n.O.e.) obtained in the rotating frame (ROESY). Conventional n.O.e. spectroscopy is of limited use for many oligosaccharides due to the particular correlation times (mobility) which they exhibit resulting in low to zero n.O.e.'s. For a non-mathematical description of such modern NMR experiments, the reader is referred to Sanders and Hunter (1987).

Arabinoxylans

Recent results (Hoffmann, 1991) suggest that ^{1}H-NMR may be the method of choice for the structure determination of the arabinoxylan family of poly-saccharides. Along a common (1→4)-β-D-Xylp backbone, single α-L-arabinofuranose units are present linked either to O-2, O-3 or both O-2 and O-3 of some

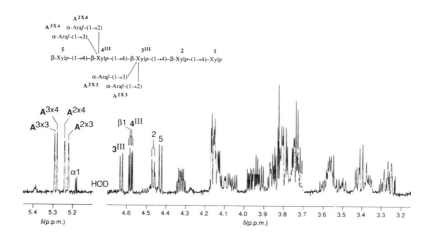

Fig.1 Resolution-enhanced 600MHz ^{1}H-NMR spectrum of the arabinoxylan nonasaccharide shown, with anomeric signal assignments. From Hoffmann et al (1992) with permission.

xyloses. Water-soluble arabinoxylan fractions from rye (Bengtsson and Åman, 1990) and wheat (Hoffmann, 1991) however contain very little mono O-2

substitution.   Endo-β-D-xylanase digestion can be
used to provide a large family of oligosaccharides
containing various combinations of branched, double-
branched and unbranched units.   Through systematic
high field ^{1}H NMR analysis, it is possible to assign
unambiguous structures for oligosaccharides
containing up to at least 14 residues.   The ^{1}H NMR
spectrum of an arabinoxylan nonasaccharide is shown
in Figure 1, in which the anomeric signals for each
residue are resolved.   Using through-bond (Figure 2)
and through-space two dimensional correlation
experiments, the complete spectrum can be assigned
(Hoffmann et al, 1992).   Similar analyses of many
oligosaccharides containing most possible

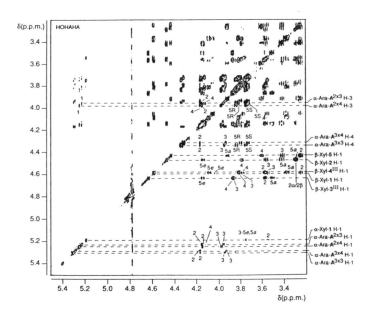

Fig.2   600MHz HOHAHA spectrum of the nonasaccharide
shown in Fig.1.   Reprinted with permission from
Hoffmann (1991).

arrangements of branches, allows anomeric ^{1}H signals

to be used as diagnostic 'structural reporter groups' (Vliegenthart et al, 1983) for the analysis of polymeric arabinoxylans (Hoffmann, 1991). ^{13}C-NMR also resolves structural features (Bengtsson and Åman, 1990; Hoffmann et al, 1991) such as branch type but the detailed micro-sequence data obtainable from ^{1}H-NMR is not achieved.

Solution Conformation

One of the main purposes of obtaining the signal assignments discussed above (particularly for samples of known strucutre) is to exploit the ability of NMR spectroscopy to provide information on solution conformation. Shapes of individual sugar rings are assumed largely invariant (not always true), so glycosidic conformation is usually the parameter of interest as this determines the overall shape of an oligo- or polysaccharide. In practice, two bond (i.e. non-(1→6) linkages) glycosidic conformation is largely determined by torsion angles ϕ and ψ (Figure 3) as tetrahedral angles at carbon and dihedral angles at the glycosidic oxygen are similar for most systems. A potentially direct

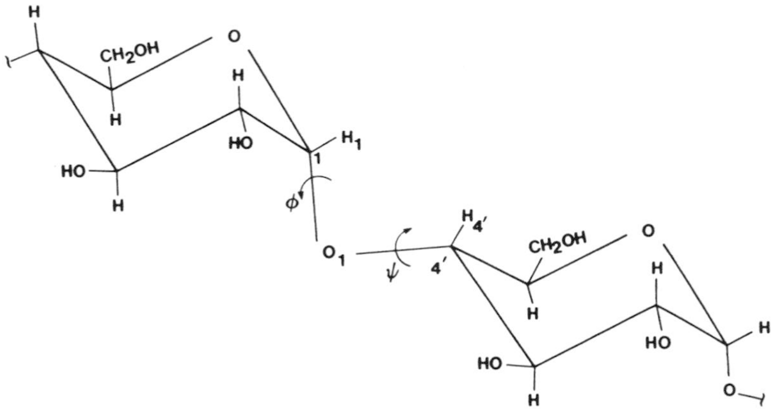

Fig.3 α-(1-4) Glucan repeating unit, illustrating the two torsion angles (ϕ and ψ) which determine overall chain conformation.

approach to determining these angles is from 3-bond coupling constants as e.g. coupling from H-1 via C-1 and O-1 to C-4' is largely determined by angle ϕ and conversely ψ is the major determinant of coupling between H-4' and C-1. Early work in this area (Parfondry et al, 1977) used ^{1}H-coupled ^{13}C-spectra and was extended (Pérez et al, 1985) to show solvent-induced conformational effects. Spectral complexity limits the measurement and assignment of heteronuclear coupling constants by direct observation of coupled spectra. However, a 2D NMR method is available in which a ^{13}C spectrum is split in the second dimension by coupling to only a selected (irradiated) proton (Bax and Freeman, 1982). This method has been shown to be applicable to carbohydrates (Gidley and Bociek, 1985) and has provided coupling constants for malto-oligomers (Gidley, 1988) and other carbohydrates useful for developing the quantitative relationship with angles ϕ and ψ (Mulloy et al, 1988). Values for starch oligomers from DP2 to 7 were particularly interesting, as differences in the ψ-determined coupling constant were observed through the series, although the ϕ-determined coupling was invariant (Gidley, 1988). Optical rotation data suggested that the higher oligomers (DP4-7) were appropriate models for the polymer, amylose. Information on glycosidic conformations in maltose and maltotriose are probably not therefore directly related to polymer behaviour.

Nuclear Overhauser enhancments provide information on the relative distances between protons and have been used (Hoffmann, 1991) to suggest that the backbone in arabinoxylan oligosaccharides adopt a conformation consistent with a left handed three-fold helix previously proposed from X-ray fibre diffraction data (Nieduszynski and Marchessault, 1972). ^{1}H coupling constant data was also used (Hoffmann, 1991) to describe the conformation of sidechain arabinofuranose units which were shown to differ slightly depending on the substituent environment. Surprisingly few n.o.e. studies have been performed on starch and related oligomers, with the exception of maltose derivatives (Shashkov et al, 1986, Alvarado et al, 1984). This may reflect the near-degeneracy of both H-2 and H-4 chemical shifts in malto-oligomers making specific enhancements difficult to resolve and quantify.

NMR can also be used to determine the presence and nature of hydrogen bonding in non-aqueous solutions. An early use of NMR data showed unusual chemical shifts for hydroxyl groups attached to C-2 and C-3 of amylose in dimethyl sulphoxide solution, suggestive of an inter-residue hydrogen bond (Casu et al, 1966). This suggestion was backed up by infrared data and was shown to occur for linear and cyclic $(1\rightarrow4)-\alpha$-D-glucans as well as for amylose. This system was later re-examined by variable temperature NMR and interpreted in terms of a hydrogen bond from OH (3^1) to OH (2) (St. Jacques et al, 1976). The same feature has more recently been illustrated using 2D NMR (Nardin et al, 1984) and partial deuteration techniques (Christofides and Davies, 1987).

^{13}C chemical shifts of starch-related materials in solution have been shown to respond to solvent composition (Peng and Perlin, 1987) and the presence of complexing agents (Jane et al, 1985). In the latter case, the observed downfield shift of C-1 and C-4 signals was interpreted in terms of a single (V-type) helix.

There are two major problems with obtaining a detailed characterisation of solution conformation of polysaccharides using NMR. Firstly, only oligomers can be used to perform the detailed conformational experiments described above and hence it needs to be shown that oligomers studied are reliable conformational models. Secondly, and probably of greater importance, there is sufficient freedom of motion about glycosidic linkages that a range of conformations should not be discounted without evidence. Any single 'conformation' derived from experimental data would thus reflect an average which may have no physical meaning. These drawbacks with solution conformation studies are overcome through the use of solid state methods as discussed later.

Starch-related spectra

The combined techniques (Schaeffer and Stejskal, 1979) of high power decoupling, rapid spinning at the 'magic angle', and cross polarisation from abundant ^1H nuclei to dilute ^{13}C nuclei (^{13}C CPMAS NMR) have revolutionised high resolution NMR spectroscopy of organic solids. Cereal grains and fractions derived from them are readily analysed by this methodology (Baianu and Förster, 1980; O'Donnell et al, 1981). In this work, characteristic spectra for starches were observed but only limited signal assignment was possible.

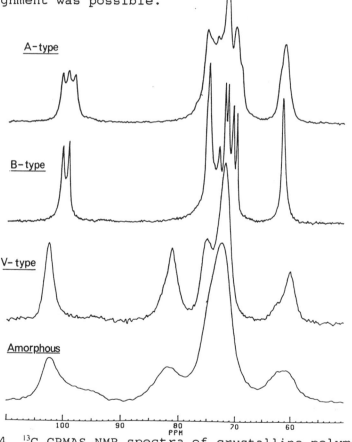

Fig.4 ^{13}C CPMAS NMR spectra of crystalline polymorphs and amorphous forms of $\alpha-(1\to4)$ glucans.

Progress in spectral interpretation came with the study of starch samples showing enhanced crystalline or amorphous character (Marchessault et al, 1985; Gidley and Bociek, 1985; Veregin et al, 1986). Crystalline fractions of both A- and B-type diffraction pattern were prepared either by controlled hydration of starches and degradation products (Veregin et al, 1986; Horii et al, 1986) or by recrystallisation of linear $(1\rightarrow4)-\alpha$-D-glucan oligomers (Gidley and Bulpin, 1987; Horii et al, 1987; Gidley and Bociek, 1985, 1988a; Tanner et al, 1987). In all cases similar spectra were obtained, as illustrated in Figure 4. The characteristic multiplicities observed for C-1 signals (ca. 100 ppm) of A and B polymorphs (Figure 4) are in line with the known unit cell contents i.e. 2 inequivalent glucoses for the A type and 3 for the B-type (Marchessault et al, 1985). The chemical shift effects which lead to these multiple peaks could arise through either intramolecular (Veregin et al, 1986) or intermolecular (Gidley and Bociek, 1985) effects. The former is usually assumed to be the case but implies glycosidic conformational variation between adjacent linkages greater than the (very small) differences deduced from X-ray fibre diffraction analyses (Imberty et al, 1988; Imberty and Perez, 1988). The latter is more unusual but is supported by the observation (Figure 4) of multiple peaks in the C-2, 3, 4, 5 spectral region (70-80 ppm) implying similar peak splittings for non-glycosidic carbons in the B-type polymorph.

Regardless of the chemical shift mechanism operative in these systems, such model crystalline spectra together with spectra of amorphous material (Figure 4) can be used (Gidley and Bociek, 1985) to simulate spectra for granular starches (Figure 5). As NMR is a probe at the atomic/sub-molecular level, it is presumed that spectra obtained from crystalline material provide 'fingerprints' for the double helices which underlie the crystalline structures. The success of the spectral simulation approach (Figure 5) allows quantitative estimates to be made

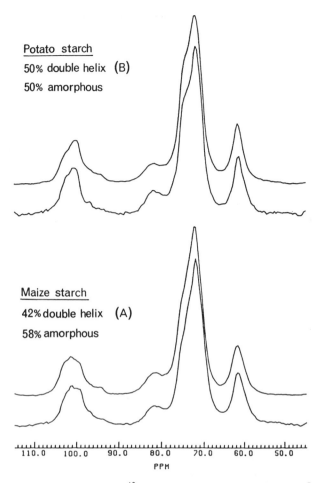

Fig.5 Simulation of ^{13}C CPMAS NMR spectra of granular starches (upper traces) by a composite spectrum of amorphous and crystalline (Fig.4) model spectra (lower traces) for potato and maize starch.

of the double helix content of starch-based materials. For a range of granular starches (Gidley and Bociek, 1985; Cooke and Gidley, 1992), double helical content was found to be usually in the range 40-50%, substantially higher than estimates of crystallinity of 25-35%. This imples that granular starches contain a significant amount of non-crystalline double helices, and provides a further probe of starch ordered structure.

The V-type single helices characteristic of amylose in the presence of complexing agents have also been widely studied by ^{13}C-CPMAS NMR (Horii et al, 1987; Veregin et al, 1987b; Gidley and Bociek, 1988a). Irrespective of the nature of the complexing agent (or the absence of any), very similar starch spectra are observed (Figure 4). Chemical shifts are distinctly different from those of double helical polymorphs and no multiplicity is evident even after resolution enhancement (Gidley and Bociek, 1988a). It is of interest to note that peak positions for C-1 and C-4 (ca. 104 and 83 ppm respectively) correspond to local intensity maxima in 'amorphous' spectra (Figure 4). As discussed later this imples that local glycosidic conformations in 'amorphous' starch are biased towards those which would be required for V-type helix formation.

Chemical shifts and glycosidic conformation

For a number of diverse organic systems (alkanes, polysaccharides, polypeptides and macrocycles), ^{13}C CPMAS NMR spectra have been interpreted in terms of a major (up to 12 ppm) effect of conformation on chemical shifts (Saitô, 1986). Although the effect is recognised, there is as yet no quantitative relationship linking chemical shift with conformation. For α-(1→4) glucans, a wide range of C-1 (and C-4) chemical shifts are apparent (Figure 4). This data set is extended if one takes into account spectra for highly crystalline cyclodextrins (Gidley and Bociek, 1986) and their inclusion complexes (Veregin et al, 1987b). Using these chemical shifts, correlations have been sought with glycosidic torsion angles ϕ and ψ (Figure 3). Analysis of (unassigned) sets of C-1 and C-4 shifts for crystalline cyclodextrin complexes whose structures were available from single crystal X-ray diffraction studies (Veregin et al, 1987b) suggested correlations of C-1 and C-4 chemical shifts with angles ψ and ϕ respectively. Using a smaller range of assigned signals, no single correlation was found with C-4 shifts, but two satisfactory correlations (with ψ and ψ+ φ) were found for C-1 shifts (Gidley and Bociek, 1988a). In order to test the validity of these latter two putative correlations, attempts were made to simulate an 'amorphous' C-1 signal by assuming that all glycosidic conformations which do not involve hard sphere interactions are equally probable. This led (Gidley and Bociek, 1988a) to the successful duplication of the chemical shift range but suggested a signal maximum to higher field than that observed (Figure 4). This points to the possibility that V-type glycosidic conformations may be preferred in amorphous starches, possibly via a similar H-bonding system to that demonstrated for amylose in dimethylsulphoxide (see earlier).

Further progress in this important field will probably require a greater theoretical appreciation of conformational effects on chemical shifts (Saitô 1986). Some progress in this area is being made (Sastry et al, 1987; Grindley et al, 1990). A further feature worthy of note is that the hydration state can affect ^{13}C chemical shifts (Furo et al, 1987; Tanner et al, 1987) although it is not clear whether this is an indirect effect via glycosidic conformation.

One class of materials which can be analysed by ^{13}C CPMAS NMR is frozen liquids (Fyfe et al, 1978). As chemical shifts (at least C-1) are considered to be responsive to conformational effects, examination of frozen oligo- and polysaccharide solutions is a potential approach to determining the range of conformations present in frozen solution (and by implication, mobile solution). First results of this method are very encouraging (Gidley and Bociek, 1988b) as they show the 'expected' results that frozen aqueous amylopectin solutions have 'amorphous' ^{13}C CPMAS spectra (Figure 4) and that frozen dimethylsulphoxide solutions of amylose (French and Zobel, 1967) show the V-type spectrum (Figure 4). It therefore appears that examination of frozen solutions is a potential test for the range of

conformations adopted in solution, and therefore
helps to determine whether the single conformational
average which could be obtained from solution state
studies of oligosaccharides (see earlier) has
physical meaning. For $(1\rightarrow4)-\alpha-D$-glucans, frozen
dimethylsulphoxide solutions give relatively narrow
^{13}C CPMAS spectra and therefore satisfy this
criterion, but frozen aqueous solutions have broad
signals and do not. The study of frozen α-
cyclodextrin solutions (Gidley and Bociek, 1988b)
suggested that the common crystalline hydrate form is
not conserved in solution, but that a less common
hydrate is an appropriate model for aqueous solution
conformation. This finding allowed a proposed model
for α-cyclodextrin complexation (based on relief of
conformational strain) to be discounted (Gidley and
Bociek, 1988b).

RELAXATION BEHAVIOUR

The preceding sections have described
'chemical' applications of NMR in which molecular
structure, organisation and conformation were the
subjects of interest. The primary concern was one of
assignment of signals to specific molecular sites.
In addition to such chemical shift (and coupling
constant) data, NMR spectroscopy can be used to probe
'physical' properties related to molecular and
segmental mobility and re-orientation. The excited
nuclear spins generated in an NMR experiment return
to equilibrium by emission of radiation of a given
frequency (chemical shift) and at a certain rate
(relaxation behaviour). A number of relaxation
pathways are possible for nuclei in different
environments and several NMR probes of relaxation
behaviour (e.g. ^{1}H and ^{13}C T_1's and T_2's) can be used to
provide information on different regimes of molecular
motion (Harris, 1983). A number of techniques have
been used in studies of starches as described below.
In solid systems, relaxation modulated by low
frequencies (e.g. T_2) is much more efficient than
intermediate frequencies (e.g. ^{1}H T_1 in the rotating

frame ($T_{1\rho}$) which is sensitive to motions of 10-100 kHz) which in turn is more efficient than higher frequencies (e.g. ^{1}H T_1 and ^{13}C T_1, which are sensitive to 10-1000 MHz motions). Thus values for ^{1}H T_2s are at the 'rigid lattice' limit of ca. $10\mu s$ and for wheat starch (Dev et al, 1987), ^{1}H $T_{1\rho}$ = 6ms, ^{1}H T_1 = 0.7s and ^{13}C T_1 = 14-28s. The ^{1}H $T_{1\rho}$ value is important as this determines the decay of magnetisation following the cross-polarisation sequence in the ^{13}C CPMAS experiment. For waxy maize and potato starches, values were found to be 4-6 ms for all signals (Gidley and Bociek, 1985) in line with the results for wheat starch. The absence of different ^{1}H $T_{1\rho}$ values for crystalline and amorphous components within starch granules suggests that domain sizes for these components are less than ca. 10nm (Havens and VanderHart, 1985). Following hydration of solid structures (e.g. hydrated flours), ^{1}H $T_{1\rho}$ values increase to 10-15ms (Garbow and Schaeffer, 1991) due to greater internal flexibility and motion.

In solutions and other highly hydrated systems, T_1 and T_2 relaxation are only ca. 1 order of magnitude different (c.f. 4-5 orders for solids) with typical ^{1}H values of 100-1000ms and 10-30ms respectively. T_1 values are important for accurate integration of solution state spectra as they govern the return to equilibrium which must be essentially complete before the next excitation pulse is applied, otherwise signals will become 'saturated' and show low intensity in the transformed spectrum. T_2 values have practical importance as, for both ^{1}H and ^{13}C, linewidths are often determined principally by the relationship:-

$$\upsilon_{1/2} = \frac{1}{\pi\, T_2}$$

where $\upsilon_{1/2}$ is the width at half height. As T_2 becomes shorter with increasing 'solidity', lines become broader and in the limit are indistinguishable from the baseline (a T_2 of $10\mu s$ corresponds to a linewidth of 32kHz or 320ppm on a 100MHz spectrometer). This effect has been monitored during the gelation of amylose (Gidley, 1989). Relaxation behaviour in polymer solution is considered to be governed by segmental mobility, and it is therefore of interest to determine the size of such segments in order to more fully appreciate the molecular origins of relaxation data. In a comprehensive study of ^{13}C relaxation parameters (T_1, T_2 and n.O.e.) for a series of oligomers of the starch-related polymer pullulan (maltotriose or maltotetraose units connected linearly by 1→6-α linkages), Benesi and Brant (1985) showed that for increasing chain length an asymptotic limit for relaxation parameters was reached at DP = 12. Furthermore the relaxation behaviour of non-terminal residues in an oligomer of DP = 15 matched those of the parent polymer (Benesi and Brant, 1985). These data therefore suggest that DP = 12 is an upper limit for a relaxation segment in pullulan and therefore presumably in structurally-related starch polymers.

The relaxation behaviour of water in the presence of starch and related systems has also been studied extensively, usually via ^1H T_2 measurements (Lechert et al, 1980; Lelievre and Mitchell, 1975; Wynne-Jones and Blanshard, 1986) but also using ^2H (Hennig and Lechert, 1977; Lechert et al, 1988) and ^{17}O (Richardson et al, 1987) spectroscopy. A number of conclusions concerning starch-water interactions have been drawn from these studies. Up to 7% H_2O in starch, the water is characterised as irrotationally 'bound'; from 7-20%, the water is considered to be rotationally restricted; anisotropic motion is observed from 15-20% up to 30% H_2O; above this water content, bulk free solvent is sugested (Guilbot and Mercier, 1985). Starch gelatinization is accompanied by an increase in ^1H T_2 values for solvent water which have been monitored directly (Lelievre and Mitchell, 1975) or indirectly _via_ high resolution linewidths (Collison and McDonald, 1960; Jaska, 1971). The relaxation properties of starch pastes (Callaghan et al, 1983) and gels (Wynne-Jones and Blanshard, 1986; Gidley, 1989) have also been probed using a range of ^1H and ^{13}C measurements.

NMR STUDIES OF STARCH FUNCTIONALITY

Many of the applications of NMR described above have attempted to address structure-function relationships and mechanisms involved in the biological and technological utilisation of cereal carbohydrates. This section seeks to summarise progress achieved in understanding gelatinization and retrogradation of starch and gelation of amylose using NMR methods. Of course, NMR is but one of many physico-chemical techniques which have been brought to bear on these important topics, but the unique capability of NMR in probing the environment of individual atomic sites has led to many insights into the molecular origins of starch functionality.

Gelatinization

Several NMR studies have demonstrated either directly or indirectly that the process of gelatinization results in decreased starch linewidths (i.e. increased T_2 values) due to enhanced segmental mobility. More detail has emerged from studies of water T_2 or linewidth (Collison and McDonald, 1960; Jaska, 1971; Lelievre and Mitchell, 1975). Data have been interpreted in terms of proton exchange between solvent water and polymer hydroxyl groups, which leads to the suggestion (Lelievre and Mitchell, 1975) that an increased hydration of the polymer chains accompanies the gelatinization process.

Gelatinization involves the loss of detectable structural order within granules and hence could be followed by ^{13}C CPMAS NMR measurements of double helix content (Gidley and Bociek, 1985). Through studies of starches isolated at various stages of the gelatinization process, it has been shown that the relative decrease in double helix content parallels the relative decrease in both crystallinity and residual gelatinization enthalpy, but occurs at higher temperature than the relative decrease in granular birefringence (Cooke and Gidley 1992). Scanning calorimetry studies of highly crystalline model materials (ca. 100% double helix content and crystallinity) and granular starches (40-50% double helix content and ~ 25% crystallinity) led to the conclusion that the characteristic enthalpy of gelatinization is due primarily to the dissociation of double helices rather than the longer range disruption of crystallinity (Cooke and Gidley, 1992).

Retrogradation

Retrogradation is generally taken to refer to all post-gelatinization changes in the state of starch-based systems. Some aspects are considered to be functionally beneficial (e.g. amylose gelation, discussed in the next section) and some aspects may be unwanted (e.g. bread staling). These processes are often associated with re-crystallisation and hence ^{13}C CPMAS NMR would be expected to provide useful information, but such studies have not yet been reported.

Changes in water T_2 values using both ^1H (Wynne-Jones and Blanshard, 1986) and ^2H (Leung et al, 1983) NMR have been reported for staling bread and model gelled systems. Storage led to a decrease in ^1H T_2, ^2H T_1 and ^2H T_2 values consistent with an overall decrease in water mobility and an increase in energetically-associated water (Leung et al, 1983; Wynne-Jones and Blanshard, 1986). This energetically-associated ('bound') water was readily displaced by D_2O (Wynne-Jones and Blanshard, 1986). Differences in timescale and temperature coefficients suggested that ^1H T_2 measurements were not simply following the same process as monitored by firming or crystallinity (Wynne-Jones and Blanshard, 1986).

Another example of retrogradation is the precipitation of amylose from aqueous solution. Using synthetic and nearly monodisperse amyloses, the physical state of retrograded material as well as the

kinetics of the transformation were found to be
highly dependent on chain length (Gidley and Bulpin),
1989). For short chains (e.g. DP ~ 40) precipitation
from aqueous solution led to a material with a sharp
X-ray diffraction pattern of the expected B-type.
With increasing chain length (e.g. DP ~ 110 and DP ~
250) the diffraction pattern became significantly
less intense (Gidley, 1989). This could be due
either to a decrease in double helix content or to
loss of crystalline register. ^{13}C CPMAS NMR data
showed the latter to be the case (Gidley, 1989) as
depicted in Figure 6. The lack of detectable

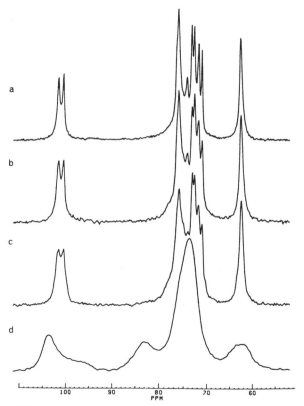

Fig.6 ^{13}C CPMAS NMR spectra of amyloses precipitated
from dilute aqueous solution having chain lengths
(DP) of (a) 40, (b) 110 and (c) 250 and (d) amorphous
α-(1→4) glucan. Reprinted, with permission, from
Gidley, 1989. Copyright 1989 American Chemical
Society.

amorphous character in the spectra of precipitated amyloses suggests nearly complete double helix formation. The reduction in NMR signal resolution with increasing chain length parallels the loss of X-ray diffraction resolution, consistent with NMR multiplicities being due to inter-helical packing effects.

Amylose gelation

A range of NMR techniques have been used in the study of amylose gels. ^1H T_2 measurements showed two ranges of relaxation behaviour of ~40μs and 3-30ms for 3% gels (Welsh et al, 1982) and ~10μs and 1-10ms for 10% gels (Gidley, 1989). Significant but incomplete loss of high resolution signal in both ^{13}C (Welsh et al, 1982) and ^1H (Gidley, 1989) spectra accompany gelation as would be expected from the T_2 measurements. High resolution ^{13}C NMR has been used to characterise the nature of segments within gelled amylose (Gidley, 1989). As shown in Figure 7, the 'solution state' spectrum of an amylose gel has broad but discernable features. The line broadening can be assigned to chemical shift anisotropy (i.e. the inability of segments to reorient rapidly enough to average the directional component of the chemical shift) as magic angle spinning at a speed corresponding to the signal widths of the solution state spectrum leads to a spectrum exhibiting narrow resonances (Figure 7b). Chemical shifts in this spectrum match those for α-(1→4) glucan solutions suggesting a solution-like (i.e. single chain) conformation. The CPMAS spectrum of the same gel (Figure 7c) shows the characteristic signals for B-type double helical amylose (Figure 7d). Two mobility-resolved conformations are therefore detected. In order to test whether these two conformational states are sufficient to describe amylose gels completely, a CPMAS spectrum of a frozen gel (Figure 7e) was obtained. The successful simulation (Figure 7f) of this spectrum with a combination of amorphous (frozen single chain) and

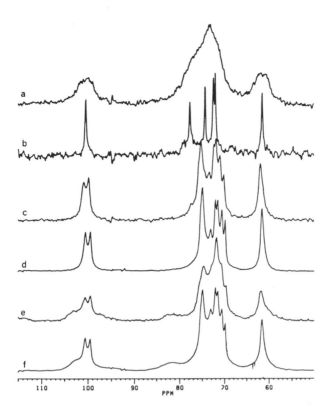

Fig.7 (a) Single pulse ^{13}C spectrum of a 10% w/v
aqueous potato amylose gel at 303K, (b) as (a) with
500Hz magic angle spinning, (c) ^{13}C CPMAS spectrum of
the same sample, (d) ^{13}C CPMAS spectrum of B-type
amylose, (e) ^{13}C CPMAS spectrum of the same gel at
233K, (f) simulation of spectrum (e) by addition of
67% of spectrum (d) and 33% of amorphous spectrum
(Fig.4). Reprinted from Gidley (1988) by permission
of Oxford University Press.

double helical features shows that the two
conformational states are sufficient. Furthermore
the ratio of amorphous (derived from relatively

mobile segments in the ambient gel) to helical features required for spectral simulation (1:2) is in good agreement with the ratio of long to short ^1H T_2 values. The combined application of relaxation time measurements and a range of high resolution experiments in such a 'mobility-resolved spectroscopy' approach promises to provide descriptions of molecular conformations within a range of biopolymer gels (Gidley et al, 1991a, b).

LITERATURE CITED

Alvarado, E., Hindsgaul, O., Lemieux, R.U., Bock, K. and Pedersen, H. 1984. The contribution of the exo-anomeric effect to the conformational preferences exhibited by methyl α-maltoside and methyl α-maltotrioside. Abstr. Int. Carbohydr. Symp., XIIth, Utrecht, p.480.

Baianu, I.C. and Förster, H. 1980. Cross-Polarisation high-field carbon-13 NMR techniques for studying physicochemical properties of wheat grain, flour, starch, gluten and wheat protein powders. J. Appl. Biochem. 2:347-354.

Bax, A. and Freeman, R. 1982. Long-range proton-carbon-13 NMR spin coupling constants. J. Am. Chem. Soc. 104:1099-1100.

Benesi, A.J. and Brant, D.A. 1985. Trends in molecular motion for a series of glucose oligomers and the corresponding polymer pullulan as measured by ^{13}C NMR relaxation. Macromolecules. 18:1109-1116.

Bengtsson, S. and Åman, P. 1990. Isolation and chemical characterisation of water-soluble arabinoxylans in rye grain. Carbohydrate Polymers. 12:267-277.

Bock, K. and Pedersen, H. 1984. Assignment of the NMR parameters of the branch-point trisaccharide of amylopectin using 2-D NMR spectroscopy at 500MHz. J. Carbohydr. Chem. 3:581-592.

Bock, K., Duus, J.Ø., Norman, B, and Pedersen, S. 1991. Assignment of structures to oligo-saccharides produced by enzymic degradation of a β-D-glucan from barley by ^{1}H and ^{13}C-n.m.r. spectroscopy. Carbohydr. Res. 211:219-233.

Callaghan, P.T., Jolley, K.W., Lelievre, J. and Wong, R.B.K. 1983. Nuclear magnetic resonance studies of wheat starch pastes. J. Coll. Interface Sci. 92:332-337.

Casu, B., Reggiani, M., Gallo, G.G. and Vigevani, A. 1966. Hydrogen bonding and conformation of glucose and polyglucoses in dimethylsulphoxide solution. Tetrahedron. 22:3061-3083.

Christofides, J.C. and Davies, D.B. 1987. Cooperative intramolecular hydrogen bonding in glucose and maltose. J. Chem. Soc. Perkin Trans. II. 97-102.

Collison, R. and McDonald, M.P. 1960. Broadening of the water proton line in high resolution nuclear magnetic resonance spectra of starch gels. Nature. 186:548-549.

Cooke, D. and Gidley, M.J. 1992. Loss of crystalline and molecular order during starch gelatinisation: origin of the enthalpic transition. Carbohydr. Res. Accepted for publication.

Dais, P. and Perlin, A.S. 1982. High-field ^{13}C-NMR spectroscopy of β-D-glucans, amylopectin and glycogen. Carbohydr. Res. 100:103-116.

Dev, S.B., Burum, D.P. and Rha, C.K. 1987. High resolution CPMAS and solution NMR studies of polysaccharides - part I: Starch. Spectroscopy Letters. 20:853-869.

French, A.D. and Zobel, H.F. 1967. X-ray diffraction of oriented amylose fibers. I. Amylose dimethylsulphoxide complex.

Furó, I, Pócsik, I., Tompa, K., Teeäär, R. and Lippmaa, E. 1987. C.P.-D.D.-M.A.S. ^{13}C-n.m.r. investigations of anhydrous and hydrated cyclomalto-oligosaccharides: the role of water of hydration. Carbohydr. Res. 166:27-33.

Fyfe, C.A., Lyerla, J.R. and Yannoni, C.S. 1978. High-resolution ^{13}C nuclear magnetic resonance spectra of frozen liquids using magic angle spinning. J. Am. Chem. Soc. 100:5635-5636.

Garbow, J.R. and Schaefer, J. 1991. Magic-angle ^{13}C NMR study of wheat flours and doughs. J. Agric. Food Chem. 39:877-800.

Gidley, M.J. 1985. Quantification of the structural features of starch polysaccharides by NMR spectroscopy. Carbohydr. Res. 139:85-94.

Gidley, M.J. and Bociek, S.M. 1985a. Long range ^{13}C ^{1}H coupling in carbohydrates by selective 2D heteronuclear J-resolved NMR spectroscopy. J. Chem. Soc. Chem. Commun. 220-222.

Gidley, M.J. and Bociek, S.M. 1985b. Molecular organisation in starches: a ^{13}C CP/MAS NMR study. J. Am. Chem. Soc. 107:7040-7044.

Gidley, M.J. and Bociek, S.M. 1986. ^{13}C CP/MAS NMR studies of α- and β- cyclodextrins: resolution of all conformationally-important sites. j. Chem. Soc. Chem. Commun. 1223-1226.

Gidley, M.J. and Bulpin, P.V. 1987. Crystallisation of malto-oligosaccharides as models of the crystalline forms of starch: minimum chain length requirement for the formation of double helices. Carbohydr. Res. 161:291-300.

Gidley, M.J. 1988. Conformational studies of α-(1→4) glucans in solid and solution states by NMR spectroscopy. Gums and Stabilisers for the Food Industry 4, IRL Press, Oxford, pp.71-80.

Gidley, M.J. and Bociek, S.M. 1988a. ^{13}C CP/MAS NMR studies of amylose inclusion complexes, cyclodextrins and the amorphous phase of starch granules: relationships between glycosidic linkage conformation and solid state ^{13}C chemical shifts. J. Am. Chem. Soc. 110:3820-3829.

Gidley, M.J. and Bociek, S.M. 1988b. ^{13}C CP/MAS NMR studies of frozen solutions of (1→4)-α-D-glucans as a probe of the range of glycosidic linkages: the conformations of cyclomaltohexaose and amylopectin in aqueous solution. Carbohydr. Res. 183:126-130.

Gidley, M.J. and Bulpin, P.V. 1989. Aggregation of amylose in aqueous systems: the effect of chain length on phase behaviour and aggregation kinetics. Macromolecules. 22:341-346.

Gidley, M.J. 1989. Molecular mechanisms underlying amylose agregation and gelation. Macromolecules. 22:351–358.

Gidley, M.J., McArthur, A.J. and Underwood, D.R. 1991. ^{13}C NMR characterisation of molecular structures in powders, hydrates and gels of galactomannans and glucomannans. Food Hydrocolloids. 5:129–140.

Gidley, M.J., Ablett, S. and Darke, A.H. 1991. Characterisation of polysaccharide networks by NMR spectroscopy. Abstr. I 15 Eurocarb. VI, Edinburgh.

Grindley, T.B., McKinnon, M.S. and Wasylishen, R.E. 1990. Towards understanding ^{13}C-NMR chemical shifts of carbohydrates in the solid state. The spectra of D-mannitol polymorphs and of DL-mannitol. Carbohydr. Res. 197:41–52.

Guilbot, A. and Mercier, C. 1985. Starch. The Polysaccharides Vol.3. G.O. Aspinall (ed.). Academic Press, Orlando. pp.209–282.

Harris, R.K. 1983. Nuclear magnetic resonance spectroscopy: a physicochemical view. Pitman, London.

Havens, J.R. and VanderHart, D.L. 1985. Morphology of poly (ethyleneterephthalate) fibers as studied by multiple-pulse ^{1}H nmr. Macromolecules. 18:1663–1676.

Hennig, H-J. and Lechert, H. 1977. DMR study of D_2O in native starches of different origins and amylose of type B. J. Coll. Interface Sci. 62:199–204.

Heyraud, A., Rinaudo, M., Vignon, M. and Vincendon, M. 1979. ^{13}C-NMR spectroscopic investigation of α- and β- 1,4-D-glucose homooligomers. Biopolymers. 18:167–185.

Hoffmann, R.A., Roza, M., Maat, J., Kamerling, J.P. and Vliegenthart, J.F.G. 1991. Structural characteristics of the cold-water-soluble arabinoxylans from the white flour of the soft wheat variety Kadet. Carbohydr. Polymers. 15:415–430.

Hoffmann, R.A. 1991. PhD Thesis. Utrecht

University, Netherlands.

Hoffmann, R.A., Geijtenbeek, T., Kamerling, J.P. and Vliegenthart, J.F.G. 1992. A ^{1}H-NMR study on enzymatically generated wheat endosperm arabinoxylan oligosaccharides. Carbohydrate. Res. 223:19-44.

Horii, F., Hirai, A. and Kitamaru, R. 1986. CP/MAS ^{13}C NMR spectroscopy of hydrated amyloses using a magic-angle spinning rotor with an O-ring seal. Macromolecules. 19:932-934.

Horii, F., Yamamoto, H., Hirai, A. and Kitamaru, R. 1987. Structural study of amylose polymorphs by cross-polarisation-magic-angle spinning, ^{13}C-NMR spectroscopy. Carbohydr. Res. 160:29-40.

Imberty, A., Chanzy, H., Pérez, S., Buléon, A. and Tran. V. 1988. The double-helical nature of the crystalline part of A-starch. J. Mol. Biol. 201:365-378.

Imberty, A. and Pérez, S. 1988. A revisit to the three-dimensional structure of B-type starch. Biopolymers. 27:1205-1221.

Jane, J-L., Robyt, J.F. and Huang, D-H. 1985. ^{13}C-NMR study of the conformation of helical complexes of amylodextrin and of amylose in solution. Carbohydr. Res. 140:21-35.

Jaska, E. 1971. Starch gelatinisation as detected by proton magnetic resonance. Cereal Chem. 48:437-444.

Lechert, H., Maiwald, W., Köthe, R. and Basler, W-D. 1980. NMR study of water in some starches and vegetables. J. Food Processing and Preservation. 3:275-299.

Lechert, H., Scoggins, W.C., Basler, W.D. and Schwier, I. 1988. New aspects of the structure of water in starches. Stärke. 40:245-250.

Lelièvre, J. and Mitchell, J. 1975. A pulsed NMR study of some aspects of starch gelatinisation. Stärke. 27:113-115.

Leung, H.K., Magnuson, J.A. and Bruinsma, B.L. 1983. Water binding of wheat flour doughs and breads as studied by deuteron relaxation. J. Food Sci.

48:95-99.

Marchessault, R.H., Taylor, M.G., Fyfe, C.A. and Veregin, R.P. 1985. Solid-state ^{13}C-CPMAS NMR of starches. Carbohyr. Res. 144:C1-C5.

Morris, G.A. and Hall, L.D. 1982. Experimental chemical shift correlation maps from heteronuclear two-dimensional nuclear magnetic resonance spectroscopy II: carbon-13 and proton chemical shifts of α-D-glucopyranose oligomers. Can. J. Chem. 60:2431-2441.

Muhrbeck, P. and Tellier, C. 1991. Determination of the phosphorylation of starch from native potato varieties by ^{31}P NMR. Stärke 43:25-27.

Mulloy, B., Frenkiel, T.A. and Davies, D.B. 1988. Long-range carbon-proton coupling constants: application to conformational studies of oligosaccahrides. Carbohydr. Res. 184:39-46.

Nardin, R., Saint-Germain, J., Vincendon, M., Taravel, F. and Vignon, M. 1984. Proton NMR conformational analysis of hydroxylated D-glucan model compounds in aprotic solvent. Nouv. J. Chim. 8:305-309.

Nieduszynski, I.A. and Marchessault, R.H. 1972. Structure of β,D(1→4')-xylan hydrate. Biopolymers. 11:1335-1344.

O'Donnell, D.J., Ackerman, J.J.H. and Maciel, G.E. 1981. Comparative study of whole seed protein and starch content via cross polarisation-magic angle spinning carbon-13 nuclear magnetic resonance spectroscopy. J. Agric. Food Chem. 29:514-518.

Parfondry, A., Cyr, N, and Perlin, A.S. 1977. ^{13}C-^{1}H Inter-residue coupling in disaccharides, and the orientations of glycosidic bonds. Carbohydr. Res. 59:299-309.

Peng, Q-J. and Perlin, A.S. 1987. Observations on NMR spectra of starches in dimethyl sulfoxide, iodine-complexing and solvation in water-dimethylsulfoxide. Carbohydr. Res. 160:57-72.

Pérez, S., Taravel, F. and Vergelati, C. 1985. Experimental evidences of solvent induced conformational changes in maltose. Nouv. J. Chim. 9:561-564.

Richardson, S.J., Baianu, I.C. and Steinberg, M.P.
1987. Mobility of water in corn starch
suspensions determined by nuclear magnetic
resonance. Stärke. 39:79-83.

St-Jacques, M., Sundararajan, P.R., Taylor, K.J. and
Marchessault, R.H. 1976. Nuclear magnetic
resonance and conformational studies on amylose
and model compounds in dimethyl sulfoxide
solution. J. Am. Chem. Soc. 98:4386-4391.

Saitô, H. 1986. Conformation-dependent ^{13}C chemical
shifts: a new means of conformational characterisation as obtained by high-resolution solid-state
^{13}C NMR. Magn. Reson. Chem. 24:835-852.

Sanders, J.K.M. and Hunter, B.K. 1987. Modern NMR
spectroscopy. University Press, Oxford.

Sastry, D.L., Takegoshi, K. and McDowell, C.A. 1987.
Determination of the ^{13}C chemical-shift tensors in
a single crystal of methyl α-D-glucopyranoside.
Carbohydr. Res. 165:161-171.

Schaefer, J. and Stejskal, E.O. 1979. High-
resolution ^{13}C NMR of solid polymers. Topics in ^{13}C
NMR spectroscopy. G.C. Levy (ed.). John Wiley,
New York. pp.283-324.

Shashkov, A.S., Lipkind, G.M. and Kochetkov, N.K.
1986. Nuclear Overhauser effects for methyl β-
maltoside and the conformational states of maltose
in aqueous solution. Carbohydr. Res. 147:175-
182.

Tanner, S.F., Ring, S.G., Whittam, M.A. and Belton,
P.S. 1987. High resolution solid state ^{13}C n.m.r.
study of some $\alpha(1\rightarrow4)$ linked glucans: the influence
of water on structure and spectra. Int. J. Bio.
Macromol. 9:219-224.

Usui, T., Yamaoka, N., Matsuda, K., Tuzimura, K.,
Sugiyama, H. and Seto, S. 1973. ^{13}C Nuclear
magnetic resonance spectra of glucobioses,
glucotrioses and glucans. J. Chem. Soc. Perkin
Trans I. 2425-2432.

Usui, T., Yokoyama, M., Yamaoka, N., Matsuda, K.,
Tuzimura, K., Sugiyama, H. and Seto, S. 1974.
Proton magnetic resonance spectra of D-gluco-
oligosaccharides and D-glucans. Carbohyr. Res.

33:105-116.

Veregin, R.P., Fyfe, C.A., Marchessault, R.H. and Taylor, M.G. 1986. Characterisation of the crystalline A and B starch polymorphs and investigation of starch crystallisation by high resolution ^{13}C CP/MAS NMR. Macromolecules. 19:1030-1034.

Veregin, R.P., Fyfe, C.A., Marchessault, R.H. and Taylor, M.G. 1987a. Correlation of ^{13}C chemical shifts with torsional angles from high resolution, ^{13}C-CPMAS NMR studies of crystalline cyclomalto-oligosaccharide complexes, and their relation to the structures of the starch polymorphs. Carbohydr. Res. 160:41-56.

Veregin, R.P., Fyfe, C.A. and Marchessault, R.H. 1987b. Investigation of the crystalline V amylose complexes by high-resolution ^{13}C CP/MAS NMR spectroscopy. Macromolecules. 20:3007-3012.

Vliegenthart, J.F.G., Dorland, L. and van Halbeek, H. 1983. High-resolution, ^{1}H-nuclear magnetic resonance spectroscopy as a tool in the structural analysis of carbohydrates related to glycoproteins. Adv. Carb. Chem. Biochem. 41:209-374.

Welsh, E.J., Bailey, J., Chandarana, R. and Norris, W.E. 1982. Physical characterisation of interchain association in starch systems. Prog. Fd. Nutr. Sci. 6:45-53.

Wynne-Jones, S. and Blanshard, J.M.V. 1986. Hydration studies of wheat starch, amylopectin, amylose gels and bread by proton magnetic resonance. Carbohydrate Polymers. 6:289-306.

CHROMATOGRAPHIC TECHNOLOGIES
FOR MACROMOLECULAR STARCH CHARACTERIZATION

Cynthia R. Sullivan
Alessandro Corona
James E. Rollings
Chemical Engineering Department
Worcester Polytechnic Institute
100 Institute Road
Worcester, MA 01606

INTRODUCTION

Starches are polymers that naturally occur in a variety of botanical sources such as corn, wheat and potatoes. Starches exhibit a wide range of physical properties making them useful for numerous applications in food, paper and biochemical industries. The physical properties of a particular starch are dependent upon its botanical source and post harvest processing. In general, utilization of starch for a specific application depends on the starch's native physical chemical properties and the alteration of these properties by various processes. Consequently, starches exhibit a wide variation in end-use functional properties.

The diversity observed in starch physical chemical properties is a result of both the macromolecular and microscopic differences in the source material and subsequent alterations caused by processing which leads to a great assortment of functional behaviors useful to the food, pharmaceutical, paper, textile and other industries. Scientists have historically examined the functional behavior of starch using a **direct** approach.

This approach generally involves processing of starch variety in some specified manner, and thereafter directly testing for its end-use performance (e.g., "mouthfeel", "binding capacity","sizing ability", "coating qualities", etc.). This approach is used in the starch industry in the process of heating and shearing (gelatinizing) a solution of a particular starch variety and evaluating the product for textural, nutritional or chemical qualities (functionality). For example, plant breeding technologies have been used to modify amylopectin solution stability and amylose gel strength and esterification techniques have been successful in lowering starch gelatinization temperatures and changing colloidal characteristics. This methodology, although successful (Whistler, et al., 1984; Davidson, 1980), is often quite expensive and laborious; consequently, alternate approaches to develop new starch applications are constantly being sought.

One such alternative approach requires understanding the fundamental relationships between processed starch physicochemical properties and the desired end-use functional attributes. The correlation of these characteristics is generally referred to as "structure-function" relationships. For example, the digestibility of a starch is improved if a pretreatment process partially hydrolyses starch macromolecules. In other words, digestibility is related to starch average molecular weight; a structure-function relationship. Similarly, macromolecular crosslinking inhibits cooking characteristics and depolymerization lowers solution viscosity; again structure-function relationships. Many starch scientists seek to understand how starch structure is related to its end use functionality. The advantage of this approach might allow investigators the ability to select and alter native starches to produce a specified physicochemical property and ultimately dictate a product's functionality.

In recent years, significant progress has been made in understanding these structure-function relationships. Slade and Levine (1987) studied thermal properties of starch systems (i.e., glass transition $[T_g]$) as a means of correlating physical chemical properties (i.e, molecular weight, MWD and moisture content) to functional properties including processability and stability. Corona (1990) related molecular weight and branching parameters to rheological properties of starch gels. He was able to discern the order in which starch molecules diffuse from gelatinized starch granules into solution (i.e., linear molecules more rapidly than branched molecules and smaller molecules more rapidly than larger molecules). Park and Rollings (in press) found a relationship between a starch's degree of branching and its susceptibility to enzymatic depolymerization, reporting that networks of $\alpha-1,6$ branch points ("clusters"), appear to restrict enzyme accessibility to $\alpha-(1,4)$ linkages. Extension of structure-function type motivated research offers the possibility of tailoring native starch varieties via genetic or hybridization techniques to meet the consumers needs and create new products.

In order to reach the goal of establishing realistic starch structure-function relationships, it will be necessary to develop new analytical tools for studying the heterogeneous nature of native and processed starch systems. Scientists have been challenged in developing new analytical tools to probe structure-function relationships; as they must design characterization strategies which are sensitive to specific starch systems' heterogeneities.

Until recently, routine analysis of starch polymer properties has relied primarily on "bulk" qualitative measurements; that is measurements which describe a starch system's average macromolecular properties. These techniques include: wet chemical methods such as iodine staining and dextrose equivalent measurements; colligative property analyses like freezing point and boiling point and osmotic and vapor pressure measurements; light scattering properties; and Brabender amylographs (viscometry). These techniques, although useful, cannot provide the necessary level of detail required to establish true macromolecular structure-function relationships for starch systems.

For heterogeneous polymer systems such as starch, unraveling the connection between physicochemical characteristics and end-use functionality has proven to be difficult. The macromolecular structure of starch is usually described in terms of its primary chemical attributes: chain configuration (bond linkages), branching, and molecular weight distributions. Starch polymers are certainly heterogeneous in all these respects; being a mixture of two fairly distinct components (the linear α-glucan amylose and the branched α-glucan amylopectin), which are both present in a broad range of molecular weights spanning several orders of magnitude; (10^2-10^8 daltons or greater). While this macromolecular information is vital to developing a clearer understanding of starch composition and primary structure, scientists have recognized that starch functionality is derived from many physical chemical properties in addition to those that are purely chemical in nature, as is the case for most polymer systems. These properties include the polymers' solution behavior (usually defined in terms of its size and conformation), thermodynamic state(s) (soluble, amorphous, crystalline), and interactions with other chemical species, including starch inter- and intra-molecular associations. All these features are dictated by the polymers' local environment (in aqueous solutions: pH, ionic strength, etc.) and processing history. Therefore to fully describe a starch system's structure-function relationships, researchers must use numerous parametric indicators, many of which are compositional and assessable by chromatographic technologies.

This paper focuses on new chromatographic techniques which can be employed to routinely describe the *distributive* macromolecular nature of processed starch system compositions and to what extent these techniques may be used to establish starch "structure-function" relationships. The operative word here is routine. Previously, distributed polymer data could only be obtained via time consuming experiments which frequently included tedious fractionation procedures. Prediction and control of functional behavior from basic starch characteristics poses a difficult challenge considering the myriad of possible physical chemical traits discussed above. New analytical techniques which combine separation schemes with advanced polymer characterization detectors are now available to expedite information acquisition. Expanded utilization of starches hinges on the researcher's ability to analyze and correlate basic starch macromolecular properties rapidly with their desired end-use applications.

POLYSACCHARIDE STRUCTURE-FUNCTION

Starch granules are composed primarily of the linear (poly)glucan, amylose, and the branched (poly)glucan, amylopectin. Amylose is composed exclusively of α-(1,4) bonded (poly)glucose as are 95% of the bond linkages of amylopectin. The remaining 5% of the amylopectin bonds are α-(1,6), which form branch points in this macromolecule. Detailed discussions of starch macromolecular architecture can be found elsewhere (Whistler, 1983). Both amylose and amylopectin are high molecular weight macromolecules with typical average molecular weights of several hundred thousand daltons for amylose and ten to one hundred times this value for amylopectin. The arrangement of these molecules within the granule continues to be a subject of great interest. Their synthesis has been discussed at length by others (Shannon and Garwood, 1984). Granules are present as a population whose composition, shape, and size distributions vary with botanical origin. It is well known that the molecular composition and polymer organization within these aggregates also varies with source (Shannon and Garwood, 1984). Furthermore, it has been shown (Corona, 1990) that the polysaccharides isolated from a single starch variety possess additional heterogeneities; being a mixture of macromolecules which vary both in molecular weight and chemical structure (i.e., branching characteristics). It is the desire to understand these heterogeneities that has stimulated research in the development of novel analytical tools capable of unraveling starch complexities and provide a logical basis for new polysaccharide derived products.

Starches exhibit diverse functional behaviors; not only because of their native botanical physicochemical heterogeneities but also because of the numerous ways in which they are processed. It is well known that certain starches function better than others in specific applications. In foods, high amylose varieties rapidly form rigid gels important in candy making. Whereas, waxy varieties are preferred as thickeners. Dent corn starch is utilized as an industrial adhesive in the manufacturing of corrugated board by a process employing mild acid or enzymatic simultaneous gelatinization and depolymerization. Starches may be treated chemically to improve their functional characteristics. Acid-modified starches are utilized in the textile industry as a warp-sizing agents to increase yarn strength, and improve fabric finishing. Starches may be chemically crosslinked in a variety of ways then gelatinized to yield high viscosity pastes with superior stability allowing starch usage in high temperature and shear applications such as those found in extrusion processes or as an ingredient in oil-well drilling muds. Starch may be blended with other polymers and processed in a variety of ways to yield biodegradable films and coatings. These are presented only as application examples.

Industrial starch processing is most often initiated by gelatinizing aqueous dispersions of granules with heat and shear. At a microscopic level, gelatinization is the process by which granules swell as they imbibe large amounts of water. A fraction of the granule polysaccharides are solubilized during gelatinization thus facilitating melting of ordered crystalline regions. The rheological behavior of gelatinized pastes and their cooled gels (important in many applications) is a result of specific starch macromolecular and microscopic properties (e.g., inter- and intra- macromolecular associations and granule- granule interactions). In addition, starch may be chemically modified (derivatized) during the gelatinization process in order to enhance thickening, coating, or adhesive behavior. When starch gelatinization is coupled with chemical (acid) or enzymatic hydrolysis (a process known as starch liquefaction) knowledge of the extent and manner of depolymerization is necessary if an understanding of the kinetic events is required or if process optimization is desired. Again, novel analytical tools capable of routinely providing macromolecular information are central to understanding these events.

Research in structure-function relationships of polysaccharides is not limited to studies of starch systems. Another example which underscores the importance of polysaccharide materials is found in biomedical research. Specific polysaccharide structures present on cell surfaces of yeasts and fungi are known to trigger desirable immunological responses in the human body. Yeasts and fungi represent a broad class of infectious agents and for the human body to fight invasion of these micro-organisms there must exist some mechanism for the body to recognize their presence. One can easily envision that leukocytes would first come into contact with the surfaces of the invading yeast or fungus. Receptor sites on the leukocytes should then "recognize" certain yeast or fungus cell wall material; i.e., polysaccharides. In order to study such biomedical processes (i.e., structure-function relationships), analytical devices such as those discussed here will play a major role.

In each case discussed above, reliable determination of polysaccharide macromolecular properties including compositions, average molecular weights, molecular weight and branching distributions, and solution conformation and size distributions is necessary in order to gain a fundamental understanding of polysaccharide functionality (i.e., structure-function relationship). Much of this information can be obtained from properly designed size exclusion chromatographic (SEC) devices coupled to various on-line detectors. Utilization of SEC and on-line detectors for understanding the molecular origins of starch functionality is discussed in other sections (below). The important "take home" message is that structure-function relationships are key to designing polysaccharides having specific utility, with physical characteristics playing a central role in gaining fundamental information.

ASSESSING STARCH MACROMOLECULAR PROPERTIES

In order to achieve the goal of establishing fundamental macromolecular relationships between starch structures and their functions, it is necessary to review which macromolecular properties are desired and how this information can be obtained. This section summarizes classical techniques which polymer scientists have used to examine starch macromolecular physical chemistry and the remaining sections of this paper build on this classical basis.

Bulk chemical assays can provide some estimates of

starch macromolecular properties in ways similar to many polymer systems (see Table I). For example, starch polymer *composition* is frequently estimated by iodine binding techniques (Sterling, 1978); as iodine complexes with linear starch polymers (amylose) in a different fashion than branched macromolecules (amylopectin). Therefore measuring the intensity of the blue colored starch iodine complex, or other associated solution physical properties, can provide some indication of the relative amount of linear and branched material. Moreover, it is known that the amylose iodine complex requires amylose to twist around the iodine molecules three full turns of the α-helix, needing 18 glucose units to accomplish this conformation (Rundle, et al., 1944). In this sense, iodine/starch complexation also provides a crude indication of the linear polymer's size. Similarly, starch polymer *concentration* in solution can be easily determined by means of a phenol sulfuric acid assay (Dubois, et al., 1956) and the average *degree of polymerization* (D.P.) of hydrolyzed starch polymers can be estimated by measuring total reducing sugars via a variety of "end-group" analyses (Miller, 1959; Dygert, et al., 1965; Somogyi, 1952).

Physical tests can also provide some information on the macromolecular properties of starches. For example *number-averaged molecular weights* of starches can be obtained from colligative property measurements (osmotic pressure, freezing point depression, vapor pressure lowering, etc.). All of these methods are

TABLE I
Starch Polymer Property Measurements

PROPERTY	TECHNIQUE
Composition	Iodine Binding, etc.
Concentration	Phenol Sulfuric
Degree of Polymerization	"End Group" Analysis
M_N	Colligative Properties
M_W	Light Scattering
$[\eta]$	Viscometry

physical chemical tests which are classical in polymer analyses. Light scattering detection and viscometry are other frequently used methods for gaining starch polymer property information: having been used to assess starch *weight-averaged molecular weights*, *intrinsic viscosity*, and other solution characteristics for over 50 years. These classical methods (commonly used by polymer scientists), are reviewed in detail elsewhere (Flory, 1953; Vollmert, 1973; Rodriques, 1982; and others), and are founded on basic physical chemical properties. Unfortunately, the macromolecular data obtained from these measurements when applied to a mixed polymer sample (such as that often present with starch systems) often does not provide the detailed description necessary to correlate starch physical chemical properties with functional performance.

Recognizing this shortcoming, researchers have developed techniques to gain *distributed* polymer property information. Distributed polymer property information can be obtained in two general ways; traditional fractional precipitation coupled to the above listed bulk chemical techniques or combined separation and detection strategies. The first methodology is discussed here and the latter method is reviewed in the remaining sections of this paper.

Classically, polymer scientists have used fractional precipitation techniques such as those which take advantage of the solvation properties of a polymer mixture. In most good solvents, polymer solubility decreases with increasing molecular weight. Fractional precipitation techniques from dilute solution are widely used which exploit this phenomena. Combining the solvent with a miscible precipitant (non-solvent), higher molecular weight fractions of most polymers can be separated from the remaining solution. Repeated applications of this procedure will result in a series of fractions varying in molecular weight. Once several fractions have been prepared, then any of the chemical and physical tests discussed above can be used on each fraction and distributed macromolecular information thereby obtained.

Other fractionation techniques can be designed which take advantage of specific physicochemical property variations within the polymer mixture. For example, the ability of polymers to form complexes with chemical agents (e.g., alcohols with linear starch polymers), polymer solubility variations with temperature, and/or the ability of the polymer to form complexes with itself (i.e., intra- or inter-macromolecular associations) have been used in some specific polymer fractionation schemes (Flory, 1953). Individual fractions of the polymer mixture can be analyzed via the same bulk measurements discussed above and thus yield the desired distributive polymer information. Utilization of these fractional precipitation techniques for starch fractionation has been reviewed by Young (1984). The basic problem with all of these fractionation techniques is that they are very time consuming and labor intensive and thus are not useful for routine analysis.

Recently, researchers have developed separation techniques which can be coupled directly to some of the physical chemical macromolecular detection methods (discussed above) and thus streamline the collection of distributed polymer information; these are the so called combined separation and detection strategies. Although this paper focuses on these new advances in starch polymer separation and detection, one can view that these techniques are simply extensions of the classical methodologies discussed above. The design basis for these new analytical tools rests firmly on the foundation of long established physical chemistry principles.

SEPARATION STRATEGIES

As previously discussed, different molecules within a given starch sample may vary considerably in physical chemical character (for example: size, solubility, charge density, etc.). Table II lists several of the heterogeneous physical chemical properties of starch and the corresponding separation techniques which can be used to understand these heterogeneities. Most techniques for starch separation involve liquid chromatography (LC) in one form or another.

High performance liquid chromatography (HPLC) is frequently used to separate starch oligosaccharides formed by hydrolytic processes. The application of HPLC to carbohydrate and oligosaccharide analysis has been reviewed in detail by Hicks (1988). Small chain starch polymers (D.P. < 25) may be separated using normal-phase partitioning (usually with amino-modified silica gel resins and aqueous acetonitrile mobile phases), and to a lesser extent, reverse phase or ion-exchange modes. In the normal-phase mode, separation is due to the preferential adsorption of polymer chains to the resin (Snyder and Kirkland, 1979). In this case, oligosaccharides elute in ascending order of degree of polymerization. The higher molecular weight oligosaccharides prefer to be in a "condensed" or precipitated state when compared to their lower molecular weight counterparts. In many ways there is a direct analogy to the solvent/non-solvent classical

TABLE II
Starch Heterogeneous Physical Chemical Properties
and Corresponding Separation Techniques

PROPERTY	SEPARATION TECHNIQUE
Solubility Hydrophilicity	HPLC–Normal Phase
Solubility Hydrophobicity	HPLC–Reverse Phase
Charge Density	HPLC–Ion Exchange
Hydrodynamic Size	SEC
Hydrodynamic Size – Diffusion Coefficients	FFF–Sedimentation
Hydrodynamic Size – Diffusion Coefficients	FFF–Flow
Thermal Diffusion Coefficients	FFF–Thermal
Electrophoretic Mobilities	FFF–Electrical
Combination of above properties	Mixed–mode

fractionation technique. Under reverse-phase LC conditions, oligosaccharides are separated by an opposite mechanism and this leads to a reversal in the order of evolution, but again the direct relationship to solvent/non-solvent fractional precipitation is evident. Ion-exchange involves exploiting the relative charge density of oligosaccharides. For example, in highly basic aqueous solutions, "neutral" polysaccharides are anionic (i.e., the pKa of the hydroxyl groups being of the order of 12). Thus higher molecular weight oligomers possess a greater charge density than smaller oligomers. The interactions with oppositely charged chromatographic resin surfaces will therefore decrease with decreasing molecular weight. Direct application of these HPLC modes to higher molecular weight polymers is not generally possible; primarily because of the limited solubility of high molecular weight polysaccharides in most resin-compatible mobile phases (usually water, buffered aqueous media, or aqueous acetonitrile mixtures), although the general principles of these separations are consistent with our basic physical chemical background and may be employed in some "mixed mode" chromatographic designs (see Table II). All of these techniques can be run using either isocratic eluents (if the partitioning forces are weak) or solvent gradient strategies (if the partitioning forces are large). In the later case the relationship of these HPLC separations and the solvent/non-solvent fractional precipitation methods is more obvious.

A promising non-LC method for resolving large polymers such as those found in most starch systems is field flow fractionation (FFF). FFF offers a family of techniques to separate polymer systems all involving the application of an external field or gradient perpendicularly applied across a flowing channel. The application of this field forces sample molecules into flow streamlines having different velocities which exist within a single flowing phase. Polymer components are separated on the basis of differences in physicochemical properties. These include variations in hydrodynamic sizes (or diffusion coefficients) with sedimentation FFF and flow FFF, electrophoretic mobilities with electrical FFF, and thermal diffusion coefficients with thermal FFF. Sedimentation FFF appears to be a most promising method for starch polymer separations but current instrument designs restrict separation of polymers to molecular weights greater than about 10^6 daltons. C. Giddings is the recognized authority on these techniques and has published extensively on the development of these methods (Giddings, 1988).

The liquid chromatographic technique, size exclusion chromatography (SEC), is the most frequently utilized separation technique for characterization of polymer mixtures. In this technique, the polymer's physical size (hydrodynamic radius) determines separation profiles. Larger molecules elute first as these materials (statistically) spend less time in the column's interstitial pores. They move along with the flowing eluent as opposed to their smaller molecular counterparts which are capable of penetrating into the non-moving solvent contained within the pores of the chromatographic resin. In principle, (in true SEC) there is little or no interaction between the polymers and the resin surface. While SEC has been the method of choice for separating starch polymers, interpretation of SEC data requires a basic understanding of its separation principles especially in light of the physical chemical diversity present in any given starch system. SEC elution behavior is often complicated by non-size effects including those arising from polymer-polymer and polymer-column interactions. SEC lends itself particularly well to the addition of on-line molecular characterization detectors downstream from the separator (more will be said of this in the section on Detection Technology) and for this reason this review will focus on advances in SEC technologies.

SIZE EXCLUSION CHROMATOGRAPHY

The retention mechanism in SEC may be described by Equation 1 (Yau, et al, 1979) which relates the elution volume of a specific polymer molecule, V_e, to the bed interstitial volume, V_o, and the bed pore volume, V_i:

$$V_e = V_o + K_D V_i \tag{1}$$

Equation 1 states that a linear relationship exists between the time at which a given polymer will elute from an SEC column and the column's pore volume. The proportionality constant (K_D) is the polymer's distribution coefficient. Fig. 1 illustrates the physical significance of K_D. K_D may be defined as the ratio of the pore volume available to a specific polymer molecule (V_a for polymer "a" or V_b for polymer "b") to the total pore volume available (V_p). For a chromatographic separation solely controlled by macromolecular size, K_D ranges from 0 to 1 and decreases proportionally with polymer hydrodynamic volume. In Fig. 1, polymer "b" sees a larger effective pore volume than polymer "a" ($V_a < V_b$) and therefore is retained longer, eluting from the column after polymer "a". In this case, $K_{Da} < K_{Db}$.

Fig. 1 displays a hypothetical SEC chromatogram

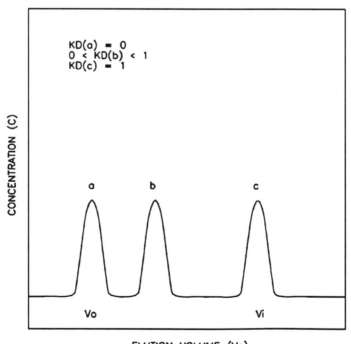

KD(a) = 0
0 < KD(b) < 1
KD(c) = 1

CONCENTRATION (C)

a   b   c

Vo   Vi

ELUTION VOLUME (Ve)

Fig. 1. SEC chromatogram depicting variation of K_D with elution volume for a polymer mixture: (A) high MW (totally excluded), (B) intermediate MW, (C) low MW (totally retained).

constructed by plotting the eluting polymer's concentration (usually obtained from differential refractometry or other spectroscopic methods) versus elution volume. For a macromolecule that is too large to diffuse into the porous matrix, $K_D=0$ and elution occurs at the interstitial bed volume; $V_e=V_o$. A value of $K_D=1$ represents a polymer which can penetrate the entire available pore volume (i.e., not sterically constrained), and hence, the polymer elutes at an elution volume equal to the total fluid volume of the bed; $V_e=V_T=V_o+V_i$. The SEC distribution coefficient (K_D) varies with the columns' pore size and shape distributions as these parameters define the range of polymer sizes which may be separated by a given column (Yau, et al., 1978). These topics are covered in more complete reviews of SEC mechanisms by Yau, et al., (1979) and Barth (1980).

The basic SEC separation mechanism discussed above is admittedly "simplistic" since actual SEC chromatographic operations are frequently complicated by the presence of other separation phenomena and leads to mixed mode chromatographic separations (see Table II). These effects, although potentially present in any SEC system, seem to be endemic to aqueous SEC systems. Recent advances in aqueous SEC have been reviewed by Barth (1986), Rollings, et al., (1982) and Dubin, (1983). Rollings, et al., (1982) pointed out that development of SEC methods for water-soluble polymers has lagged behind SEC techniques when compared to polymers soluble in organic solvents due to the lack of readily available monodispersed calibration standards and chromatographic supports suitable for characterizing water-soluble polymers via a size separation process. The presence of non-size exclusion separation effects caused by intramolecular interactions, polymer-support interactions (most of which are "ionic" in nature due to the polyelectrolytic character of many water-soluble polymers), and flow related anomalies (viscous fingering, frit binding, etc.) further complicate chromatography operation. These non-size effects in aqueous SEC have been reviewed in detail elsewhere (Corona and Rollings, 1988). In order to extract meaningful macromolecular information from an SEC experiment either; (1) some means of relating elution profiles to macromolecular data must be established (i.e., column calibration) or (2) some absolute macromolecular detector must be placed down stream from the chromatographic separator. Both options are discussed below.

SEC CALIBRATION TECHNIQUES

As noted above, in "true" SEC separation, the distribution coefficient, K_D, is directly related to the polymer's elution volume via Equation 1 and as a consequence, a direct calibration technique can be employed to determine molecular weight distributions (MWD) and average molecular weights of the polymer sample under study. The technique requires the polymer's molecular weight to scale directly with the polymer's hydrodynamic volume; a condition that is usually satisfied with a homopolymer system chromatographed on inert supports. This "direct" calibration scheme obligates the use of compositionally identical and well-characterized polymer standards (i.e., with known average molecular weights and MWDs) to establish the relationship between K_D or V_e and molecular weight for a specific set of chromatographic conditions. Direct calibration is performed by plotting the log molecular weights of a set of known polymer standards verses elution volume at peak height, establishing the functionality,

$$\log M_v = (V_e) \qquad\qquad (2)$$

where M_v is the polymer's molecular weight at elution volume, V_e, and then relating the elution behavior of the unknown sample to this calibration relationship. One can establish the unknown sample's molecular weight distribution by obtaining the concentration of the unknown sample as a function of molecular weight via Equation 2 and summing over all parts of the chromatogram. Average molecular weights of the distribution may be established using equations given in Table III. Standard computer software packages are readily available to perform these calculations. The main problem with this approach is that primary standards are generally unavailable for most polymers

TABLE III
Average Molecular Weight Calculations

PARAMETER	FORMULA
Weight-average molecular weight	$M_w = \Sigma\ C_v M_v\ /\ \Sigma\ C_v$
Number-average molecular weight	$M_n = \Sigma\ C_v M_v\ /\ \Sigma\ C_v$
Z-average molecular weight	$M_z = \Sigma\ C_v M_v^2\ /\ \Sigma\ C_v M_v$
Polydispersity	$P.D. = M_w\ /\ M_n$

(including starch), thus direct calibration is generally not possible for starch studies.

When primary standards are unavailable, secondary standards are often employed for SEC calibration. This technique provides polymer molecular information relative to a set of standards which usually differs significantly in chemical nature (e.g., primary structure and ionic behavior in aqueous applications); thus sacrificing accuracy of analysis. If accurate polymer molecular weight information is desired and suitable calibration standards are not available, a correlation between molecular weight and solution size (the controlling SEC separation parameter) is required.

A column calibration technique more commonly employed in SEC studies than the direct calibration scheme is an indirect method often referred to as "universal" calibration. This method was first proposed over twenty years ago by Grubsic, Rempp, and Benoit (1967). Their experiments revealed that plots of the log of the product of the polymer's intrinsic viscosity and molecular weight, $[\eta]M$, versus elution volume yielded a common calibration curve for several sets of polymers varying in chemical composition and configuration; all soluble in organic solvents. The theoretical framework for the universal calibration concept was attributed to Einstein's viscosity law which states that for solid spheres, the product, $[\eta]M$, is proportional to the particle's hydrodynamic radius (R_G).

Extending this relationship to macromolecules, Flory (Fox and Flory, 1951) derived Equation 3 which relates the polymer's intrinsic viscosity to its size in solution and molecular weight for linear polymers with Gaussian segment distributions.

$$[\eta] = \Phi <r_{rms}^2>^{3/2}/M \qquad (3)$$

Here, M is the polymer molecular weight, $<r_{rms}^2>$ is the polymer root-mean-square end-to-end distance, and Φ is Flory's constant. A generalized equation for both linear and branched polymers exhibiting Gaussian chain statistics can be written in which the root-mean-square radius of gyration $<R_G>$ is proportional to the product, $[\eta]M$. (Fox and Flory, 1951).

$$[\eta]M = \Phi <R_G^2>^{3/2} \qquad (4)$$

Universal calibration in SEC implies that Equation (4) holds at each elution volume (v) over the separation range of the column.

$$([\eta]_u M_u = [\eta]_k M_k)_v \qquad (5)$$

Universal column calibration is carried out by determining the intrinsic viscosity $[\eta]_k$ of a set of standards of known molecular weight (M_k) and narrow molecular weight distribution and plotting the log of the product $[\eta]M$ versus elution volume. The molecular weight of the unknown polymer (M_u) can be determined at each elution volume with knowledge of its intrinsic viscosity $[\eta]_u$ at each elution volume. Only in a few cases, have the relationships between intrinsic viscosity and molecular weight for both known and an unknown polymers in the solvent of interest been established. This is the primary limitation of universal calibration. Others have reviewed this topic in greater detail and the reader is referred to these works (Yau, et al. 1979; Dubin, 1983). In addition, the validity of scaling SEC elution behavior with the "universal" parameter has only been clearly established for <u>flexible</u> linear polymers. Branched polymers dissolved in aqueous solvents (Kuge, et al., 1984; Kato, et al, 1983; Yu and Rollings, 1987) appear to follow universal calibration in some cases (but not generally) whereas asymmetric rod-like macromolecules (Dubin and Principi, 1989) do not follow universal trends. These macromolecules elute earlier than radially symmetric polymers (spheres and flexible coils) of equal $[\eta]M$, thus demonstrating the failure of universal calibration in some cases. Failure of universal calibration in some aqueous systems can arise from polyelectrolytic behavior which can result in intramolecular polymer-polymer and polymer-support interactions (Dubin and Spec, 1986; Bose, et al., 1982) and alteration of elution behavior.

In summary, inherent limitations exist for all the proposed calibration techniques and are most severely limited for heterogeneous polymer systems solubilized in polar solvent systems such as highly basic aqueous media. Researchers interested in the study of starch macromolecular systems should employ absolute molecular detectors such as those discussed below.

DETECTION TECHNOLOGY

All of the absolute macromolecular detectors commercially available today are built on fundamental physical chemical principles; i.e., tools that have been available for several decades. The fundamental differences between the tools scientists used thirty years ago and those available today is scale. The newly designed tools are capable of using extremely small samples and have been designed to be combined with modern polymer separation equipment such as chromatography. This section reviews the two principle types of tools available today; light scattering and intrinsic viscosity.

On-Line Laser Light Scattering

With the development of on-line, laser light scattering (LS) detection in the late 1970's, molecular weights, molecular weight distributions, and polymer branching can now be determined rapidly during chromatographic elution thereby eliminating the need for direct and secondary calibration techniques. Light scattering is an absolute molecular weight determination technique and is in this sense the most direct method of establishing starch sample molecular weight and MWD.

The scattering of light through dilute polymer solutions is a well-understood phenomena (Flory, 1953) and has been the subject of many reviews (Strazelle, 1972; Yamakawa, 1971). The intensity of scattered light is dependent upon the polymer's concentration and the angle from the incident beam at which the scattered light intensity is measured. The relationship between this intensity and a polymer's molecular parameters was originally formulated by Rayleigh and Debye (see Katime and Quintana, 1989) and applied by Zimm (see Strazelle, 1972).

$$Kc/R(\theta,c) = 1/M_w P(\theta) + 2A_2 c + O(c^2) + \ldots \qquad (6)$$

where: $K = (2\pi^2 n^2/\lambda^4 N)(dn/dc)^2(1 + \cos^2\theta)$

In Equation 6, the weight-average molecular weight of the polymer (M_w) can be determined from the solution concentration (c), the solute's excess Rayleigh ratio (the solution's scattered intensity less the solvent's contribution) for polarized incident radiation at the scattering angle $R(\theta,c)$, the eluent refractive index (n), the wavelength of light in vacuo (λ), Avogadro's number (N), and the second and higher order viral coefficients (A_2, A_3, . .). The term $P(\theta)$ is the form factor, which is a function of the size and shape of the macromolecule in solution, and represents the modulation of light due to the polymer's finite size and deviation from sphericity ($P(\theta)->1$ as $\theta->0$). The term dn/dc is the specific refractive index increment and is defined as the change in solution refractive index with polymer concentration. In dilute conditions, such as those existing in SEC experiments, terms containing $O(c^2)$ and higher can be neglected.

Equation 6 serves as the starting equation for describing light scattering photometry of polymers eluting from chromatographic separation devices. Equation 6 can be simplified if light scattering measurements are collected at a single low angle (< 10°) and under dilute solution conditions to yield Eqn. 7.

$$Kc_v/R(c_v) = 1/M_v + 2A_2 c_v \qquad (7)$$

where the subscript v is used to denote constant elution volume comparison. M_v and c_v represent the weight-average molecular weight and concentration of a sample within the detector cell at a specific SEC elution volume. Equation 7 is the working equation for application of low-angle laser light scattering (LALLS). Here, the molecular weight distribution of a polymer in solution as a function of elution volume (molecular size) can be determined using Equation 7. Use of Equation 7 requires knowledge of the macromolecule's solvent-specific and temperature-specific second viral coefficient (A_2) and specific refractive index increment (dn/dc). The first commercially available LALLS instrument was developed by Chromatix Corp. (Sunnyvale, CA) and this company was later purchased by LDC Milton Roy (Riveria Beach, FL). Two different LALLS instruments were marketed (CMX 100 and the KMX-6). The KMX-6 was the instrument most conveniently designed for research studies as the design of the KMX-6 allows for adjustments of optical measurements not available with the other instrument.

Recently, Wyatt Corporation (Santa Barbara, CA) has introduced a Multi-angle Laser Light Scattering Photometer (MALLS) also suitable for on-line use with SEC. Unlike the LALLS instrument, scattering intensities are measured at as many as 15 separate angles for each elution volume slice. Determination of molecular information follows again from Zimm's work (Strazelle, 1972). Equation 6 can be reciprocated and expressed as:

$$R(\theta,c)/Kc = M_wP(\theta) - 2A_2M_w^2P(\theta)^2c + O(c^2) + \ldots \quad (8)$$

where all variables and parameters are the same as in Equation 6. The molecular weight at each elution volume is determined by extrapolating the Debye plot [$R(\theta)/Kc$ vs. $\sin^2(\theta/2)$] to zero angle ($P(\theta) = 1$) to obtain the intercept (M_w). Again, truncation of higher ordered terms in c is valid for concentrations encountered during SEC experiments. In addition, polymer size and shape information may be gained from the Debye plots, i.e., utilization of multiple angle data will provide polymer hydrodynamic size and conformation information.

On-line light scattering holds a great advantage

213

over competing analytical tools in that it provides direct measurement of eluting polymer molecular weights. Furthermore, light scattering has the added versatility of light source wave length adjustment permitting some optical parametric (K) flexibility. However, on-line light scattering does have certain operating limitations. Molecules of MW < 50,000 daltons do not scatter enough light to be discernable from the base line signal to noise ratio in many of the experimental conditions best suited for starch. Also, if a polymer has a very low refractive index (dn/dc) the optical constant (Equation 6) approaches zero and the polymer will be undetectable. Luckily this is not the case for starch which has a refractive index increment of 0.146 in 0.5N NaOH (Vink, et al., 1967). Another method of obtaining on-line polymer macromolecular information is by viscometry. Although on-line viscometry is an indirect method of obtaining molecular weight information it has the advantage of being sensitive to the lower molecular weight polymers which can not be detected by on-line laser light scattering.

On-Line Viscometry

On-line intrinsic viscosity is an other technique capable of providing molecular weight information on starch and other polysaccharide samples. Intrinsic viscosity is phenomenologically related to a polymers molecular weight through the well known Mark-Houwink-Sukarada power law formula.

$$[\eta] = KM^a \qquad (9)$$

where K, a are parameters specific to a given polymer in a given solvent at a fixed temperature, and $[\eta]$ and M are the polymers intrinsic viscosity and molecular weight. Essentially there are three ways in which on-line intrinsic viscometers can be used to provide molecular weight information from a polymer sample eluting from a chromatographic separator. Firstly, if the Mark-Houwink-Sukarada parameters for a particular polymer are known, then equation 9 can be rearranged so that molecular weight at each elution volume can be obtained directly from the measured intrinsic viscosity data.

$$M_v = [\eta]_v^{1/a}/K \qquad\qquad (10)$$

Similarly molecular weight distributions can be obtained by appropriately averaging this data over the entire chromatogram. A secondary way in which molecular weight information can be obtained from SEC and on-line intrinsic viscometry detection is to assume (or independently verify) that universal calibration is appropriate for the polymers under study with the chosen separation system. Using a well characterized standard polymeric reference material (i.e., a known (k) polymer), the molecular weight of an unknown (u) polymer can be determined at each elution volume as;

$$M_v^u = ([\eta]_v^k/[\eta]_v^u)M_v^k = (K^k M_v^{a+1})/[\eta]_v^u \qquad (11)$$

A third method of characterizing starch samples using SEC and on-line intrinsic viscosity measurements is to use on-line light scattering in conjunction with the viscometer. In one sense, this might seem unnecessary, redundant or otherwise overkill, but on the other hand, this may be a preferable method considering the complexity of starch samples. Combined instrumentation generates more data allowing for a more complete analysis (e.g., branching analysis). More will be said of this in the next section.

Instruments designed to collect intrinsic viscometry data on-line with chromatographic separations are available from Viscotech Corp. (Porter, TX) and Waters Division of Millipore Corp. (Milford, MA). These instruments measure viscosity by monitoring gage pressure drops across a calibrated capillary and employ a Hagen-Poiseuille Law applicable for laminar flows present in SEC operations. The Water's instrument is available only as an add-on to the Water's 150C instrument and operates with a single capillary design. The operating principles have been discussed by Lesec, et al. (1988). Viscotech Corporation markets two instruments, one a duel capillary design originally developed by Yau (1990) and coworkers at Dupont and the other a four capillary design arranged as the hydrodynamic equivalent of a wheatstone bridge. Haney (1985) has described the details this later instrument's functioning.

One final note about the use of modern intrinsic viscometry instrumentation is that the instruments can be employed as a "batch" instrument, i.e., not connected to an SEC separator. Used in this mode, the on-line intrinsic viscometer can provide molecular information of dynamic processes such as might occur in liquefaction, saccharification, or other chemical or physical processes. The authors are unaware of any published reports which have employed the viscometer in this capacity.

The SEC/VIS coupled technique has some advantages and disadvantages when compared to SEC/LS. One advantage is that often times the SEC/VIS technique is more sensitive to low molecular weight polysaccharides (i.e., below 50,000 daltons) and thus more useful than SEC/LS in this molecular weight range. A second advantage is that viscosity is often times the functional property of interest to the starch researcher and in this sense SEC/VIS is more specifically targeted at an industrial parameter of interest. This also implies the fact that SEC/VIS is a disadvantage in that the SEC/VIS technique is really an indirect measurement of starch molecular weight. Therefore if absolute starch molecular weights are required, the SEC/VIS technique posses an inherent limitation.

THE SEC/LS/VIS APPROACH

Starch polymers possess a vast array of useful functional properties that derive from their fundamental molecular characteristics. Establishment of the relationship between these fundamental physical chemical properties and functional behavior is highly desirable and necessitates accurate and reliable determination of key macromolecular properties such as composition, MW and MWD. SEC is routinely used for homopolymer analysis, but starch samples are typically mixtures of amylose and amylopectin molecules and their derivatives. SEC combined with on-line detectors may be utilized in a variety of ways to obtain macromolecular information even for mixed polymer samples.

216

SEC separates by polymer molecular size not molecular weight. Starch is composed primarily of amylose and amylopectin, two macromolecules which posses significant differences in their primary, secondary, and tertiary molecular structure and hence solution properties. Amylose is an essentially linear polyglucan displaying a wide range of hydrodynamic conformations depending on solvent conditions (e.g., non-free draining coil in water to helix in DMSO [Ring, et al., 1985; Jordan and Brant, 1980]). Amylopectin possesses about 4-5% branched linkages (Banks and Greenwood, 1975) and its solution conformation is believed to be planar-like in DMSO and spherical (perhaps due to aggregation) in water (Callaghan and Lelievre, 1985). Polymer molecular size varies with MW in a unique fashion depending on polymer structure and its solution conformation. In SEC of starch, amylose and amylopectin molecules with the same hydrodynamic size will co-elute, but these two materials (at the same elution volume) will have substantially different molecular weights, amylopectin molecular weights being approximately 5 times greater than the same sized amylose macromolecules (Yu and Rollings, 1988). This fact has been overlooked by some researchers (Jackson, et al., 1989; Hizukuri, 1985 and 1986). In fact, for mixtures of macromolecules, like starch, traditional SEC will not be very useful. Mixed polymer samples cannot be quantitatively analyzed without the use of some absolute detection device.

Several research groups (Yu and Rollings, 1987, 1988; Jordan and McConnell, 1980; Jordan, et al., 1984; Alexson and Knapp, 1980 and Grinshpun, et al., 1986) have extended basic light scattering theory and experimentation to obtain mixed polymer sample molecular information. The work of Yu and Rollings was specifically directed at starch analysis using combined SEC, light scattering and intrinsic viscosity. Yu and Rollings were able to analyze starch polymer branching as a function of polymer column separation (i.e., polymer size) and/or polymer molecular weight. An example of their branching results is illustrated in Fig. 2., for the three homopolymers (amylose, amylopectin and glycogen). Recall that 4-5% of amylopectin bonds are branch point bonds and that 10% of glycogen bonds are branch points. SEC/LS branching analysis indicates that amylopectin has roughly 4.5 times the molecular weight of amylose at the same elution volume; glycogen is 15 to 20 times heavier than an equivalent hydrodynamic volume amylose. With this information, compositions of mixed linear and branched starch samples were obtained directly with a single SEC/LS or SEC/VIS experiment. These methods have a number of interesting applications which provide new insights on starch processing.

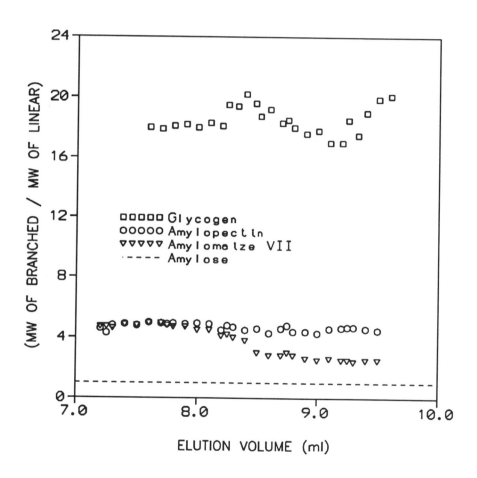

Fig. 2. Branching distribution of polysaccharides: M_b/M_l vs. elution volume for three branched polysaccharides as compared to amylose (linear). M_b/M_l is the ratio of branched to linear molecular weights.

The capability to determine relative compositions of mixed polymer samples of each SEC elution slice makes it possible for researchers to observe preferential chemical and enzymatic hydrolysis of polysaccharide macromolecules. As was mentioned in the introduction, Park and Rollings (in press) found a relationship between a starch's degree of branching and its susceptibility to enzymatic depolymerization, reporting that networks of α-(1,6) branch points (i.e., "clusters"), appear to restrict enzyme accessibility to α-(1,4) linkages. They also observed that the initial rate of hydrolysis was more rapid for amylopectin than amylose, attributing this to the conformational differences of these two molecules in solution (amylose-helical and amylopectin-random coil). Beri, et al. (submitted for publication) found that a critical minimum injected mass of dextran polymer was required for accurate analysis. This value is system specific as, the absolute molecular weight detector was found to give consistently the same macromolecular information only if the total sample injected mass exceeds this critical value, (see Fig. 3.).

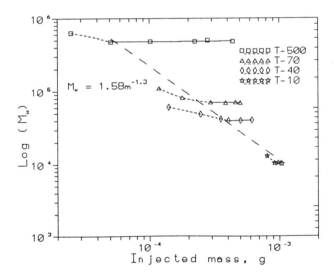

Fig. 3. Detected dextran molecular weight (LS) vs. total injected sample mass. Beri, et al. (Submitted for publication). Courtesy of Carbohydrate Research.

In these experiments the elution volume was observed to change significantly due to non-size separation effects (10-15% variation in elution behavior); but the calculated molecular weight distributions (using LS detector) were invariant. This is a good example of how on-line light scattering can circumvent inaccuracies inherent in SEC calibration methods. The experimental considerations for application of SEC/LS/VIS technology are discussed in the following section.

EXPERIMENTAL CONSIDERATIONS

The components of a typical SEC/on-line detector system are shown in Fig. 4. The system includes a solvent reservoir, pump, sample injector, SEC column(s), a concentration detector, a light scattering detector and/or an intrinsic viscosity detector. Data acquisition and processing is carried out by computer using software supplied by the photometer or viscometer manufacturer.

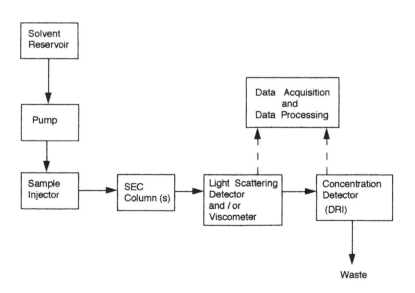

Fig. 4. Typical SEC/on-line detector system.

A major consideration in starch SEC is mobile phase selection. Starch polymers have limited solubility in water even at high temperatures and are insoluble (due to amylose complexation) in most organic solvents (eg. alcohols, esters, ketones) (Young, 1984). Dimethyl sulfoxide (DMSO) readily dissolves starch as does aqueous alkali. Unfortunately, DMSO is quite viscous at room temperature thus requiring SEC at elevated temperatures (80°C) (Chuang and Sydor, 1987). Use of alkali solvents is limited by their compatibility with relatively few commercial column resins. Starch polymers degrade with prolonged exposure to aqueous alkali. In some cases, starch polymers may contain polyelectrolytic functionality (e.g., phosphate groups in potato) and are therefore susceptible to polymer-solvent-support interactions. The addition of simple salts to the eluting media usually eliminates this complication. When pure water or buffer is chosen as the eluent, small amounts of sodium azide (0.02%) may be added to prevent bacterial growth (Takagi and Hizukuri, 1984). Mobile phase and column resins for starch SEC analysis are summarized in Table IV.

Regardless of the choice, solvents for combined SEC/light scattering analysis must be of the highest quality; free from dust and other contaminates. This requirement is met by a combination of solvent distillation, filtration and/or centrifugation steps prior to use. For solvents with high air solubility (eg. water), degassing is essential. Filter selection must be based on solvent compatibility, and if the filter is to be placed downstream from the column(s), filter pore size.

Accurate determination of the polymer-solvent specific refractive index increment (dn/dc) is necessary since it appears as a squared term in Equation 6. Values of this increment for starch polymers in various solvents (at various wavelengths) have been collected by Huglin (1972) and others and are listed in Table V. They may be obtain directly through separate measurements with a refractometer or interferometer. The increment may also be estimated by the ratio technique proposed by Yu and Rollings (1987). The only

TABLE IV
Solvents and Columns Utilized in Starch SEC

MOBILE PHASE	COLUMN RESIN	STARCH POLYMER	REFERENCES
10 mM phosphate buffer (pH=6.1)	Pharmacia Sephadex G-50	Enzymatic hydrolysates of manioc starches	Hood and Mercier (1978)
		Native and lintnerized potato, waxy and their hydrolysates	Robin, et al. (1974 and 1975)
50mM phosphate buffer (pH=6.1)	Toyo-Soda TSK SW PW series 1	Various amylose and debranched amylopectin fractions	Hizukuri and Takagi (1984)* Takeda, et al.,(1986, 1988 A&B)* Hizukuri (1985-6)* Takagi and Hizukuri (1984)*
Water	Pharmacia Sepharose CL	Potato and amylomaize VII	Eberman and Swartz (1975)
	Bio-Gel A-series	Potato and waxy-maize amylose and amylopectin	Ma and Robyt (1987)
	Pharmacia Superose	Amylose enzyme hydrolysates	Praznik, et al. (1987)

1 Now manufactured by Toso-Hass
* On-line light scattering measurements employed

TABLE IV (Continued)
Solvents and Columns Utilized in Starch SEC

MOBILE PHASE	COLUMN RESIN	STARCH POLYMER	REFERENCES
Water	Shodex S-series Ionpak	Water sol. fractions of corn & sorghum	Jackson, et al., (1988 and 1989)
Aqueous NaOH (0.5N)	Pharmacia Sepharose CL	Corn starch enzyme hydrolysates	Rollings, et al. (1982&84)
	Toyo-Soda TSK HW series[1]	Amylose, amylopectin & amylomaize VII enz.hyd.	Yu and Rollings (1987&88)* Park, et al (1988&89)*
Aqueous KOH (0.5N)	Pharmacia Sepharose CL	Drum dried & extrusion cooked starches	Schweizer & Reimann (1986)
	Pharmacia Sepharose CL	Native starches and oxidation products	Henriksnas and Brunn (1978)
DMSO	Waters E-series Bondagel	Native wheat amylose and amylopectin fractions	Kobayashi, et al., (1985)
DMSO with 0.03M NaNO$_3$	Waters E-series Bondagel, Styragel Brownlee Aquapore OH	Potato, high amylose corn and their acid and enzyme hydrolysates	Chuang and Sydor (1987)

1 Now manufactured by Toso-Hass
* On-line light scattering measurements employed

Table V
Specific Refractive Index Increments for
Starch Components in Various Solvents

	SOLVENT	REF. INDEX INC. (k nm)	REFERENCE
AMYLOPECTIN	DMSO	0.066 (546)	Erlander, 1968
	DMSO/water 90/10 vol.	0.074 (546)	Erlander, 1968
	80/20 vol.	0.083 (546)	Erl., 1968
	60/40 vol.	0.101 (546)	Erl., 1968
	50/50 vol.	0.110 (546)	Erl., 1968
	Water	0.152 (546)	Erl., 1968
	1N KOH	0.142 (546)	Stacy, 1956
	0.2N NaOH	0.142 (436)	Witnauer, 1955
	0.5N NaOH	0.146	Vink, 1967
AMYLOSE	DMSO	0.066 (546)	Everett, 1959
	DMSO/water 30%-90% vol	0.07 - 0.13 (436)	Jordan, 1980
	0%-100% vol	0.06 - 0.15 (546)	Dintzis, 1978
	Water	0.152 (546)	Kodma, 1978
	1N KOH	0.146	Paschall, 1952
	0.5N NaOH	0.146	Vink, 1967

other unknown in Equation 6 is the second viral coefficient A_2 of the polymer in the eluting solvent. Very little data for starch polymers have been collected. Kodama, et al., (1978) and Ring and coworkers (1985) analyzed fractions of amylose from potato and pea (MW 1 x 10^6) in water and found A_2 of 1.88 x 10^{-4} and 4.8 x 10^{-4} mol•ml/g^2 respectively. Jordan and Brant(1980) studied amylose in mixed DMSO/water systems and found A_2 to increase from 0.5 x 10^{-4} to about 2.5 x 10^{-4} mol•ml/g^2 as DMSO volume fraction was increased from 0.30 to 0.9. Fujii, Honda and Fujita (1973) carried out light scattering experiments on a number of amylose fractions in DMSO and found A_2 to vary between 2.9 and 6.1 x 10^{-4} mol•ml/g^2 depending on MW. Second viral coefficients may be obtained for the polymer and solvent of interest through measurement of (or for MALLS extrapolation to) low angle scattering intensities at several dilute polymer concentrations and use of Equation 7 or Equation 8.

Column Selection

The selection of columns for SEC of starch depends on the separation range desired as well as the aforementioned solvent considerations. Native starch varieties have polymer molecular weight distributions spanning almost four orders of magnitude. Full resolution may require use of two or more columns of limited separation range. Selection criteria for column lengths, diameters and flow rates is discussed in general by Yau et al. (1979) and by Rollings and co-workers (1982) for aqueous starch SEC. Columns employed in starch SEC analysis are listed in Table IV. Native starch polymers have molecular weights exceeding 10^7 (primarily amylopectin). In most cases these high molecular weight materials elute at the column exclusion volume (V_o) thus making MW determination of the entire distribution by any calibration or on-line technique impossible. In general, higher MW exclusion limits and greater single column separation ranges are possible with silica and styrene-divinyl benzene based resins (eg. Bondagel, Styragel) with DMSO eluents than carbohydrate gel based resins and aqueous solvents. Resin pore size can be controlled and mechanical stability is significantly better. High MW polymer adsorption and shearing by column end fittings is also possible. SEC separation of starch hydrolysis products (SHP) is more easily undertaken owing to their smaller MW (size) distributions. It should be noted that on-line light scattering detection is particularly useful in studying column separation performance, since molecular weights are determined at each elution volume directly.

Sample Preparation

The starch sample is usually prepared in the mobile phase solvent. Adequate care must be taken to insure complete dissolution. For SEC in pure water or buffer, starch preparation requires dispersion in a good solvent and/or heating. When aqueous alkali is used, dissolution frequently takes several hours and must be followed by neutralized with acid prior to injection (Takagi and Hizukuri, 1984). DMSO and DMSO-water mixtures (Chuang and Sydor, 1987; Park, 1987) are sometimes used. When mobile phase and sample prep solvents are not matched, a large solvent (DRI) peak is seen at the column(s) void volume due to differences in refractive index. This may interfere with detection of low MW polymer peaks. Use of aqueous alkali or DMSO as both the eluent and the sample prep solvent alleviates this problem. Aqueous alkali has the added advantage of allowing sample preparations directly from aqueous processing conditions. In enzymatic hydrolysis reactions, addition of a low pH aqueous alkali solution deactivates depolymerases as well as solubilizes SHPs (Rollings, 1981). Heating and agitation in any of these solvents aids the dissolution process but may have deleterious depolymerizing and/or ring opening effects. For SEC preparation of native granules, a defatting step is frequently added prior to solubilization. Despite these pitfalls, systematic optimization of starch SEC sample preparation procedures have not been studied.

Guidelines for determining sample amount and concentration in SEC analysis have been set forth by Yau et al., (1979). In general, increases in sample size increase column band broadening effects. Sample volumes also affect apparent solute retention times. When ever possible, sample volumes should be kept as small as possible and constant if comparative data is desired. For high MW starch polymers, sample concentration is limited by sample viscosity. Viscous streaming or fingering may result in tailing on the trailing side of the solute peak. Proper choice of starch sample size also depends on restrictions imposed by light scattering analysis. Scattered light intensity at any elution volume depends on both starch concentration and MW. For light scattering of high MW starch polymers, sample size must be restricted to keep the LS signal within the dynamic range of the photometer. SHP analysis becomes increasingly difficult at low molecular weights since scattering intensity decreases with both concentration and molecular weight (Equation 6). Here, larger sample sizes are required to maintain adequate signal to noise (see Beri, et al., in press). Native starch samples with broad molecular weight distributions require some intermediate sample mass amount. In this case, MW of the distribution is likely overestimated as signals from small molecular weight polymers vanish (see Fig. 3.). Some comments on future research directions are discussed below.

RESEARCH DIRECTIONS

Lesec and Volet (1990) have done research to evaluate the triple coupling of aqueous SEC with light scattering and viscometry. They concluded that this method yields precise polymer characterization data because of the absolute molecular weights detected by light scattering combined with the excellent solvent flow rate control provided by the viscometer. Use of the latest SEC/on-line detector analysis techniques will allow starch researchers to gain detailed starch structural information for use in gathering new structure-function data and providing new insights on past research. For example, Patil and Kale's (1973) work on preferential adsorption of amylose on cellulose resin, could be repeated using an SEC/LS/VIS experiment designed to determine accurate MW or branching specificity of solvent effects. Future research studies might include accessment of starch macromolecular structural effects on mixed enzyme hydrolysis. SEC/LS/VIS should also be extended to study starch complex formations (e.g., starch -starch, $-I_2$, -lipid, - hydrocolloid and -particle). Armed with new analytical tools to determine starch structure-function relationships, scientists now have the ability to select and alter native starches to produce specified physical chemical properties and ultimately "engineer" a product's functionality. Applications to the starch industry include, novel product development (e.g., starches for use in drug delivery systems (Bjork and Edman, 1990)), starch hybridization and genetic manipulation, and reactor - separator designs for specific hydrolysates or derivatives.

ACKNOWLEDGEMENTS

This paper is based upon work supported by the National Science Foundation, grant # CBT-8451013 and Novo Nordisk Bioindustrials, Inc. (Danbury, CT).

LITERATURE CITED

Axelson, D. E. and Knapp, W. C. 1980. Size exclusion chromatography and low-angle laser light scattering. Application to the study of long chain - branched polyethylene. J. Appl. Polym. Sci. 25, 119.

Banks, W. and Greenwood, C. T. 1975. Pages 38-43 in: Starch and its Components. Edinburgh University Press, Edinburgh, Scotland.

Barth, H. G. 1980. A practical approach to stearic exclusion chromatography of water soluble polymers. J. of Chromatogr. Sci. 18, 409.

Barth, H. G. 1986. Characterization of water-soluble polymers using size-exclusion chromatography. Amer. Chem. Soc. Adv. Chem. 213, 31.

Beri, R. G., Walker, J., Reese, E. T. and Rollings, J. E. (submitted for publication) Characterization of chitosans via coupled size exclusion chromatography and multiple angle laser light scattering technique. Carbohydrate Research.

Bjork, E. and Edman, P. 1990. Characterization of degradable starch microspheres as a nasal delivery system for drugs. International J. of Pharmaceutics. 62, 187.

Bose, A., Rollings, J. E., Caruthers, J. M., Okos, M. R. and Tsao, G.T. 1982. Polyelectrolytes as secondary calibration standards for aqueous size-exclusion chromatography. J. Appl. Polym. Sci. 27, 795-810.

Callaghan, P. T. and Lelievre, J. 1985. The size and shape of amylopectin: a study using pulsed-field gradient nuclear magnetic resonance. Biopolymers. 24, 441-460.

Chuang, J.-Y. and Sydor, R. J. 1987. High performance size exclusion chromatography of starch with dimethyl sulfoxide as the mobile phase. J. Appl. Polym. Sci. 34, 1739.

Corona, A., Rollings, J. E., 1988. Polysaccharide characterization by SEC/LALLS: a review. Separation Sci. and Tech. 23, 855.

Corona, A. 1990. M.S. Thesis. Worcester Polytechnic Institute.

Davidson, R. L., 1980. Handbook of Water Soluble Gums and Resins. McGraw Hill, New York.

Dintzis, F. R. and Tobin, R. 1978. Optical rotation of some α-1,4-linked glycopyranosides in the system H_2O-DMSO and solution conformation of amylose. Carbohydr. Res. 66, 71.

Dubin, P. L. 1983. Exclusion chromatography of water soluble polymers. Am. Lab. (Fairfield, CT). 15, 62.

Dubin, P. L. and Speck, C. M. 1986. Electrostatic effects in aqueous size exclusion chromatography. Proc. Amer. Chem. Soc. Div. PSME. 54, 194.

Dubin, P. L. and Principi J. M. 1989. Failure of universal calibration for size exclusion chromatography of rodlike macromolecules versus random coils and globular proteins. Macromolecules. 22, 1891.

Dubois, M., Gilles, K. A., Hamilton, J. K., Rebers, P. A. and Smith, F. 1956. Colormetric method for determination of sugars and related substances. Analytical Chemistry. 28(3), 350.

Dygert, S., Li, L. H., Florida, D. and Thoma, J. A. 1965. Determination of reducing sugar with improved precision. Analytical Biochemistry. 13(3), 367.

Ebermann, R. and Schwartz, R. 1975. Fractionation of starch by gel filtration on agarose beads. Starch. 27, 361.

Erlander, S. R.and Tobin 1968. The stability of the helix of amylose and amylopectin in DMSO and water solutions. Makromol.Chem. 111, 194.

Everett, W. B. and Foster, J. F. 1959. The subfractionation of amylose and characterization of the subfractions by light scattering. J. Amer. Chem. Soc. 81, 3459.

Flory, P. J. 1953. Principles of Polymer Chemistry. Cornell University Press, Ithaca, New York.

Fox, T. G. and Flory, P. J. 1951. Treatment of intrinsic viscosities. J. Amer. Chem. Soc. 73, 1904.

Fujii, M., Honda, K. and Fujita, H. 1973. Dilute solution of amylose in dimethylsulfoxide. Biopolymers. 12, 1177.

Giddings, J. C. 1988. Field-flow fractionation. Chem. and Eng. News. 66(41), 34.

Grinshpun, V., Rudin, A., Russell, K. E. and Scammell, M. V. 1986. Long chain branching indices from size-exclusion chromatography of polyethylenes. J. Polym. Sci., Polym. Phys. Ed. 24, 1171.

Grubisic, Z., Rempp, R. and Benoit, H. 1967. A universal calibration for gel permeation chromatography. J. Polym. Sci., Part B. 5, 763.

Haney, M. A. 1985. The differential viscometer. II. On-line viscosity detector for size exclusion chromatography. J. Appl. Polym. Sci. 30, 3037.

Henriksnas, H. and Bruun, H. 1978. Molecular weight distribution in starch solutions when hydrolysed with alpha-amylase and when oxidized with sodium hydrochlorite. Starch. 30, 233.

Hicks, K. B. 1988. High-performance liquid chromatography of carbohydrates. Advances in Carbohydrate Chem. and Biochem. 46, 17-72.

Hizukuri, S. and Takagi, T. 1984. Estimation of the distribution of molecular weight for amylose by low angle laser light scattering technique combined with high performance gel chromatography. Carbohydrate Research. 134, 1-10.

Hizukuri, S. 1985. Relationship between the distribution of the chain length of amylopectin and crystalline structure of starch granules. Carbohydrate Research. 141, 295.

Hizukuri, S. 1986. Polymodal distribution of the chain lengths of amylopectins, and its significance. Carbohydrate Research. 147, 342-347.

Hood, L. F. and Mercier, C. 1978. Molecular structure of unmodified and chemically modified manioc starches. Carbohydrate Research. 61, 53-66.

Huglin, M. B. 1972. Page 165 in: Light Scattering from Polymer Solutions. M. B. Huglin, Ed. Academic Press, New York.

Jackson, D. S., Choto-Owen, C., Waniska, R. D. and Rooney, L. W. 1988. Characterization of starch cooked in alkali by aqueous high-performance size exclusion chromatography. Cereal Chem. 65(6), 493-496.

Jackson, D. S., Waniska, R. D. and Rooney, L. W. 1989. Differential water solubility of corn and sorghum starches as characterized by high performance size -exclusion. Cereal Chemistry. 66(3), 228.

Jordan, R. C. and Brant, D. A. 1980. Unperturbed dimensions of amylose in binary water/dimethyl sulfoxide mixtures. Macromolecules. 13, 491.

Jordan, R. C. 1980. Size exclusion chromatography with low angle laser light scattering detection. J. Liq. Chromatography. 3(3), 439.

Jordan, R. C. and McConnell, M. L. 1980. Page 107 in: Size Exclusion Chromatography (GPC). T. Provder, ed. (ASC Symposium Series No. 138). American Chemical Society, Washington.

Jordan, R. C., Silver, S. F., Sehon, R. D. and Rivard, R. J. 1984. Page 295 in: Size Exclusion Chromatography: Methodology and Characterization of Polymer Related Materials. T. Provder, ed. (ACS Symposium Series No. 245). American Chemical Society, Washington.

Katime, I. S. and Quintana, J. R. 1989. Page 103 in: Comprehensive Polymer Science, Volume 1. Sir G. Allen, J. C. Bevington, C. Booth and C. Price, eds. Pergamon Press. Oxford.

Kato, T., Tokuya, T. and Takahashi A. 1983. Comparison of poly(ethylene oxide), pullulan and dextran as polymer standards in aqueous gel chromatography. J. Chromatogr. 256, 61.

Kobayashi, S., Schwartz, S. J. and Lineback, D. R. 1985. Raped analysis of starch. Amylose and amylopectin by high-performance size-exclusion chromatography. J. Chromatography. 319, 205-214.

Kodama, M., Noda, H. and Kamata, T. 1978. Conformation of amylose in water. I. Light scattering and sedimentation-equilibrium measurements. Biopolymers. 17, 985-1002.

Kuge, T., Kobayashi, K., Tanahashi, H., Igushi, T. and Kitamura, S. 1984. Gel permeation chromatography of polysaccharides: universal calibration curve. Agric. Biol. Chem. 48, 2375.

Lansky, S., Kooi, M. and Schoch, T. J. 1949. Properties of the fractions and linear subfractions from various starches. J. Amer. Chem. Soc. 71, 4066.

Lesec, J., Lecacheux, D., Marot, G. 1988. The continuous viscometric detector in size exclusion chromatography. J. Liq. Chromatography. 11, 2571.

Lesec, J. and Volet, G. 1990. Data treatment in aqueous GPC with on-line viscometer and light scattering detectors. J. Liq. Chromatography. 13(5), 831-849.

Ma, W. -P. and Robyt, J. F. 1987. Preparation and characterization of soluble starches having different molecular sizes. Carbohydrate Research. 166, 283.

Miller, G. L. 1959. Use of Dinitrosalicylic acid reagent for determination of reducing sugar. Analytical Chemistry. 31(3), 426.

Park, J. T. 1987. M.S. Thesis, Worcester Polytechnic Institute.

Park, J. T., Yu, L. -P., Corona, A. and Rollings, J. E. 1988. The effects of polysaccharide branching on enzymatic depolymerization reactions. ASC Division of Polymer Chem., Polym. Preprints. 29(1), 626-628.

Park, J. T. and Rollings, J. E. 1989. Biopolymetric Structural effects of α-amylase catalyzed amylose depolymerization. Enzyme Microbiology Technology. 11, 334-340.

Park, J. T. and Rollings, J. E. (submitted for publication) A kinetic model of mixed linear and branched polysaccharide hydrolysis reactions; I: amylose/amylopectin α-amylosis. Biotech. Bioeng.

Paschall, E. F.and Foster, J. F. 1952. An investigation by light scattering of the state of aggregation of amylose in aqueous solutions. J. Polym. Sci. 9, 85.

Patil, N. B. and Kale, N. R. 1973. Quantitative separation of starch components on a cellulose column. J. of Chromatogr. 84, 75-85.

Praznik, W., Beck, R. H. F. and Berghofer, E. 1987. Molecular characterization of hydrolytically degraded starches with high performance GPC analysis. Starch. 39, 397.

Ring, S. G., I'Anson, K. J. and Morris, V. J. 1985. Static and dynamic light scattering studies of amylose solutions. Macromolecules. 18, 182.

Robin, J.P., Mercier, C., Charbonniere, R. and Guilbot, A. 1974. Linterized starches. Gel filtration and enzymatic studies of insoluble residues from prolonged acid treatment of potato starch. Cereal Chemistry. 51, 389-406.

Robin, J.P., Mercier, C., Duprat, F., Charbonniere, R. and Guilbot, A. 1975. Linterized starches. Chromatographic Enzymatic Studies of insoluble residues from acid hydrolysis of various cereal starches, particularly waxy maize starch.Starch. 27(2), 36-45.

Rodriguez, F. 1982. Principles of Polymer Systems, Second Edition. McGraw-Hill Book Co., New York.

Rollings, J. E. 1981. Ph. D. Thesis, Purdue University.

Rollings, J. E., Bose, A., Okos, M. R. and Tsao, G.T. 1982. System development for size exclusion chromatography of starch hydrolysates. J. Appl. Polym. Sci. 27, 2281.

Rollings, J. E. and Thompson, R. W. 1984. Kinetics of enzymatic starch liquefaction: simulation of the high-molecular-weight product distribution. Biotech. Bioeng. 26, 1475-1484.

Rundle, R. E., Foster, J. F. and Baldwin, R. R. 1944. On the nature of the starch iodine complex. J. Am. Chem. Soc. 66, 2116-2120.

Schweizer, T. F. and Reimann, S. 1986. Influence of drum drying and twin-screw extrusion cooking on wheat carbohydrates, II. effects of lipids on physical properties, degradation and complex formation of starch in wheat flour. J. Cereal Sci. 4, 193-203.

Shanon, J. C. and Garwood, D. L. 1984. Genetics and
    Physiology of Starch Development. Pages 26-86 in:
    Starch: Chemistry and Technology. R. L. Whistler, J.
    N. Bemiller, and E. F. Paschall, eds. Academic
    Press, Orlando.

Slade, L. and Levine, H. 1987. Pages 387-430 in: Recent
    Developments in Industrial Polysaccharides. S. S.
    Stivala, V. Crescenzi and I. C. M. Dea eds. Gordon
    and Breach Science, New York.

Snyder, L. R. and Kirkland, J. J. 1979. Introduction to
    Modern Liquid Chromatography. John Wiley and Sons,
    Inc. New York.

Somogyi, M. 1952. Notes on sugar determination. J. Biol.
    Chem. 195, 19.

Stacy, C. and Foster, J. F. 1956. A light scattering
    study of corn amylopectin and its beta-amylase limit
    dextrin. J. Polym. Sci. 20, 57.

Sterling, C. 1978. Textural qualities and molecular
    structure of starch products. J. of Textural
    Studies. 9, 225-255.

Strazielle, C. 1972. Page 663 in: Light Scattering from
    Polymer Solutions. M. B. Huglin, ed. Academic Press,
    New York.

Takagi, T. and Hizukuri, S. 1984. Molecular weight and
    related properties of lily amylose determined by
    monitoring of elution from TSK-GEL PW high
    -performance gel chromatography columns by low-angle
    laser light scattering technique and precision
    differential refractometry. J. Biochem. 95, 1459-
    1467.

Takeda, Y., Hizukuri, S. and Juliano, B. O. 1986.
    Purification and structure of amylose from rice
    starch.Carbohyd. Res. 148, 299-308.

Takeda, Y., Suzuki, A. and Hizukuri, S. 1988A. Influence of steeping conditions for kernels of corn starch. Starch. 40(4), 132-135.

Takeda, Y., Shitaozono, T. and Hizukuri, S. 1988B. Molecular structure of corn starch. Starch. 40(2), 51-54.

Vink, H. and Dahlstrom, G., 1967. Refractive Index Increments for polymers in solution multicomponent systems. Makromol. Chem. 109, 249.

Vollmert, B. 1973. Polymer Chemistry. Spring-Verlag. New York.

Whistler, R. L., BeMiller, J. N., Paschall, E. F. 1984. Starch: Chemistry and Technology. Academic Press, Orlando, FL.

Witnauer, L. P., Senti, F. R. and Stern, M. D. 1955. Light scattering investigation of potato amylopectin. J. Polymer Sci. 16, 1.

Yamakawa, H., 1971. Modern Theory of Polymer Solutions. Harper and Row, New York.

Yau, W. W., Kirkland, J. J. and Bly, D. D. 1979. Modern Size Exclusion Liquid Chromatography. Wiley, New York.

Yau, W. W., Ginnard, C. R., Kirkland, J. J., 1978. Broad-range linear calibration in high-performance size-exclusion chromatography using column packing with bimodal pores. Journal of Chromatography. 149, 465.

Yau, W. W. 1990. New polymer characterization capabilities using SEC with on-line MW-specific detectors. Chem. Tracts - Macromolecular Chemistry. 1(1), 1-36.

Young, A. H. 1984. Fractionation of Starch. Pages 249-284 in: Starch: Chemistry and Technology. R. L. Whistler, J. N. Bemiller, and E. F. Paschall, eds. Academic Press, Orlando.

Yu, L. -P. and Rollings, J. E., 1987. Low angle laser light scattering - aqueous size exclusion chromatography of polysaccharides: Molecular weight distribution and polymer branching determination. J. Appl. Polym. Sci. 33, 1909-1921.

Yu, L. -P. and Rollings, J. E., 1988. Quantitative branching of linear and branched polysaccharide mixtures by size exclusion chromatography and on -line low angle laser light scattering detection. J. Appl. Polym. Sci. 35, 1085-1102.

HIGH PERFORMANCE LIQUID CHROMATOGRAPHY
OF CEREAL-DERIVED CARBOHYDRATES

Kevin B. Hicks and Arland T. Hotchkiss, Jr.

U.S. Department of Agriculture
Agricultural Research Service
Eastern Regional Research Center
600 East Mermaid Lane
Philadelphia, PA 19118

INTRODUCTION

During the last 20 years, High Performance Liquid Chromatography (HPLC) has developed from a mild curiosity into a powerful method for carbohydrate analysis. In this chapter, we will describe current state-of-the-art methods for HPLC analysis of cereal-derived carbohydrates. Separations of simple sugars, sugar alcohols, and cyclic- and linear- oligosaccharides up to a dp of approximately 85 will be described. High performance separations of polysaccharides will be covered in another chapter in this book. Some information will be given here on stationary phases, instruments, and other equipment useful for HPLC analyses. If more details in this area are required, the reader will be referred to a number of extensive recent reviews.

It is the aim of this chapter to present the most recent carbohydrate HPLC methods that are available and to indicate which of these methods are most frequently used for specific applications in carbohydrate-related industries. In order to obtain the latter information, a voluntary survey was completed by skilled practitioners of HPLC at 20 major companies in the wet milling, ingredient manufacturing, food, alcoholic- and non-alcoholic beverage, and candy manufacturing industries. Companies which participated are included in Table I.

INSTRUMENTATION

Instrumentation for carbohydrate HPLC analysis has recently been reviewed (Hicks, 1988) so only critical aspects will be discussed.

Solvent Delivery Systems

Almost all modern, commercially available HPLC solvent delivery systems are useable for carbohydrate analysis. However, since detectors commonly used with carbohydrate analyses are quite sensitive to changes in solvent flow, pressure, and composition, those pumping systems which provide the most pulse-free and precisely blended mobile phases are desirable. Most recent single- and multi-pump commercial systems can provide acceptable performance. Since HPLC detectors are also sensitive to dissolved gasses in mobile phases, the use of commercial degassing units or those systems which operate under a positive pressure of an inert gas such as helium are also recommended. Because many users of analytical HPLC techniques will want to scale up to preparative levels, it is recommended that researchers purchase HPLC systems capable of delivering flow rates useful for both analytical (0.1-5 ml/min) and preparative (5-50 ml/min) applications. Such systems which provide these flow rates without compromising performance are commercially available. Because many methods require the use of buffers and alkaline mobile phases, systems configured with inert non-metallic components, such as polyetheretherketone (PEEK), and self washing pump heads, can avoid corrosion of components and destruction of pump seals and pistons, saving considerable maintenance cost and time.

TABLE I
Participants of HPLC Survey

Type of company	Participants in Survey
Milling Companies	American Maize-Products Co.; A. E. Staley Manufacturing Co; Cargill Analytical Services; Corn Products, a Unit of CPC International, Inc.; 2 more requesting anonymity.
Alcoholic and Non-alcoholic Beverage Manufacturers	Miller Brewing Co.; The Seven Up Co.; Adolph Coors Co.; The Coca-Cola Company.
Food Ingredient Manufacturers	Biospherics Inc.; The NutraSweet Co.; Kelco, a Division of Merck & Co., Inc.; Suomen Xyrofin Oy, Quest International.
Food and Confectionery Manufacturers	Hershey Foods Corp.; Kraft General Foods, Inc. (2 responses); Nabisco Brands; 1 requesting anonymity

Fittings, Valves, Tubing, and Injectors

One of the most common causes of poor resolution in HPLC is a poorly "plumbed" system. Columns, injectors, and other components, connected with excess tubing and poorly constructed unions lead to band broadening and poor resolution in the final chromatogram. The use of PEEK tubing and plastic ferrules and nuts has greatly simplified the "plumbing" of HPLC instruments. These components, which operate well at pressures typically encountered in HPLC analyses, can easily be used to plumb a system that contains a minimum amount of dead-volume. Plastic fittings also provide "universal" connections for fittings from various manufacturers. However, some plastic fitting may fail at elevated temperatures.

Fixed loop injectors are the most useful type for HPLC analysis. When loops are properly overfilled, reproducibility of the injected amount is excellent and one injector can serve for both analytical and preparative applications by simply installing a loop with the desired volume (available from 5 microliters to over 10 ml in volume).

Filters and Pre- or Guard-columns

HPLC column life can be greatly extended by installing a pre-column between the injector and the analytical column. Those pre-columns that contain replaceable 0.2 micron filters and replaceable cartridge-type precolumns are especially convenient and effective. It is not easy to predict how often a pre-column should be changed. This is usually determined after acquiring some experience with the particular HPLC method being used. Build-up of sample particulate material on the frits/filters on a system will result in a gradual rise in pressure. Whenever pressure rises more than a few psi, the cause should be determined and corrected. Properly working systems should operate at a consistent pressure.

Column Hardware

The most useful feature developed in recent years are cartridge-type columns in which the stationary phase, housed in a glass, metal, or plastic cartridge is used in conjunction with a device which may provide radial or axial pressure, thereby eliminating voids which may develop in the packed stationary phase. Such configurations can greatly extend the useful "lifetime" of a stationary phase and also result in a lower replacement cost (only the cartridge is replaced).

Detection Systems

A detailed analysis of HPLC detection systems has been published (Hicks, 1988). The most commonly used detectors for carbohydrate analysis are refractive index (RI) detectors which were used by 95% of our survey respondents. The sensitivity and characteristics of commonly used detectors are given in Table II. In general, one should use the most sensitive and stable detector available since this permits injection of small samples (and corresponding impurities), resulting in longer column life. The majority of survey respondents listed RI detectors as the most stable, sensitive, trouble free, and useful for quantitative purposes. Since homologous series of oligosaccharides all give approximately the same peak area per unit weight when RI detectors are used (Scobell et al., 1977), one can quantify a whole series of oligosaccharides by using only one pure oligosaccharide of that series as an external standard. Pulsed amperometric detectors (PAD) programmed with optimized settings (LaCourse and Johnson, 1991), give reproducible and predictable responses for simple sugars, di-, and trisaccharides (Paskach et al., 1991). For higher oligosaccharides, the situation is not so clear and researchers who use High Performance Anion Exchange (HPAE) chromatography columns for their unequalled ability to separate oligosaccharides up to an extremely high DP level, are frequently frustrated by their inability to quantify all the corresponding chromatographic peaks with the PAD. A number of investigators have studied this detection system in order to predict response factors for oligosaccharides for which standards are not available. Koizumi, et al. (1989 and 1991a), for instance, demonstrated that the PAD response for DP 2-13 malto-oligosaccharides increased directly with the number of HCOH groups per oligosaccharide. The relative detector response continued to increase with larger oligosaccharides (DP 14-17) but it was not quite proportional to the number of HCOH groups per molecule. Barsuhn and Kotarski (1991) using the same PAD detector settings, showed that the PAD response per micromole of malto-oligosaccharide steadily increased from DP 2-7.

TABLE II
Characteristics of Detectors Used for Carbohydrate HPLC

Feature	RI	PAD	UV
Detectability	"Universal"	"Selective"	"Selective"
Lower limit of detection	20-40 ng	10-30 ng	1-12 μg
Advantages	Predictable response; useful for quantitative analysis; stable; easy to maintain; non-destructive; analytical and preparative use.	Can be used with solvent gradients; sensitive; monitors only analyte of interest.	High sensitivity for some analytes; monitors only selected analytes.
Disadvantages	Sensitive to changes in temperature, pressure, and solvent; not useable with gradients.	Response factors are not well understood; Must have standards for all analytes; requires maintenance.	Poor sensitivity for many carbohydrates; many interfering compounds.

Ammeraal et al. (1991) observed that the malto-oligosaccharide response per micromole increased from DP 2 to -14. Beyond that point, response was variable, but the standards used contained more than one DP oligosaccharide. If pure oligosaccharides of higher DP values were available, this would facilitate quantification of these oligosaccharides by PAD. Unfortunately methods do not currently exist to isolate malto-oligosaccharides larger than DP 14-17 (Koizumi et al., 1991a).

HPAE methods provide excellent separations of oligosaccharides in part, because of the ability to apply mobile phase gradients to the column. However, the changing solvent composition effects the response of the PAD detector, further complicating the accurate quantification of eluting oligosaccharides. An innovative approach to compensate for these detector response variations was proposed by Larew and Johnson (1988), who sent the effluent from the HPAE column into an immobilized glucoamylase reactor, which completely hydrolyzed the separated oligosaccharides to produce glucose, prior to entering the PAD detector. By this method, a single calibration curve for glucose based on peak area was used to accurately quantify each malto-oligosaccharide from DP 2-7.

Three other types of detectors, flame ionization, polarimetric, and light scattering (mass) detectors have recently been reviewed (Hicks, 1988). These detectors have a number of features that make them ideal for detection of carbohydrates. For some reason, they have not been extensively utilized by workers in the field.

Respondents to our survey indicated that the majority (95%) used refractive index detectors with 27% using amperometric types and 11% reporting the use of UV detectors at low wavelengths. Of these individuals, 80% chose modern RI detectors as being the most sensitive, with amperometric detectors being next in sensitivity which is in contrast to the literature values in Table II. Refractive index detectors were considered to be the most stable (92%); most quantitative (83%) and most trouble free (75%). The most trouble-prone was indicated by 71% of respondents to be amperometric detectors. Fortunately, newer versions of the PAD detectors, such as the pulsed electrochemical detector (PED), and those from at least two other manufacturers, are addressing some of these problems. The above discussion of carbohydrate amperometric detection was applicable to gold electrodes. Luo et al. (1991) compared gold, platinum, copper, nickel, silver and cobalt as materials for constant potential amperometric detection of glucose and found the copper electrodes had the best range of linear response, detection limits, and stability. The use of various types of electrochemical detectors is likely to become more reliable and commonplace in the cereal carbohydrate field.

HPLC STATIONARY- AND MOBILE PHASES FOR CARBOHYDRATE ANALYSIS

Because this area has been covered in detail in recent reviews (Hanai, 1986; Hicks, 1988; Ball, 1990; Churms, 1990; Ben-Bassat and Grushka, 1991) only information pertinent to cereal-derived carbohydrates will be given here. At present there are four main types of stationary phases for HPLC of carbohydrates. These are given in Table III.

Normal Phases

Polar amino-bonded silica gels were introduced in the mid-1970's (Conrad and Fallick, 1974; Linden and Lawhead, 1976; Schwarzenbach, 1976, and Rabel et al., 1976) and were immediately applied to numerous carbohydrate separations. Initially, these columns were relatively unstable, but improvements in manufacturing technology have resulted in phases with very reasonable column "life" (from several months to a year, depending on applications). In addition to these amino-bonded phases, silica gels with covalently attached diol (Brons and Olieman, 1983), carbamoyl or polyamine (Koizumi et al., 1987 and 1991a) and cyclodextrin (Armstrong and Jin, 1989), functionalities have now been developed. Some early literature mistakenly described the use of these columns with acetonitrile/water mobile phases as "reversed-phase" chromatography, however, it is clearly a normal-phase system because increasing the amount of water, the "strong" solvent, in the mobile phase leads to decreased retention times for all sugars. These columns provide excellent separation of mono-, di- (Nikolov et al., 1985a), tri- (Nikolov et al., 1985b), and higher (Kainuma et al., 1981) oligosaccharides. In addition, their large sample capacity makes them quite useful for preparative applications (Hicks and Sondey, 1987).

TABLE III

Stationary Phases Used for HPLC of Neutral, Cereal-Derived Carbohydrates

Classification	Examples
Normal Phase	Silica gel with bonded aminopropyl-, diol-, or cyclodextrin groups; organic polymer-bound amino functionality.
Reversed-phase	Silica gel alkylated with C_8, C_{18}, or Phenyl groups; graphite; polystyrene-divinylbenzene-based polymers.
Cation Exchange Phases	Sulfonated polystyrene-divinylbenzene polymers in H^+, Ca^{++}, Pb^{++}, Ag^+, etc. form.
Anion Exchange Phases	Pellicular, sulfonated polystyrene-divinylbenzene strongly basic anion exchange resins.

Their disadvantages include: the need for expensive and thoroughly-degassed solvent (acetonitrile), shorter lifetime than ion exchange columns (see section below), and the insolubility of larger oligosaccharides in the mobile phase. Despite these factors, almost half (47%) of survey respondents reported using these phases. Further details about these phases are given in the reviews noted above.

Reversed-Phases

These alkylated silica gels are extremely useful in other areas of chemistry but receive minor use by carbohydrate analysts (32% of survey respondents). Underivatived mono- and di-saccharides have very little retention on these columns and reducing oligosaccharides generally elute in anomeric pairs. Situations in which these phases may be useful for carbohydrate separations may be found in articles by Vratny et al., (1983), Porsch (1985), McGinnis, et al., (1986) and Hicks (1988), in which their uses for separating mono-, linear-, and cyclic oligosaccharides as well as sugar derivatives are described. The most used phase of this type, C_{18} bonded silica gel, is usually eluted with pure water, or water containing 5-10% methanol. An innovative aminopropyl "doped" reversed-phase material that provided excellent separation of oligosaccharides was developed by Porsch (1985) but unfortunately was never made commercially available.

Cation Exchange Resin Phases

Since the late 1970's, these phases have been the "workhorse" of the analytical carbohydrate chemist. This was confirmed by our survey in which 95% of the respondents used such columns. This is not surprising since they are not only versatile, but durable (some have lasted for up to 10 years, with only occasional regeneration). Such phases, in a variety of ionic forms, including Ag^+, Ca^{++}, Pb^{++}, Na^+, and H^+, can be used with a simple mobile phase of water (or dilute acid for the H^+-form) for the separation of practically every class of carbohydrate. In the survey, the frequency of use was $Ca^{++} > Ag^+ > H^+ > Pb^{++} > Na^+$-forms. These columns are usually run at elevated temperatures (70-85°C) to prevent the separation of anomeric forms of reducing sugars. The historical advances which led to the commercialization of these phases has been described elsewhere (Hicks, 1988) but it is important to note that researchers in the corn refining industry played a major role in these developments (see Brobst et al., 1973; Scobell et al., 1977; Fitt, 1978; Fitt et al., 1980; and Scobell and Brobst, 1981). For information on the care, handling, mechanisms of separation, and general applications, see Hicks (1988).

Anion Exchange Resin Phases

Since their introduction (Rocklin and Pohl, 1983), pellicular, high-performance anion exchange resin phases have steadily gained popularity. About 32% of our survey respondents use them. When they are used with alkaline eluents (pH 10-12), sugars act as anions and separate according to their pKa values and additional steric factors, resulting in powerful separation capability for numerous sugars, small oligosaccharides and sugar alcohols (Paskach et al., 1991; van Riel and Olieman, 1991) and starch-derived oligosaccharides as large as DP 85 (Koizumi et al., 1989; Koizumi et al., 1991a; Ammeraal et al., 1991). No other HPLC stationary phase can match the versatility, selectivity, and efficiency of these relatively new pellicular resins.

In our survey, respondents ranked the following phases as the most durable: cation exchange resins (67%), high-performance anion exchange resins (17%), reversed-phases (8%) and polar amino-bonded phases (8%). The low durability ranking of high-performance anion exchange resins may be misleading. Most investigators have used these phases for a short period, and in many cases have not yet learned how to effectively prolong the "life" of the column. In the laboratory of one of the authors, a high-performance anion-exchange column (Dionex HPIC-AS6) has been used for 4 years with little change in selectivity and resolution. When resolution begins to fail, the column is washed with 3 M NaOH solution, which cleans the phase and restores nearly original performance.

METHOD SELECTION FOR SPECIFIC APPLICATIONS

In our survey we asked respondents to give the source of the HPLC methods used in their laboratories. The results were: 37% from the scientific literature, 33% developed at their location, and 30% official methods from a group such as AOAC, CRA, AACC, etc. It was interesting to note that less than one third of the respondents used officially sanctioned HPLC methods for carbohydrate analysis, even though a number of these methods are available. For instance, the 1990 edition of the AOAC Official Methods of Analysis provides official HPLC methods for major saccharides in corn syrup (method 979.23), minor saccharides in corn sugar (983.22), sugars in milk chocolate (980.13), and sugars in pre-sweetened cereals (982.14). The AACC has two official HPLC methods, 80-04 for simple sugars in cereal products, and 80-05 for saccharides in corn syrup. The sixth edition of the Corn Refiner's Association's Standard Analytical Methods manual lists HPLC method E-61 for saccharides in corn syrup, and F-51 for minor saccharides in corn sugar.

Survey respondents were also asked to provide specific stationary phases, mobile phases, and detectors they used for separating specific carbohydrates. The data were collected, analyzed and are given in Table IV. Most of the methods given by respondents were readily identified with known literature references, which are given in the Table. In most cases, original references are given for a particular separation, rather than the most recent reference that simply uses that technique. This was done because such early references generally have a wealth of information that is not always conveyed in later publications. The original references of Scobell et al. (1977) and Scobell and Brobst (1981) are given for separation of dextrose, maltodextrins, corn syrup solids, HFCS, and fructose because those references provide details of resin conversion, packing, effects of different metal ion ligands, effects of temperature, and other fundamental parameters on sugar separations. More recent references such as that of Warthesen (1984) are given to indicate more recent refinements of methods and to indicate sources of commercially available columns that are pre-packed with resins of optimum size and counter ion for specific applications.

TABLE IV
Survey Results: Methods Used for Specific Applications

Carbohydrate	Stationary Phase	Eluent/Detector	Reference
dextrose:	43% Ca^{++}-CER[a]	H_2O, 85°, RI	Scobell et al., 1977
	21% NH$_2$ phase[b]	CH$_3$CN/H$_2$O, RI	Verhaar and Kuster, 1981
	7% Ag$^+$-CER	H_2O, 85°, RI	Scobell and Brobst, 1981
	7% Na$^+$-CER	.001 $\underline{M}$ Na$_2$SO$_4$, RI	
	7% HPAE[c]	NaOH, PAD, RI	Paskach et al., 1991
other	50% Ca^{++}-CER	H_2O, RI	Richmond et al., 1983
monosacch's:	21% HPAE	NaOH, PAD	Paskach et al., 1991
	14% Pb^{++}-CER	H_2O, RI	Petersen et al. 1984
	14% NH$_2$ phase	CH$_3$CN/H$_2$O, RI	Verhaar and Kuster, 1981
malto-	56% Ag$^+$-CER	H_2O, RI	Warthesen, 1984
dextrins:	22% NH$_2$ phase	CH$_3$CN/H$_2$O, RI	Kainuma et al., 1981
	11% Ca^{++}-CER	H_2O, RI	Scobell and Brobst, 1981
	11% HPAE	NaOH/NaOAc, PAD	Koizumi et al., 1989
			Ammeraal et al., 1991
di- and	31% NH$_2$ phase	CH$_3$CN/H$_2$O, RI	Nikolov et al., 1985a, 1985b
tri-sacch's:	23% HPAE	NaOH/NaAc, PAD	Paskach et al., 1991
	15% RP[d]	H_2O, RI	Rajakyla, 1986
	15% Ag$^+$-CER	H_2O, RI	van Riehl and Olieman, 1986
	8% Ca^{++}-CER	H_2O, RI	Brons and Olieman, 1983
	8% Pb^{++}-CER	H_2O, RI	Voragen et al., 1986
corn syrup	36% Ca^{++}-CER	H_2O, RI	Scobell and Brobst, 1981
solids:	27% Ag$^+$-CER	H_2O, RI	Warthesen, 1984; Scobell and Brobst, 1981.
	9% Pb^{++}-CER	H_2O, RI	Voragen et al., 1986
	9% Na$^+$-CER	.001$\underline{M}$ Na$_2$SO$_4$, RI	
	9% NH$_2$ phase	CH$_3$CN/H$_2$O, RI	Kainuma et al., 1981
	9% HPAE	NaOH/NaAc, PAD	Koizumi et al., 1989
HFCS:	67% Ca^{++}-CER	H_2O, RI	Scobell, et al., 1977
	22% NH$_2$ phase	CH$_3$CN/H$_2$O, RI	Verhaar and Kuster, 1981
	11% HPAE	NaOH, PAD	

Survey Results: Methods Used for Specific Applications

Carbohydrate	Stationary Phase	Eluent/Detector	Reference
fructose:	44% Ca^{++}-CER	H_2O, RI	Scobell et al., 1977
	33% NH_2 phase	CH_3CN/H_2O, RI	Iverson and Bueno, 1981
	11% Na^+-CER	.001M Na_2SO_4, RI	
	11% HPAE	NaOH, PAD	Rocklin and Pohl, 1983
Poly-	50% HPGPC[e]	NaAc buffer, RI	
dextrose®:	50% Na^+-CER	H_2O, RI	
pyro-	50% Ag^+-CER	H_2O, RI	
dextrins:	50% HPGPC	H_2O, RI	
non-starch	20% Ca^{++}-CER	H_2O, RI	Schmidt et al., 1981
syrups:	20% Ag^+-CER	H_2O, RI	Wood et al., 1991
	20% RP	H_2O, RI	Vratny et al., 1983
	20% NH_2 phase	CH_3CN/H_2O, RI	Smiley et al., 1982
	20% HPAE	NaAc/NaOH, PAD	Ammeraal et al., 1991
cyclo-	50% Ca^{++}-CER	H_2O, RI	Hokse, 1980
dextrins:	50% other		Koizumi et al., 1988 and 1991b
sugar	80% Ca^{++}-CER	H_2O, RI	Wood et al., 1980
alcohols:	20% HPAE	NaOH, RI,PAD	Paskach et al., 1991
hydro-	50% Ca^{++}-CER	H_2O, RI	
genated	50% Ag^+-CER	H_2O, RI	
corn syrup:			
Sucralose:	100% RP	25% CH_3CN/ H_2O pH 2.7, RI	
Saccharides in wort and beer	NH_2-phase	CH_3CN/H_2O, RI	Uchida et al., 1990

[a]43% Ca^{++}-CER = 43% of respondents used a Ca^{++}-form Cation Exchange Resin for this application.
[b]Polar, amino-type normal phase, such as aminopropyl silica gel.
[c]High performance anion exchange resin.
[d]Reversed-phase, such as C_{18}.
[e]High-performance gel permeation chromatography.

In addition to the information in Table IV for carbohydrate-specific methods, it may be useful to provide some general guidelines on method selection. These are given below.

General Guidelines for HPLC Method Selection

1. Use the simplest method that provides an acceptable separation.
2. Use cation exchange resin-based phases whenever possible:

For analysis of:	First try:
- one or two monosaccharides	Ca^{++}-form resins
- mixtures of several monosaccharides	Pb^{++}-form resins
- one or two disaccharides	Ca^{++}-form resins
- mixtures of several disaccharides	Ca^{++}- or Pb^{++}-form resins
- mixtures of mono-, di-, and higher oligosacch's	Ag^{+}-form resins
- mixtures of sugar alcohols	Ca^{++}-form resins.
- cyclodextrins	Ca^{++}-form resins

3. When choosing a cation exchange resin phase, use 8% crosslinked resins for monosaccharides and 4% crosslinked resins for di- and larger oligosaccharides. For separation of larger than DP-8 and less than DP-13 oligosaccharides, 2% crosslinked resins may be used (Hicks and Hotchkiss, 1988)

4. If cation exchange resin-based phases will not provide the needed separations, use a pellicular, high performance anion exchange (HPAE) phase.

5. If an HPAE phase is not effective, or is not available, try a normal phase column with an acetonitrile/water mobile phase. In selecting a normal phase column, the use of cyclodextrin-bonded phase (Armstrong and Jin, 1989) and diol-bonded phase (Brons and Olieman, 1983) columns should be examined as well as amino-bonded phases because of the durability of the former two.

6. Use RI detection whenever possible for isocratic separations. If it is not possible to use RI detection (gradient separations, etc.), use a UV detector if an appropriate UV absorbing group is present on the analyte. Otherwise, use a pulsed amperometric detector if available.

7. Consider using an evaporative light scattering ("mass") or flame ionization detector if available. Unlike RI detectors, these can accommodate solvent gradients.

8. Always use pre-filters and pre-columns containing a stationary phase similar to that in the analytical column.

9. Use mobile phase to dissolve samples prior to injection. This may not be necessary with HPAE-PAD, (Paskach et al., 1991).

TECHNOLOGY BREAKTHROUGHS

Survey participants were asked to list the recent techniques, stationary phases, detectors, etc. that have had the greatest positive impact on their ability to analyze carbohydrates. Those breakthroughs are listed in Table V. The most important breakthrough was found to be the development of recent RI detectors which are capable of providing extremely sensitive and stable detection of carbohydrates. These detectors were first introduced in the mid 1980's and are now available from at least six manufacturers. These detectors are commonly used with the second ranked breakthrough, the cation exchange resin based stationary phases, which generally use pure water as a mobile phase. This simple eluent also contributes significantly to the sensitivity obtained on these detectors. The third ranking breakthrough, high-performance anion exchange chromatography, coupled with pulsed amperometric detection, is a relatively new technique, and may eventually replace many of the currently popular HPLC methods. Conversely, the fourth ranked breakthrough, the development of aminopropyl bonded silica gel stationary phases would have ranked much higher in surveys 5 to 10 years ago and may continue to receive less attention in the coming years.

TABLE V
Greatest Technology Breakthroughs in HPLC of Carbohydrates

Rank (% of responses)	Breakthrough
1 (41%)	Ultra-sensitive and -stable RI detectors
2 (35%)	Cation exchange resin-based HPLC columns
3 (18%)	High-performance anion exchange columns coupled with pulsed amperometric detection
4 (6%)	Aminopropyl silica gel bonded "carbohydrate" HPLC columns

PREPARATIVE HPLC OF CEREAL-DERIVED CARBOHYDRATES

The combination of relatively large sample injections and non-destructive detection makes HPLC an ideal preparative technique. During the late 1970's and early 1980's, commercial preparative HPLC systems were introduced. These initial systems were far from ideal since they operated at extremely high flow rates and therefore, used large volumes of solvents. The preparative columns available were packed with relatively inefficient particles, leading to poor resolution. In many cases, better, and certainly less expensive results were obtained on low-pressure LC equipment than on those "first generation" preparative HPLC instruments. Currently, however, highly advanced systems are available which consist of automated, microprocessor controlled solvent delivery systems, extremely efficient columns, and "intelligent" fraction collectors which can be programmed to collect using a drop, time, or peak mode. Such systems have been used for the efficient preparative isolation of mono- and di-saccharides (Hicks et al., 1983) and malto-oligosaccharides (Hicks and Sondey, 1987). For a complete review of preparative HPLC of carbohydrates, please see Hicks (1988).

CURRENT NEEDS AND POTENTIAL SOLUTIONS

Survey respondents provided a list of several important needs in the HPLC area. Some of these and some potential solutions are given below.

Need: Ability to quantify lactose in the presence of fructose/sucrose/maltose.

Solution: Relatively recent methods are available for this separation using cation exchange resin- (van Riel and Olieman, 1986) and high-performance anion exchange resin- (van Riel and Olieman, 1991) phases.

Need: Ability to separate mixtures of saccharides of the same DP value.

Solution: Separation on high-performance anion exchange resin phases (Paskach et al., 1991; Koizumi et al., 1989).

Need: Need more effective methods to analyze Polydextrose®.

Solution: When Polydextrose® is present in matrices free from other larger oligosaccharides, a method based on a Ca^{++}-form cation exchange resin phase may be appropriate (Noffsinger et al., 1990). Alternatively, HPAE-PAD, with its excellent selectivity toward oligosaccharides should be applied to this analysis.

Need: Need methods to quantify higher oligosaccharides (>DP 20) in mixtures.

Solution: For malto-oligosaccharides for which no standards exist, use of an immobilized glucoamylase reactor (Larew and Johnson, 1988), prior to the detector, can provide quantitative information.

CONCLUSIONS

HPLC methods exist for analysis of most cereal-derived carbohydrates. Breakthroughs in technology over the last 10-15 years have resulted in methods which allow accurate analysis of mixtures that were impossible to analyze a generation ago. Technology continues to develop in this field and additional breakthroughs in separation and detection methods will likely continue. Industrial laboratories surveyed in this work had a relatively sophisticated working knowledge of the state-of-the-art in

HPLC methods and used them as their methods of choice for numerous analyses of cereal-derived carbohydrates.

LITERATURE CITED

Ammeraal, R. N., Delgado, G. A., Tenbarge, F. L., and Friedman, R. B. 1991. High-performance anion-exchange chromatography with pulsed amperometric detection of linear and branched glucose oligosaccharides. Carbohydr. Res. 215:179.

Armstrong, D. W., and Jin, H. L. 1989. Evaluation of the liquid chromatographic separation of monosaccharides, disaccharides, trisaccharides, tetrasaccharides, deoxysaccharides and sugar alcohols with stable cyclodextrin bonded phase columns. J. Chromatogr. 462:219.

Ball, G. F. M. 1990. The application of HPLC to the determination of low molecular weight sugars and polyhydric alcohols in foods: a review. Food Chem. 35:117.

Barsuhn, K., and Kotarski, S. F. 1991. Ion chromatographic methods for the detection of starch hydrolysis products in ruminal digesta. J. Chromatogr. 546:273.

Ben-Bassat, A. A., and Grushka, E. 1991. High performance liquid chromatography of mono- and oligosaccharides. J. Liq. Chromatogr. 14:1051.

Brobst, K. M., Scobell, H. D., and Steele, E. M. 1973. Analysis of carbohydrate mixtures by liquid chromatography. Proc. Ann. Meet. Am. Soc. Brewing Chem. 43.

Brons, C., and Olieman, C. 1983. Study of the high-performance liquid chromatographic separation of reducing sugars, applied to the determination of lactose in milk. J. Chromatogr. 259:79.

Churms, S. C. 1990. Recent developments in the chromatographic analysis of carbohydrates. J. Chromatogr. 500:555.

Conrad, E. C., and Fallick, G. J. 1974. High pressure liquid chromatography and its application in the brewing laboratory. Brewers Dig. 49:72.

Fitt, L. E. 1978. Convenient, in-line purification of saccharide mixtures in automated high-performance liquid chromatography. J. Chromatogr. 152:243.

Fitt, L. E., Hassler, W., and Just, D. E. 1980. A rapid and high-resolution method to determine the composition of corn syrups by liquid chromatography. J. Chromatogr. 187:381.

Hanai, T. 1986. Liquid Chromatography of Carbohydrates. Advances in Chromatogr. 25:279.

Hicks, K. B., Symanski, E. V., and Pfeffer, P. E. 1983. Synthesis and high-performance liquid chromatography of maltulose and cellobiulose. Carbohydr. Res. 112:37.

Hicks, K. B., and Sondey, S. M. 1987. Preparative high-performance liquid chromatography of malto-oligosaccharides. J. Chromatogr. 389:183.

Hicks, K. B. 1988. High-performance liquid chromatography of carbohydrates. Advances in Carbohydrate Chemistry and Biochemistry. 46:17.

Hicks, K. B., and Hotchkiss, A. T., Jr. 1988. High-performance liquid chromatography of plant-derived oligosaccharides on a new cation-exchange resin stationary phase: HPX-22H. J. Chromatogr. 441:382.

Hokse, H. 1980. Analysis of cyclodextrins by high-performance liquid chromatography. J. Chromatogr. 189:98.

Iverson, J. L., and Bueno, M. P. 1981. Evaluation of high pressure liquid chromatography and gas-liquid chromatography for quantitative determination of sugars in foods. J. Assoc. Off. Anal. Chem. 64:139.

Kainuma, K., Nakakuki, T., and Ogawa, T. 1981. High-performance liquid chromatography of maltosaccharides. J. Chromatogr. 212:126.

Koizumi, K., Utamura, T., and Kubota, Y. 1987. Two high-performance liquid chromatographic columns for analyses of malto-oligosaccharides. J. Chromatogr. 409:396.

Koizumi, K., Kubota, Y., Okada, Y., and Utamura, T. 1988. Retention behavior of cyclodextrins and branched cyclodextrins of reversed-phase columns in high-performance liquid chromatography. J. Chromatogr. 437:47.

Koizumi, K., Kubota, Y., Tanimoto, T., and Okada, Y. 1989. High-performance anion-exchange chromatography of homogenous D-gluco-oligosaccharides and -polysaccharides (polymerization degree >50) with pulsed amperometric detection. J. Chromatogr. 464:365.

Koizumi, K., Fukuda, M., and Hizukuri, S. 1991a. Estimation of distributions of chain length of amylopectins by high-performance liquid chromatography with pulsed amperometric detection. J. Chromatogr. 585:233.

Koizumi, K., Taimoto, T., Okada, Y., Nakanishi, N., Kato, N., Takagi, Y., and Hashimoto, H. 1991b. Characterization of five isomers of branched cyclomaltoheptaose (β CD) having degree of polymerization (d.p.) = 9: Reinvestigation of three positional isomers of diglucosyl-β CD. Carbohydr. Res. 215:127.

LaCourse, W. R., and Johnson, D. C. 1991. Optimization of waveforms for pulsed amperometric detection (p.a.d.) of carbohydrates following separation by liquid chromatography. Carbohydr. Res. 215:159.

Larew, L. A., and Johnson, D. C. 1988. Quantitation of chromatographically separated maltooligosaccharides with a single calibration curve using a postcolumn enzyme reactor and pulsed amperometric detection. Anal. Chem. 60:1867.

Linden, J. C., and Lawhead, C. L. 1975. Liquid chromatography of saccharides. J. Chromatogr. 105:125.

Luo, P., Zhang, F., and Baldwin, R. P. 1991. Comparison of metallic electrodes for constant-potential amperometric detection of carbohydrates, amino acids, and related compounds in flow systems. Analytica Chimica Acta 244:169.

McGinnis, G. D., Prince, S., and Lowrimore, J. 1986. The use of reverse-phase columns for separation of unsubstituted carbohydrates. J. Carbohydr. Chem. 5:83.

Nikolov, Z. L., Meagher, M. M., and Reilly, P. J. 1985a. High-performance liquid chromatography of disaccharides on amine-bonded silica columns. J. Chromatogr. 319:51.

Nikolov, Z. L., Meagher, M. M., and Reilly, P. J. 1985b. High-performance liquid chromatography of trisaccharides on amine-bonded silica columns. J. Chromatogr. 321:393.

Noffsinger, J. B., Emery, M., Hoch, D. J., and Dokladalova, J. 1990. Liquid chromatographic determination of Polydextrose® in food matrixes. J. Assoc. Off. Anal. Chem. 73:51.

Paskach, T. J., Lieker, H-P., Reilly, P. J., Thielecke, K. 1991. High-performance anion-exchange chromatography of sugars and sugar alcohols on quaternary ammonium resins under alkaline conditions. Carbohydr. Res. 215:1.

Petersen, R. C., Schwandt, V. H., and Effland, M. J. 1984. An analysis of the wood sugar assay using HPLC: A comparison with paper chromatography. J. Chromatogr. Sci. 22:478.

Porsch, B. 1985. High-performance liquid chromatography of oligosaccharides in water on a reversed phase doped with primary amino groups. J. Chromatogr. 320:408.

Rabel, F. M., Caputo, A. G., Butts, E. T. 1976. Separation of carbohydrates on a new polar bonded phase material. J. Chromatogr. 126:731.

Rajakyla, E. 1986. Use of reversed-phase chromatography in carbohydrate analysis. J. Chromatogr. 353:1.

Richmond, M. L., Barfuss, D. L., Harte, B. R., Gray, J. I., and Stine, C. M. 1982. Separation of carbohydrates in dairy

products by high performance liquid chromatography.
J. Dairy Sci. 65:1394.

Rocklin, R. D., and Pohl, C. A. 1983. Determination of carbohydrates by anion exchange chromatography with pulsed amperometric detection. J. Liquid Chromatogr. 6:1577.

Schmidt, J., John, M., and Wandrey, C. 1981. Rapid separation of malto-, xylo-, and cello-oligosaccharides (dp 2-9) on cation-exchange resin using water as eluent.
J. Chromatogr. 213:151.

Schwarzenbach, R. 1976. A chemically bonded stationary phase for carbohydrate analysis in liquid chromatography.
J. Chromatogr. 117:206.

Scobell, H. D., Brobst, K. M., and Steele, E. M. 1977. Automated liquid chromatographic system for analysis of carbohydrate mixtures. Cereal Chem. 54:905.

Scobell, H. D., and Brobst, K. M. 1981. Rapid high-performance separation of oligosaccharides on silver form cation-exchange resins. J. Chromatogr. 212:51.

Smiley, K. L., Slodki, M. E., Boundy, J. A., and Plattner, R. D. 1982. A simplified method for preparing linear isomalto-oligosaccharides. Carbohydr. Res. 108:279.

Uchida, M., Nakatani, K., Ono, M., and Nagami, K. 1991. Carbohydrates in brewing. I. Determination of fermentable sugars and oligosaccharides in wort and beer by partition high-performance liquid chromatography.
J. Am. Soc. Brew. Chem. 49:65.

Van Riel, J. A. M., and Olieman, C. 1986. High-performance liquid chromatography of sugars on a mixed cation-exchange resin column. J. Chromatogr. 362:235.

Van Riel, J. A. M., and Olieman, C. 1991. Selectivity control in the anion-exchange chromatographic determination of saccharides in dairy products using pulsed amperometric detection. Carbohydr. Res. 215:39.

Verhaar, L. A. Th., and Kuster, B. F. M. 1981. Liquid chromatography of sugars on silica-based stationary phases.
J. Chromatogr. 220:313.

Voragen, A. G. J., Schols, H. A., Searle-Van Leeuwen, M. F., Belden, G., and Rombouts, F. M. 1986. Analysis of oligomeric and monomeric saccharides from enzymatically degraded polysaccharides by high-performance liquid chromatography.
J. Chromatogr. 370:113.

Vratny, P., Coupek, J., Vozka, S., and Hostomska, Z. 1983. Accelerated reversed-phase chromatography of carbohydrate oligomers. J. Chromatogr. 254:143.

Warthesen, J. J. 1984. Analysis of saccharides in low-dextrose equivalent starch hydrolysates using high-performance liquid chromatography. Cereal Chem. 61:194.

Wood, P. J., Weisz, J., and Blackwell, B. A. 1991. Molecular characterization of cereal β-D-Glucans. Structural analysis of oat β-D-glucans and rapid structural evaluation of β-D-glucans from different sources by high-performance liquid chromatography of oligosaccharides released by lichenase. Cereal Chem. 68:31.

Wood, R., Cummings, L., and Jupille, T. 1980. Recent developments in ion-exchange chromatography. J. Chromatogr. Sci. 18:551.

ACKNOWLEDGMENTS

The authors thank Drs. Landis Doner and Jeffrey Brewster and Ms. Rebecca Haines for assistance in preparing this chapter. We also thank the following individuals and their companies for participating in this survey: James R. Beadle, Biospherics, Inc.; Raffaele Bernetti, Corn Products, a Unit of CPC International, Inc.; Mike Blin, Cargill Analytical Services; Allan G. W. Bradbury, Robert Engel, Kraft General Foods Inc.; Stuart A. S. Craig, Nabisco Brands; Robert B. Friedman, American Maize-Products Company; Amelia L. Grace, The Seven-Up Company; W. J. Hurst, Hershey Foods Corporation; Blair S. Kunka, Quest International Bioproducts Group; R. Maruyama, Adolph Coors Company; Zohar M. Merchant and Keith A. Meyer, Kraft General Foods Inc.; Linda J. Marinelli, Miller Brewing Company; Brian E. Mueller, Kelco Division of Merck & Co., Inc.; Christine M. Quinn, The Coca-Cola Company; Eero Rajakyla, Suomen Xyrofin Oy; Steven Schroeder, The NutraSweet Company; Henry D. Scobell, A. E. Staley Manufacturing Co.; and four other individuals from companies requesting anonymity.

STRUCTURE, BIOSYNTHESIS, AND USES OF NONSTARCH POLY-SACCHARIDES: DEXTRAN, ALTERNAN, PULLULAN, AND ALGIN

John F. Robyt
Department of Biochemistry and Biophysics
Iowa State University
Ames, IA 50011

DEXTRANS AND RELATED POLYSACCHARIDES

Jeanes *et al.* (1954) surveyed 96 strains of *Leuconostoc* and *Streptococci* for the formation of polysaccharides from sucrose. They found that the polysaccharides exclusively contained D-glucose and had a relatively high amount of α-1$\rightarrow$6 glycosidic linkages. The polysaccharides were branched by α-1$\rightarrow$2, α-1$\rightarrow$3, or α-1$\rightarrow$4 linkages to varying degrees, depending on the strain of the organism. The alcohol precipi-tates were described in various qualitative terms such as *pasty, fluid, stringy, tough, long, short, floccu-lent, etc.* These somewhat fanciful descriptions, however, were forerunners of the differences in their structures. The definitive natures of the structures in terms of the types of linkages, the nature of the branch linkages, the amount of branching, and their arrangement was later determined by methylation and ^{13}C-NMR (Van Cleve, 1956; Seymour *et al.*, 1976, 1977, 1979a, 1979b, 1980; Jeanes and Seymour, 1979; Seymour and Knapp, 1980a, 1980b). Table 1 gives a selection of *Leuconostoc* and *Streptococci* that have enzymes that synthesize glucans from sucrose, the type and percent-ageage of linkages in the glucans, and a qualitative description of the ethanol precipitates.

It was also found that some of the strains were producing two types of polysaccharides. These poly-saccharides could be separated by a differential alcohol fractionation (Wilham *et al.*, 1955). In many cases, the first polysaccharide was precipitated by 36-37% ethanol and was designated as the *L-fraction* for <u>less</u> <u>soluble</u>, *i.e.*, it could be precipitated by a lower alcohol concentration than the second fraction.

Table 1 Glucans synthesized from sucrose by selected *Leuconostoc* and *Streptococci*

Species and Strain No.[a]	Solub. Class[b]	Linkages					Description of ethanol precipitate
		1→6	1→3	1→3Br[c]	1→2Br	1→4Br	
L.m. B-512F	L	95		5			translucent gel
L.m. B-742	L	87	13				heavy, opaque
L.m. B-742	S	50	50				fine[d]
L.m. B-1299	L	66		1	27		flocculent[d]
L.m. B-1299	S	65			35		fine[d]
L.m. B-1355	L	95		5			translucent gel
L.m. B-1355	S	54	35	11			heavy, opaque
L.m. B-1211	L	95				5	short, smooth[d]
L.m. B-1119	L	94		2		4	cohesive, stringy[d]
L.m. B-1308	L	95				5	pasty, crumbly[d]
L.m. B-1525	S	83				17	fluid, stringy[d]
L.m. B-523	I		100				water-insoluble
L.m. B-1149	I		100				water-insoluble
S.m. 6715	S	64	36				heavy, opaque
S.m. 6715	I	4	94	2			water-insoluble
S.v. B-1351	S	89		11			short[d]

[a]L.m. = *Leuconostoc mesenteroides*; S.m. = *Streptococcus mutans*; S.v. = *Strep. viridans*
B-numbers refer to the strain number in the Northern Regional Research Laboratory (NRRL)
culture collection of the USDA, Peoria, IL.
[b]L = less soluble (ppted by 36-37% ethanol); S = more soluble (ppted by 40-44% ethanol);
I = water-insoluble; [c]Br = branch linkage; [d]Description taken from Jeanes et al. (1954).

The second fraction was precipitated by 40–44% ethanol and hence was designated the *S-fraction* for <u>soluble</u>, *i.e.*, more soluble than the L-fraction. The differential alcohol-precipitation curves for three different *Leuconostoc mesenteroides* strains are given in Fig. 1. This figure shows that the B-512F strain gave a single polysaccharide, whereas strains B-1355 and B-742 each gave two polysaccharides. The alcohol precipitates of the two polysaccharides that were obtained from a single strain had different appearances (see Fig. 2 for a comparison of the S- and L-polysaccharides of B-1355) and different kinds and arrangements of linkages (see Table 1). In some cases, a strain produced a very water-insoluble polysaccharide.

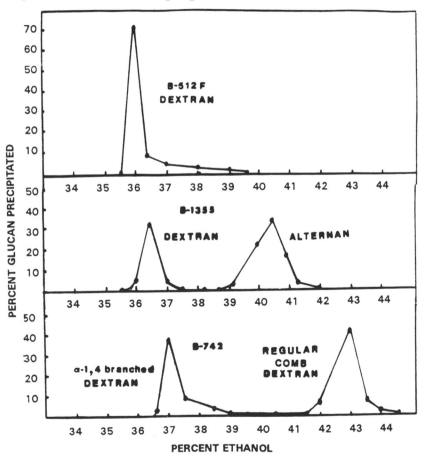

Fig. 1 Differential precipitation of glucans with ethanol.

263

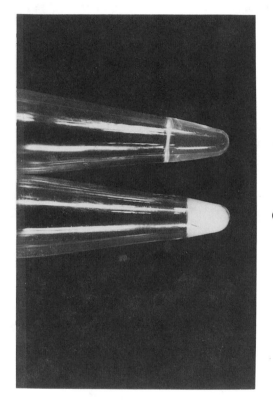

S L
B-1355 Glucans

Fig. 2 Appearance of the ethanol precipitates of alternan (fraction-S) and dextran (fraction-L) from *Leuconostoc mesenteroides* B-1355. **S** = alternan, **L** = dextran. From Cote and Robyt, 1982a.

The two polysaccharides produced by *L. mesenteroides* B-1355 had very obvious differences in their appearance when precipitated by ethanol. The L-fraction was a translucent gel and the S-fraction appeared white and opaque (Cote and Robyt, 1982a). The S-fraction was quite resistant to the action of endo-dextranase, whereas the L-fraction was readily hydrolyzed and converted into a mixture of glucose and isomaltodextrins. This also was true for several of the other organisms that produced two polysaccharides. The L-fraction was hydrolyzed by dextranases and the S-fraction was usually resistant to hydrolysis (Cote and Robyt, 1982a, 1983).

Structures of the Polysaccharides

A dextran is now defined as a glucan that has contiguous α-1$\rightarrow$6 linked glucose residues in the main chains with a varying percentage of α-1$\rightarrow$2, α-1$\rightarrow$3, and α-1$\rightarrow$4 branch linkages. A large percentage of the dextrans have α-1$\rightarrow$3 branch linkages. *L. mesenteroides* B-512F dextran is the classical dextran containing a high percentage (95%) of consecutive α-1$\rightarrow$6 linkages and a relatively low percentage (5%) of α-1$\rightarrow$3 branch linkages (Robyt, 1986a). B-512F dextran contains both single glucose residues linked by an α-1$\rightarrow$3 branch linkage to the main α-1$\rightarrow$6 linked chains and a relatively long chain of α-1$\rightarrow$6 linked glucose residues attached by an α-1$\rightarrow$3 branch linkage to α-1$\rightarrow$6 linked main chains (Larm *et al.*, 1971; Senti *et al.*, 1959; Bovey, 1959). This is consistent with the mechanism of synthesis of the branch linkages (see the section on biosynthesis below).
The L-fraction of *L. mesenteroides* B-1355 has a structure very similar to that of B-512F dextran (Cote and Robyt, 1982a). A structure related to these dextrans is the L-fraction of *L. mesenteroides* B-742. It contains 87% α-1$\rightarrow$6 and 13% α-1$\rightarrow$4 branch linkages instead of α-1$\rightarrow$3 branch linkages (Seymour *et al.*, 1979a). The α-1$\rightarrow$4 branch linkages produce a structure that differs from B-512F dextran with α-1$\rightarrow$3 branch linkages. The B-742 L-dextran is much more resistant to endo-dextranase hydrolysis than it should be for only having 13% branch linkages (Cote and Robyt,

1983). It has been suggested that the α-1$\rightarrow$4 branch linkage imparts a different conformation to the dextran molecule than does the α-1$\rightarrow$3 branch linkage (Torii et al., 1966). This possible difference in conformation may account for its difference in susceptibility to endo-dextranase.

The S-fraction of L. mesenteroides B-1355 contain 50% α-1$\rightarrow$6 linkages and 50% α-1$\rightarrow$3 linkages, but it is not a dextran as it does not have a contiguous sequence of α-1$\rightarrow$6 linkages. The main chains have an alternating sequence of α-1$\rightarrow$6 and α-1$\rightarrow$3 linked glucopyranosyl residues (Seymour et al., 1979b) with 11% α-1$\rightarrow$3 branch linkages (Seymour et al., 1977). Misaki, et al. 1980 have reported that an unspecified amount of endodextranase hydrolyzed B-1355 S-glucan to 7.3% isomaltose in 24 hours. Without knowing the amount of enzyme used, one can only conjecture that the glucan is within 1/100th to 1/10,000th as susceptible to endodextranase hydrolysis as is B-512F dextran. These differences are significant enough that the glucan can be considered not to be a dextran. It has consequently been named alternan (Cote and Robyt, 1982b).

L. mesenteroides B-742 S-fraction contains 50% α-1$\rightarrow$6 linkages and 50% α-1$\rightarrow$3 linkages. Unlike alternan, however, the α-1$\rightarrow$3 linkages are all branch linkages (Seymour et al., 1979a). Like B-512F dextran its main chains are composed of consecutive α-1$\rightarrow$6 linked glucose residues with long α-1$\rightarrow$6 linked branch chains attached to an α-1$\rightarrow$6 linked chain by an α-1$\rightarrow$3 branch linkage. It differs from B-512F dextran, however, by having single glucose residues attached to each of the glucose residues in the α-1$\rightarrow$6 linked chain. This fulfills the definition of a dextran by having contiguous α-1$\rightarrow$6 linked glucose chains, but is a very highly branched dextran by virtue of having single glucose residues branched onto every glucose residue in the main chains. The structure of this dextran can be considered a bifurcated comb in which the single glucose residues are the "teeth" on the backbone of the comb, the contiguous α-1$\rightarrow$6 linked chain.

The degree of branching for this glucan does seem to be dependent upon the conditions of synthesis (Cote and Robyt, 1983). Apparently under some condi-

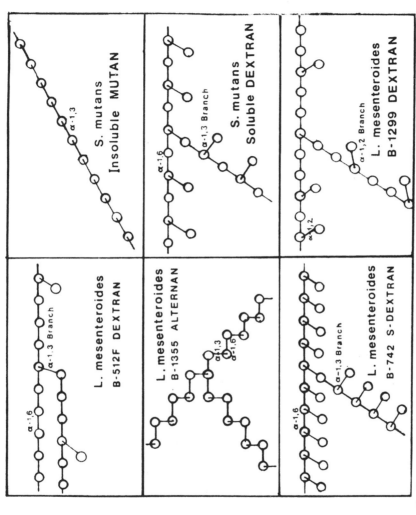

Fig. 3 Structures of the various glucans synthesized from sucrose by glucansucrases. Symbols according to Robyt (1986).

267

tions, the degree of branching can be less than 50%. Slodki *et al.*, 1986 have reported the degree of branching to be as low as 28-33%. This is still a substantial amount of branching, but it does render the dextran more susceptible to endodextranase hydrolysis. Because endodextranase requires a minimum of 6-7 contiguous, unsubstituted α-1→6 linked glucose residues for hydrolysis to occur (Bouren *et al.*, 1962; Richards and Streamer, 1978), the increased susceptibility to endodextranase hydrolysis suggests that the dextran has a run of at least six glucose residues that are not substituted by α-1→3 linked glucose residues. The biosynthetic mechanism for the formation of these gaps in the structure, and the consequent decrease in the degree of branching, have not, however, been examined.

A dextran related in structure to the B-742 S-dextran is the soluble dextran produced by *Streptococcus mutans* strains. These dextrans also contain a relatively high percentage (35%) of α-1→3 branch linkages (Shimamura *et al.*, 1982). Like B-742 S-dextran, the branch chains are primarily single glucose residues attached to α-1→6 linked main chains. If the α-1→3 branch linkages are regularly spaced, the resulting structure is an alternating comb dextran in which single glucose residues are linked to every other glucose residue in the main chains.

In addition to the soluble comb dextran, *S. mutans* also produces a very water-insoluble glucan that contains mostly consecutive α-1→3 linked glucose residues (Hare *et al.*, 1978). Two *L. mesenteroides* strains, B-523 and B-1149, also produce water-insoluble, α-1→3 linked glucans. Because of the differences in the structures and physical properties (the water-insoluble glucans are totally resistant to endo-dextranase hydrolysis), they are called *mutans*.

Several strains of *Neisseria perflava* synthesize glucans from sucrose that stain with triiodide. Some strains produce glucans that have glycogen-like structures (Tao *et al.*, 1988) and give a brown stain with triiodide. Other strains produce glucans that have amylopectin-like structures and stain wine-red to purple with triiodide, and some strains produce glucans that stain dark blue with triiodide, indicating an amylose-like structure (Simet and Robyt, 1990). Figure 3 is a summary of the structures of the various glucans synthesized from sucrose.

Biosynthesis of Dextran and Related Polysaccharides

In the biosynthesis of these glucans by dextransucrase or glucansucrases, sometimes called sucrose-glucosyltransferases, the sucrose serves as a high-energy donor of glucose. Sucrose has a glycosidic bond energy equivalent to that of UDP-glucose (6-7 kcal/mole). Mechanism studies of the action of these enzymes have shown that the B-512F dextransucrase (Robyt et al., 1974) and the S. mutans dextransucrase and mutansucrase (Robyt and Martin, 1983) form covalent, glucosyl intermediates. The glucosyl intermediates are transferred to the reducing-end of a growing glucanosyl chain, which is covalently linked to the active-site of the enzymes . The proposed mechanism for the synthesis of a sequence of α-1$\rightarrow$6 linked glucose residues involves two nucleophiles at the active-site that attack sucrose and displace fructose to give two ß-glucosyl enzyme intermediates (Robyt et al., 1974). The C-6-hydroxyl group of one of the glucosyl residues attacks C-1 of the other glucosyl-intermediate to form an α-1$\rightarrow$6 linkage. The freed, enzyme nucleophile then attacks another sucrose molecule forming a new glucosyl intermediate. The C-6-hydroxyl of this new glucosyl intermediate then attacks C-1 of the isomaltosyl unit (i.e., attacks the growing dextran chain), which is in effect transferred to the glucosyl intermediate. The glucosyl and dextanosyl units alternately go back and forth between the two nucleophiles as the dextran chain is elongated at the reducing-end (see Fig. 4A). The elongation is terminated and the dextran chain is released by acceptor reactions (25), one of which may be with an exogenous dextran chain to give a branch linkage (see below for a discussion of the formation of branch linkages).

A similar mechanism may be proposed for the synthesis of the α-1$\rightarrow$3 glucan (mutan), with the difference that the two nucleophiles orient the glucosyl-intermediates so the C-3-hydroxyl is in position to make the attack onto C-1 of the growing chain to form α-1→3 linkages instead of α-1→6 linkages. The chain is elongated like dextran by the transfer of glucosyl residues to the reducing-end of the growing mutan chain (Robyt and Martin, 1983) (see Fig. 4B). It is likely that alternan is synthesized in a similar manner in which one of the glucosyl-intermediates is oriented so the C-6-hydroxyl makes an attack onto the C-1 of the other glucosyl-intermediate, and the other intermediate is oriented so the C-3-hydroxyl makes an attack onto C-1 of the other glucosyl-intermediate. The elongation, thus, would go back and forth alternately synthesizing α-1→6 and α-1→3 glycosidic linkages and the formation of the alternan chain (see Fig. 4C).

The glucansucrases catalyze another reaction when other carbohydrates in addition to sucrose are added or are present in the enzyme digests. In these reactions the glucosyl-intermediate is transferred to the carbohydrate and is diverted from the synthesis of glucan (Robyt and Walseth, 1978). The carbohydrates are called *acceptors*. When the acceptors are low molecular weight monosaccharides or di- or tri-saccharides, the products are low molecular weight oligosaccharides (see Fig. 5). The acceptor-products themselves can be acceptors and a series of homologous acceptor products result (Robyt and Walseth, 1978; Robyt and Eklund, 1983). For some acceptors, however, only a single acceptor-product results as the acceptor-product is not an acceptor. Table 2 lists some of the common acceptors and the number of products they form along with their relative efficiencies as acceptors.

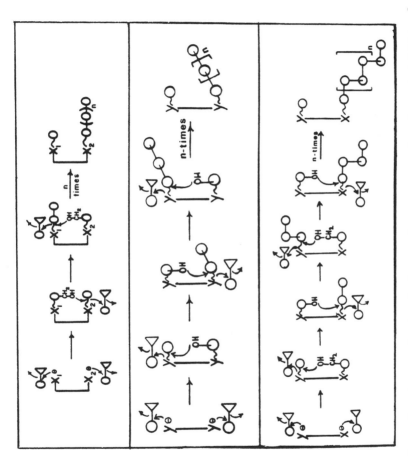

Fig. 4 Mechanisms for the synthesis of glucans from sucrose by glucansucrases. A synthesis of dextran by dextransucrase; B, synthesis of mutan by mutansucrase; C, synthesis of alternan by alternansucrase. X and Y are nucleophiles at the active-site of the enzymes. O = glucose, O–◁ = sucrose.

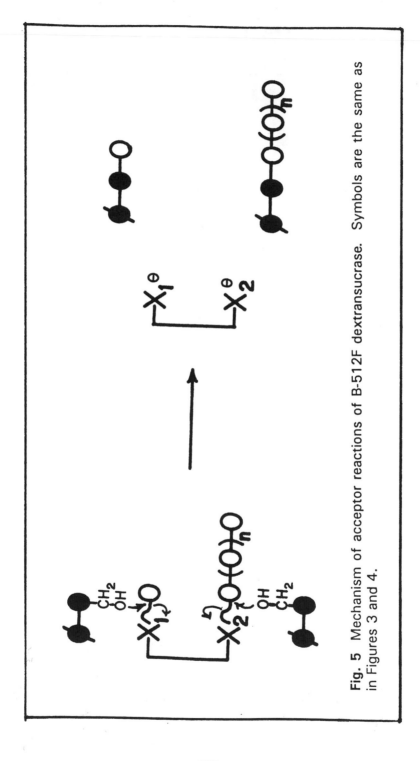

Fig. 5 Mechanism of acceptor reactions of B-512F dextransucrase. Symbols are the same as in Figures 3 and 4.

Table 2 Some Acceptors for Dextransucrase, the Number of Products, and the Relative Acceptor Efficiencies for the Acceptors[a]

Acceptor	% Glc in acceptors	% Glc in dextran	No. of acceptor products	Relative acceptor efficiency
Maltose	80.7	19.3	5	100
Isomaltose	71.9	28.1	6	89
Nigerose	47.4	52.6	6	58
α-Me-Glc	42.3	57.7	7	52
Glucose	14.2	85.8	7	17
Turanose	11.2	88.8	7	13
Lactose	9.0	91.0	1	11
Cellobiose	7.5	92.5	4	9
Fructose	5.3	94.7	1	6
Raffinose	3.7	96.3	1	4
Melibiose	3.5	96.5	1	4
Mannose	2.4	97.6	1	3
Xylose	0.5	99.6	1	.5

[a]Products were measured using a ratio of acceptor to sucrose of 1:1 at 80 mM. After Robyt and Eklund, 1983.

Kinetic studies have shown that the acceptor binds at an acceptor binding-site (Tanriseven and Robyt, 1992) and attacks C-1 of the glucosyl- or glucanosyl-intermediates to release them from the active-site of the enzyme, with the formation of a covalent linkage between the acceptor and the glucose or glucan chain. When the acceptor displaces the glucan from the active-site, the polymerization of the glucan chain is terminated (Robyt and Walseth, 1978).

Exogenous dextran chains can act as acceptors. In doing so, the C-3-hydroxyl of one of the glucose residues of dextran makes an attack onto C-1 of either the glucosyl-intermediate or the dextranosyl-intermediate. The exogenous dextran, thus, becomes branched with a single glucose residue or with a dextran chain (Robyt and Taniguchi, 1976) (see Fig. 6). The acceptor reactions terminate glucan elongation. This can occur during synthesis, when no acceptors are added, by the fructose by-product acting as an acceptor to give the disaccharide, leucrose, or by a dextran chain acting as an acceptor to give the synthesis of branch linkages (Robyt and Taniguchi, 1976).

The synthesis of the 5% branch linkages of B-512F dextran is apparently diffusion controlled, and consequently the branch linkages are randomly distributed. It would seem very unlikely, however, that the synthesis of the more highly branched dextrans, such as the B-742 regular comb dextran (50% branch linkages) and the *S. mutans* alternating comb dextran (33% branch linkages), are diffusion controlled and therefore random. In the case of the former glucan, it is meaningless to speak of a random distribution of branch linkages as every glucose residue in the main chains is substituted by an α-1$\rightarrow$3 linked glucose residue. In the case of the latter glucan, it is also nearly meaningless to describe the distribution of branch linkages as random as one out of three glucose residues in the main chains is substituted by an α-1$\rightarrow$3 linked glucose residue. The biosynthetic details, for the attachment of these single glucose residues to the α-1$\rightarrow$6 linked glucose residues in the main chains by the enzymes that synthesize these highly branched dextrans, have yet to be worked out.

Uses of Dextran and Related Polysaccharides

These polysaccharides can be a nuisance when they are found as contaminants in food preparations containing sucrose, *e.g.*, in wine, soft drinks, canned fruit juices, sugar-cured hams, *etc.* (Fabian and Henderson, 1950; Deibel and Niven, 1959). They also present a problem in the sugar refining industry, where they plug-up process pipes and filters and inhibit the crystallization of sucrose. The two polysaccharides produced by enzymes from *Streptococcus mutans*, alternating comb-dextran and mutan, form the principal components of dental plaque and are involved in the development of dental caries (Krasse, 1965; Gibbons and Banghart, 1967; Hamada and Slade, 1980). Nevertheless, one of the polysaccharides synthesized from sucrose, *viz.*, B-512F dextran, has been produced commercially as it was found that it could be used as a therapeutic agent in restoring blood volume in mass

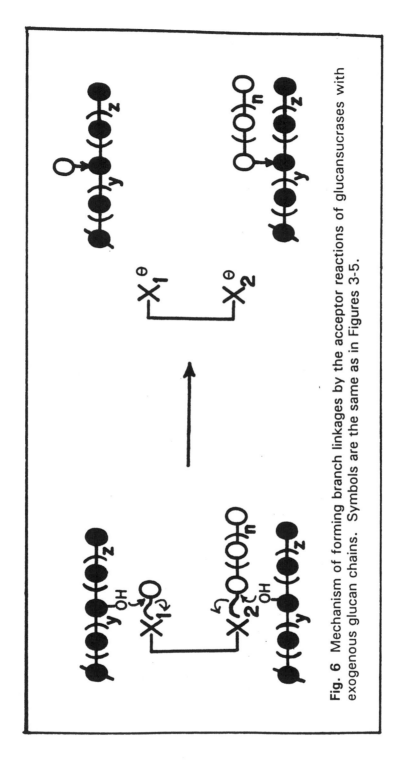

Fig. 6 Mechanism of forming branch linkages by the acceptor reactions of glucansucrases with exogenous glucan chains. Symbols are the same as in Figures 3-5.

casualties (Gronwall and Ingleman, 1944, 1945).

Many different esters and ethers of B-512F dextran have been synthesized to give macromolecules having diverse properties with negative, positive, or hydrophobic groups. The most widely used dextran derivative is the epichlorohydrin, cross-linked dextran that has become the commercial material known as *Sephadex*. This is a gel-filtration or molecular sieving agent, manufactured by Pharmacia Fine Chemicals, Ltd. of Sweden. The introduction of these materials with varying degrees of cross-linking in the 1960's, revolutionized the purification and separation of biochemically important macromolecules (Porath and Flodin, 1959; Whitaker, 1963; Porath, 1968). Another derivative, dextran sulfate, acts as a blood anticoagulant and a heparin substitute (Ricketts, 1959). Recently, it also has been reported to have inhibitory effects on virus replication (Takemoto and Liebhaber, 1962), including the HIV virus (Ito *et al.*, 1987).

Although B-512F dextran currently is the only dextran or related glucan that is produced on a commercial scale, alternan and the other highly branched dextrans offer possibilities for uses as highly soluble, noncaloric or low caloric bulking agents, fillers, binders, and extenders. Cote and colleagues at the USDA Northern Regional Research Center, Peoria, Illinois are developing a series of branched limit dextrans that are synthesized from sucrose. The resulting glucans are treated with endodextranases to give products with molecular weights of 2-10K. These dextranase-treated glucans give solutions with lower viscosities than do the much higher molecular weight native-glucans, and they are apparently resistant to further enzymatic attack. The solution and gel properties are similar to the commercial bulking agent, Polydextrose, but have the advantage of having a distinct, relatively uniform, and reproducible structure. They are also similar to commercial maltodextrins, D.P. 10-50, but they are not susceptible to hydrolytic cleavage by α-amylases and are only very slowly hydrolyzed by dextranases. They, therefore, are of low caloric value. The use of alternan and the highly branched dextrans in other areas other than as bulking agents needs further exploration. Further, the production of the enzymes and their efficient use to produce the glucans at a cost-effective price also needs investigation and development to make them commercially feasible.

PULLULAN

Pullulan is a neutral polysaccharide produced by the black yeast, *Pullularia pullulans* (now known as *Aureobasidium pullulans*). Pullulan is a water soluble, linear polysaccharide of D-glucose, containing α-1→4 and α-1→6 linkages in the ratio of 2:1 (Bender *et al.*, 1959; Wallenfels *et al.*, 1961, 1965; Bouveng *et al.*, 1962, 1963). It has been reported to have a range of molecular weights from 10^4 to 10^6. Pullulan can be thought to be essentially a polymer of maltotriose joined end to end by an α-1→6 linkage (see Fig. 7A). A careful analysis of pullulan has shown that in addition to maltotriose there is a small amount (5-7%) of maltotetraose (Catley and Whelan, 1971). The maltotetraose is located in the interior part of the polysaccharide (Catley and Whelan, 1971) (see Fig. 7B).

Two different endohydrolases have been found to hydrolyze pullulan with two different specificities. Pullulanase from *Aerobacter aerogenes* catalyzes the hydrolysis of the α-1→6 linkage to give maltotriose and small amounts of maltotetraose (Catley *et al.*, 1966; Catley and Whelan, 1971; Bender and Wallenfels, 1961). Isopullulanase specifically hydrolyzes the second α-1→4 linkage of the maltotriose unit to give isopanose (Sakano *et al.*, 1971, 1972). See Figure 7C for a summary of the action of pullulanase and isopullulanase on pullulan.

Very little is known about the biosynthesis of pullulan. A lipid carrier intermediate (presumably a polyisoprenoid), associated with the cell membrane, has been implicated (Taguchi *et al.*, 1973; Catley and McDowell, 1982), although no specific details of the biosynthetic reactions involved in the biosynthesis of the pullulan chain have been reported. The very regular structure, however, suggests a highly specific synthetic mechanism.

The presence of the α-1→6 bond on every third

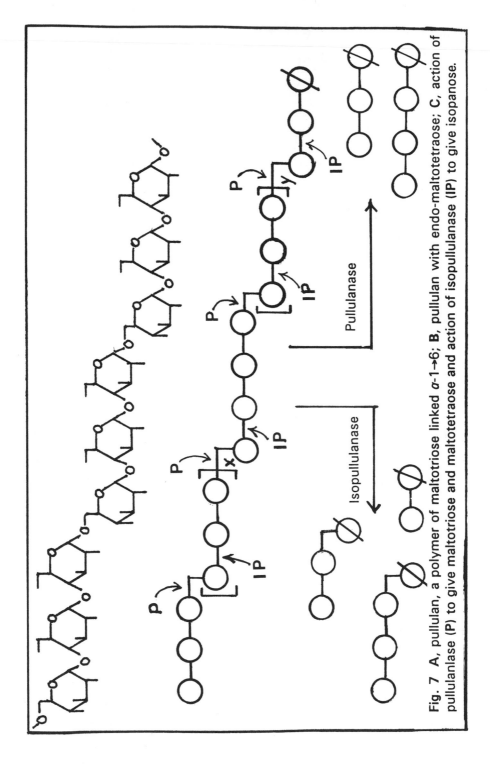

Fig. 7 **A**, pullulan, a polymer of maltotriose linked α-1→6; **B**, pullulan with endo-maltotetraose; **C**, action of pullulanase (P) to give maltotriose and maltotetraose and action of isopullulanase (IP) to give isopanose.

glucose residue introduces a flexibility that produces a highly water-soluble linear polymer. This is in contrast with other linear polymers such as amylose (an α-1$\rightarrow$4 linked glucan) and mutan (an α-1$\rightarrow$3 linked glucan). Although amylose can be solubilized, it readily undergoes retrogradation to give a very water-insoluble material. But, even B-512F dextran, with contiguous α-1$\rightarrow$6 linkages becomes water-insoluble when the degree of α-1$\rightarrow$3 branch linkages is decreased. So, the water solubility of pullulan is more than just the presence of the α-1$\rightarrow$6 linkage, it is its presence interspersed with two α-1$\rightarrow$4 linkages that gives it water-solubility.

The combination of linkages in pullulan readily gives the formation of fibers and films. Films formed from pullulan, without plasticizers, are water-soluble, impervious to oxygen, and a suitable packaging material for foods and pharmaceuticals (Yuen, 1974a, 1974b). These films, therefore, have particular potential as edible packaging agents. Further, ingestion of pullulan results in calorie reduction and when incorporated into foods produces a sensory improvement in taste, flavor, and texture. It has been suggested that its incorporation in foods will prevent drying of the food and retrogradation of starch. Pullulan fibers have a shiny gloss that resembles rayon, and after stretching, it has a tensile strength that is comparable to nylon (Yuen, 1974a, 1974b). Pullulan is also suitable as a sizing agent in the production of speciality papers and it can be used as an adhesive or as a coating or laminating agent. It has also been investigated as a material that can be molded by compression into solid objects and is reported to have properties similar to polyvinyl alcohol or polystyrene (Yuen, 1974a), but with the added value of being completely biodegradable under usual waste disposal conditions.

ALGIN

Algin is a polysaccharide composed of β-D-mannuronic acid and α-L-guluronic acid linked 1$\rightarrow$4 (see Fig. 8). It is found in many species of sea weed or brown algae as an important constituent of the cell walls.

The polysaccharide actually exists as the salts of the carboxylate group and as such is called *alginate*. The most common algin is sodium alginate. The thickening, suspending, emulsifying, stabilizing, gel-forming, film-forming, and encapsulating properties have re-sulted in a large number of commercial uses.

Several species of brown algae serve to produce commercial alginate: *Macrocystis pyrifera*, *Laminaria hyperborea*, *L. digitate*, and *Ascophyllum nodosum*. *M. pyrifera* is a giant kelp found in the temperate zones of the Pacific Ocean. The commercial beds are found off of Southern California and Australia. Other species of brown algae are harvested off the coasts of England, Norway, France, and Japan.

At least three bacterial species produce extra-cellular algins: *Pseudomonas aeruginosa*, *P. syringae*, and *Azotobacter vinelandii*. Although none of these bacterial algins have, to date, been developed for commercial use, they hold out the future possibility for a stable source of algins as the sea beds become contaminated and seaweed becomes unavailable because of depletion.

Algin Structure

Structural studies have shown that algin is linear and consists of ß-1→4 linked D-mannopyranosyluronic acid and α-1→4 linked L-gulopyranosyluronic acid (Hirst and Rees, 1965; Rees and Samuel, 1967; Atkins *et al.*, 1963, 1970; Hirst *et al.*, 1964). Saccharides obtained by partial acid hydrolysis contain both uronic acid residues, showing that algin is a copolymer and not a mixture of two polysaccharides, each containing one of the uronic acids. The amounts of D-mannuronic acid and L-guluronic acid are in the ratio of approximately 2:1 for many algins, with the exception of the algin from *L. hyperborea* stipes, which has a ratio of 1:2.3 (McNeely and Pettitt, 1973). The ratio varies with the algal species, the type of tissue, and the age of the tissue. *A. nodosum* receptacles, for example, contains essentially pure polymannuronic acid. The bacterial algins are rich in D-mannuronic acid, but the ratio can be made to vary, depending on the bacte-rial species and the culturing conditions (Larsen and

Haug, 1971). For example, the ratio has been observed to vary between 0.4 to 2.4 for *A. vinelandii* algin.

The alginate molecule is a copolymer containing blocks of D-mannopyranosyluronic acid residues and blocks of L-gulopyranosyluronic acid residues (see Fig. 8). These blocks are linked together by segments having predominantly alternating monomer composition (McNeely and Pettitt, 1973) (see Fig. 8). X-ray diffraction data have indicated that the D-mannuronic acid units are in the C1 conformation and the L-gulu-ronic acid residues are in the 1C conformation (Atkins *et al.*, 1970). These somewhat unusual structural features give unique properties to the alginates.

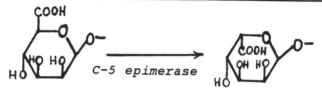

β−D−mannopyranosyl   α−L−gulopyranosyl
uronic acid   uronic acid

-MMM-GG-MMMMM-GGG-MM-GGGG-MM-

block copolymer

-M-G-M-G-M-G-M-G-M-G-M-G-M-G-M-G-

alternating β-D-mannopyranosyl uronic acid and α-L-gulopyranosyl uronic acid that link the block copolymers together.

Fig. 8 **Structure of Algin**

Algin Biosynthesis

Very few hard facts are known about the biosynthesis of algin. A possible pathway has been postulated for *Azotobacter vinelandii* alginate biosynthesis based on detectable enzymes (Pindar and Bucke, 1975) (see Fig. 9). The biosynthetic scheme is proposed to proceed from hexose-6-phosphate formed by hexokinase and interconverted by hexose-6-phosphate isomerases. Based on other systems, it is postulated that GDP-mannuronic acid is the high energy donor of the D-mannuronic acid monomer, although no real evidence for this has been given. Polymannuronic acid is the initial polysaccharide formed. The mechanism and enzymes involved in its synthesis have not been elaborated. A polymannuronic acid C-5-epimerase has been found and purified from the culture supernatants of *A. vinelandii*. This enzyme catalyzes the conversion of ß-D-mannuronic acid into α-L-guluronic acid by inversion of the carboxyl group attached to C-5 (Haug and Larsen 1971). A similar enzyme has been found in brown algae (Madgwick *et al.*, 1973). The alginate molecule is, thus, synthesized as poly-ß-1→4-D-mannuronic acid and post-synthetically modified by a C-5-epimerase to give blocks of α-1→4 linked L-guluronic acid. The amount of α-L-guluronic acid being dependent on the activity of this epimerase.

Algin Properties

Alginates will form gels by interacting with divalent metal ions. The affinity of alginates for divalent ions decreases in the following order: Pb>Cu>Cd>Ba>Sr>Ca>Co, Ni, Zn, Mn>Mg (McNeely and Pettitt, 1973). Gel strength increases with increasing concentration and/or degree of polymerization of the alginate. The structure of the alginate is also important. High L-guluronic acid alginates readily form gels, whereas polymannuronic acid will not gel at all (Smidsrød and Haug, 1972).

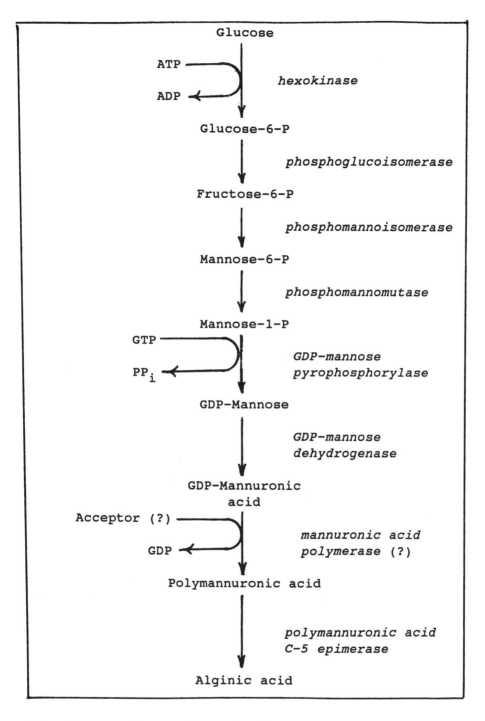

Fig. 9 Postulated Biosynthesis of Algin from hexoses

Gels are readily formed by mixing a solution of sodium alginate with a solution of divalent metal ion. For example, calcium-alginate beads are formed by dripping a 3-5% solution of sodium alginate into a 2% solution of calcium chloride. The gel beads form instantly.

Water-soluble alginate films can be cast from sodium alginate solutions. Water-insoluble films can be obtained by extruding sodium alginate solutions through a slit into a divalent metal ion solution. Another method is the treatment of a sodium alginate film with a divalent metal ion solution. These films are clear, tough, and flexible.

Algin Uses

Alginate appears today as a constituent in many food preparations. It appears on the GRAS (*generally recognized as safe*) list of the U. S. Food and Drug Administration.

Alginates have physical properties of holding water, gelling, emulsifying, and stabilizing food preparations. For example, it is used in ice cream to produce a smooth texture and prevent the formation of ice crystals. Similarly, it is found in various frozen foods, syrups, bakery icings, pastry fillings, cottage cheese, and spread cheese. Its gelling properties are used in instant and cooked puddings. Its emulsifying properties are used in salad dressings, and its stabilizing effects are utilized in beer foams, whipped cream, fountain syrups, and fruit juices. An interesting application appears in "pimento" centered olives. The red material in these olives is primarily calcium alginate gel to which some red dye and pimento flavoring has been added. The alginate gives stability to the material. Because of the β-D-1$\rightarrow$4 and α-L-1$\rightarrow$4 linkages, the alginates are not metabolized and are completely noncaloric.

A significant market for microbial polysaccharides is the petroleum industry where their viscous solutions are used to produce enhanced oil recovery as drilling muds. In enhanced oil recovery, the polysaccharides are used to improve water flooding techniques by increasing the efficiency of contact with oil and its displacement.

Another interesting use of alginate has developed in biotechnology and biochemical engineering in which divalent metal ion alginate gels are used to encapsulate whole cells (Smidsrød and Skjak-Braek, 1990; Vorlop and Steinert, 1987) that are used in biochemical reactors to produce various kinds of products by the continuous action of the cells' biochemical machinery. Calcium alginate also can be used to immobilize enzymes by encapsulating them into gel-beads. This must be combined with some additional technique of forming membranes around the beads (Tanaka et al., 1984; Bajpai and Margaritis, 1985) or by increasing the molecular weight of the enzyme by chemically coupling it to starch, dextran, or polyvinyl alcohol (Svensson, 1973). Without one of these modifications, the enzymes slowly diffuse out of the alginate gels due to the relatively large pore sizes in the gels. The use of calcium-alginate gels is particularly efficacious in that the gels are easily formed under mild conditions that are conducive to relatively fragile living cells and/or enzymes. The half-lives of the cells or the enzymes are usually greatly increased, making their use in bioreactors very efficient and feasible.

LITERATURE CITED

Atkins, E. D. T., Mackie, W., and Smolko, E. E., 1970, Crystalline structures of alginic acids, Nature (London) 225:626.

Atkins, E. D. T., Mackie, W., Parker, K. D., and Smolko, E. E., 1963, Structural studies of alginic acid, Chem. Ind. (London), 257.

Bajpai, P., and Margaritis, A., 1985, Improvement of inulinase stability of calcium alginate immobilized *Kluyveromyces marxianus* cells by treatment with hardening agents, Enzyme Microb. Technol. 7:34.

Bender, H., Lehmann, J., and Wallenfels, K., 1959, Pullulan, an extracellular glucan of *Pullularia pullulans*, Biochim. Biophys. Acta 36:309.

Bender, H., and Wallenfels, K., 1961, Pullulan. II. specific degradation by a bacterial enzyme, Biochem. Z. 334:79.

Bourne, E. J., Hutson, D. H., and Weigel, H., 1962, Studies on Dextrans and Dextranases: 2. The action of mould dextranases on modified isomaltodextrins and the effect of anomalous linkages on dextran hydrolysis, Biochem. J. 85:158.

Bouveng, H. O., Kiessling, H., Lindberg, B., and McKay, J., 1962, Polysaccharides elaborated by *Pullularia pullulans*. I. The neutral glucan synthesized from sucrose solutions, Acta Chem. Scand. 16:615.

Bouveng, H. O., Kiessling, H., Lindberg, B., and McKay, J., 1963, Polysaccharides elaborated by *Pullularia pullulans*. II. The partial acid hydrolysis of the neutral glucan synthesized from sucrose solutions, Acta Chem. Scand. 17:797.

Bovey, F. A., 1959, Enzymatic polymerization: I. Molecular weight and branching during the formation of dextran, J. Polym. Sci. 35:167.

Catley, B. J., Robyt, J. F., and Whelan, W. J., 1966, Maltotetraose, a minor component of pullulan, Biochem. J., 100:5p.

Catley, B. J., and Whelan, W. J., 1971, Observations on the structure of pullulan, Arch. Biochem. Biophys. 143:138.

Catley, B. J., and McDowell, W., 1982, Lipid linked saccharides formed during pullulan biosynthesis in *Aureobasidium pullulans*, Carbohydr. Res. 103:65.

Cote, G. L. and Robyt, 1982a, Isolation and partial characterization of an extracellular glucansucrase from *Leuconostoc mesenteroides* NRRL B-1355 that synthesizes an alternating (1→6), (1→3)-α-D-glucan, Carbohydr. Res. 101:57.

Cote, G. L. and Robyt, 1982b, Acceptor reactions of alternansucrase from *Leuconostoc mesenteroides* B-1355, Carbohydr. Res. 111:127.

Cote, G. L. and Robyt, 1983, The formation of α-D-(1→3) branch linkages by an extracellular glucansucrase from *Leuconostoc mesenteroides* NRRL B-742, Carbohydr. Res. 119:141.

Deibel, R. H., and Niven, C. F., 1959, Contaminating *Leuconostoc mesenteroides* synthesis of dextran in foods containing sucrose, Appl. Microbiol. 7:138.

Fabian, F. W. and Henderson, R. H., 1950, *Leuconostoc mesenteroides* as the cause of ropiness in canned peaches, Food Res. 15:415.

Gibbons, R. J., and Banghart, S. B., 1967, Synthesis of extracellular dextran by cariogenic bacteria and its presence in human dental plaque, Arch. Oral Biol. 12:11.

Gronwall, and Ingleman, B. G. A., 1944, Untersuchungen über Dextran und sein Verhalten bei parenteraler Zufuhr I., Acta Physiol. Scand. 7:97.

Gronwall, and Ingleman, B. G. A., 1945, Untersuchungen über Dextran und sein Verhalten bei parenteraler Zufuhr II., Acta Physiol. Scand. 9:1.

Hamada, S., and Slade, H. D., 1980, Biology, immunology, and cariogenicity, Microbiol. Revs. 44:331.

Hare, M. D., Svensson, S., and Walker, G. J., 1978, Characterization of the extracellular, water-insoluble α-D-glucans of oral *Streptococci* by methylation analysis, and by enzymic synthesis and degradation, Carbohydr. Res. 66:245.

Haug, A., and Larsen, B., 1971, Biosynthesis of alginate: II. Polymannuronic acid C-5-epimerase from *Azotobacter vinelandii* (Lipman), Carbohydr. Res. 17:297.

Hirst, E. L., Percival, E., and Wold, J. K., 1964, The structure of alginic acids: IV. Partial hydrolysis of the reduced polysaccharide, J. Chem. Soc. 1493.

Hirst, E. L., and Rees, D. A., 1965, The structure of alginic acid: V. Isolation and unambiguous characterization of some hydrolysis products of the methylated polysaccharide, J. Chem. Soc., 1182.

Ito, M., Baba, M., Sator, A. Pauwels, R. DeClercq, E., and Shigeta, S., 1987, Inhibitory effect of dextran sulfate and heparin on the replication of humandeficiency virus (HIV) *in vitro*, Antiviral Res. 7:361.

Jeanes, A., Haynes, W. C., Wilham, C. A., Rankin, J. C., Melvin, E. H., Austin, M. J., Cluskey, J. E., Fisher, B. E., Tsuchiya, H. M., and Rist, C. E., 1954, Characterization and classification of dextrans from ninety-six strains of bacteria, J. Amer. Chem. Soc. 76:5041.

Jeanes, A. and Seymour, F. R., 1979, The α-D-glucopyranosidic linkages of Dextrans: comparison of percentages from structural analysis by periodate oxidation and by methylation, Carbohydr. Res. 74:31.

Krasse, B., 1965, The effect of caries-inducing streptococci in hamsters fed diets with sucrose or glucose, Arch. Oral Biol. 10:223.

Larm, O., Lindberg, B., and Svensson, S. 1971, Studies on the length of the side chains of the dextran elaborated by *Leuconostoc mesenteroides* B-512F, Carbohydr. Res. 20:39.

Larsen, B., and Haug, A., 1971, Biosynthesis of alginate: I. Composition and structure of alginate produced by *Azotobacter vinelandii* (Lipman), Carbohydr. Res. 17:287.

Madgwick, J., Haug, A., and Larsen, B., 1973, Polymannuronic acid 5-epimerase from the marine alga, *Pelvetia canaliculata*, Acta Chem. Scand. 27:3592.

McNeely, W. H., and Pettitt, D. J., 1973, Algin, Chap. IV in Industrial Gums, 2nd ed. (R. L. Whistler and J. N. BeMiller, eds) Academic Press, 49-81.

Misaki, A., Torii, M., Sawai, T., and Goldstein, I. J., 1980, Structure of the dextran of *Leuconostoc mesenteroides* B-1355, *Carbohydr. Res.* 84:273.

Pindar, D. F., and Bucke, C., 1975, The biosynthesis of alginic acid by *Azotobacter vinelandii*, Biochem. J. 152:617.

Porath, J., and Flodin, P., 1959, Gel filtration: a method for desalting and group separation, Nature (London) 183:1657.

Porath, J., 1968, Molecular sieving and adsorption, Nature (London) 218:834.

Rees, D. A., and Samuel, J. W. B., 1967, The structure of alginic acid: VI. Minor features and structural variations, J. Chem. Soc. C., 2295.

Richards, G. N., and Streamer, M., 1978, Mode of
    action of dextranase D$_2$ from *Pseudomonas* UQM 733
    on oligosaccharides, *Carbohydr. Res.*, 62:191.
Ricketts, C. R., 1959, Dextran sulfate a synthetic
    analogue of heparin, Biochem. J. 51:129.
Robyt, J. F., Kimble, B. K., and Walseth, T. F., 1974,
    The Mechanism of dextransucrase action: direction
    of dextran biosynthesis, Arch. Biochem. Biophys.
    165:634.
Robyt, J. F., and Taniguchi, H., 1976, The mechanism
    of dextransucrase action: biosynthesis of branch
    linkages by acceptor reactions with dextran,
    Arch. Biochem. Biophys. 174:129.
Robyt, J. F., and Walseth, T. F., 1978, The mechanism
    of acceptor reactions of *Leuconostoc mesente-*
    roides B-512F dextransucrase, Carbohydr. Res.
    61:433.
Robyt, J. F., and Eklund, S. H., 1983, Relative,
    quantitative effects of acceptors in the reaction
    of *Leuconostoc mesenteroides* B-512F dextransu-
    crase, Carbohydr. Res. 121:279.
Robyt, J. F., and Martin, P. J., 1983, Mechanism of
    synthesis of D-glucans by glucosyltransferases
    from *Streptococcus mutans* 6715, Carbohydr. Res.
    113:301.
Robyt, J. F., 1986a, "Dextran" in Encyclopedia of
    Polymer Science and Engineering, 2nd ed., vol. 4
    (H. F. Mark, N. M. Bikales, C. G. Overberger, and
    G. Menges, eds.) pp. 752-767. New York: John
    Wiley & Sons.
Robyt, J. F., 1986, Graphic Representation of Oligo-
    saccharide and Polysaccharide Structures Contain-
    ing Hexopyranose Units, J. Chem. Educ. 63:560.
Sakano, Y., Matsuda, N., and Kobayashi, T., 1971,
    Hydrolysis of pullulan by a novel enzyme from
    Aspergillus niger, Agric. Biol. Chem. 35:971.
Sakano, Y., Higuchi, M., and Kobayashi, T., 1972,
    Pullulan 4-glucanohydrolase from *Aspergillus*
    niger, Arch. Biochem. Biophys. 153:180.
Senti, F. R., Hellman, N. N., Ludwig, N. H., Babcock,
    G. E., Robin, R., Glass, C. A., and Lamberts, B.,
    1959, Wiscosity, sedimentation, and light-
    scattering properties of fractions of an acid-
    hydrolyzed dextran, J. Polym. Sci. 17:527.

Seymour, F. R., Knapp, R. D., and Bishop, S. H., 1976, Determination of the structure of dextran by ^{13}C-nuclear magnetic resonance, Carbohydr. Res. 51:179.

Seymour, F. R., Slodki, M. E., Plattner, R. D., and Jeanes, A., 1977, Six unusual dextrans: methylation structural analysis by combined g.l.c.m.s. of per-O-acetyl-aldononitriles, Carbohydr. Res. 53:153.

Seymour, F. R. , Chem, E. C. M., and Bishop, S. H., 1979a, Methylation structural analysis of unusual dextrans by combined gas-liquid chromatography-mass spectrometry, Carbohydr. Res. 68:113.

Seymour, F. R., Knapp, R. D., Chen, E. C. M., Bishop, S. H., and Jeanes, A., 1979b, Structural analysis of *Leuconostoc* dextrans containing 3-O-α-D-glucosylated α-D-glucosyl residues in both linear-chain and branch-point positions, or only in branch-point positions by methylation and by ^{13}C-n.m.r. spectroscopy, Carbohydr. Res. 74:41.

Seymour, F. R., Julian, R. L., Jeanes, A., and Lamberts, B. L., 1980, Structural analysis of insoluble D-glucans by fourier-transform, infrared difference-spectrometry: correlation between structures of dextrans from strains of *Leuconostoc mesenteroides* and of D-glucans from strains of *Streptococcus mutans*, Carbohydr. Res. 86:227.

Seymour, F. R., and Knapp, R. D., 1980a, structural analysis of α-D-glucans by ^{13}C-nuclear magnetic resonance, spin-lattice relaxation studies, Carbohydr. Res. 81:67.

Seymour, F. R., and Knapp, R. D., 1980b, Structural analysis of dextrans, from strains of *Leuconostoc* and related genera, that contain 3-O-α-D-glucosylated α-D-glucopyranosyl residues at the branch points, or in consecutive, linear positions, Carbohydr. Res. 81:105.

Shimamura, A., Tsumori, H., and Mukasa, H., 1982, Purification and properties of *Streptococcus mutans* extracellular glucosyltransferase, Biochim. Biophys. Acta, 702:72.

Slodki, M. E., England, R. E., Plattner, R. D., and Dick, Jr., W. E., 1986, Methylation analyses of NRRL dextrans by capillary gas-liquid chromatography, *Carbohydr. Res.* 156:199.

Simet, I., and Robyt, J. F., 1990, Isolation of *Neisseria perflava* from dental plaque that strain brown, purple, wine, and blue with triiodide when grown on sucrose medium, unpulished results.

Smidsrφd, O., and Skjak-Braek, G., 1990, Alginate as immobilization matrix for cells, Trends Biotech. 8:71.

Svensson, B., 1973, Covalent coupling of amylodextrin to subtilisin type Novo by means of cyanogen bromide activation, Compt. Rend. Trav. Lab. Carlsberg, 39:339.

Taguchi, R., Sakano, Y., Kikuchi, Y., Sakuma, M., and Kobayashi, T., 1973, synthesis of pullulan by acetone-dried cells and cell-free enzyme from *Pullularia pullulans*, and the participation of lipid intermediate, Agric. Biol. Chem. 37:1635.

Takemoto, K. K., and Liebhaber, H., 1962, Virus-polysaccharide interactions: II. Enhancement of plaque formation and the detection of variants of poliovirus with dextran sulfate, Virology, 17:499.

Tanriseven, A., and Robyt, J. F., 1992, Inhibition of dextran synthesis by acceptor reactions of dextransucrase, and the demonstration of a separate acceptor binding site, Carbohydr. Res. 225:321.

Tanaka, H., Kurosawa, H., Kokufuta, E., and Veliky, I. A., 1984, Preparation of immobilized glucoamylase using Ca-alginate gel coated with partially quaternized poly(ethyleneamine), Biotech. Bioeng. 26:1393.

Tao, B. Y., Reilly, P. J., and Robyt, J. F., 1988, *Neisseria perflava* amylosucrase: characterization of its product polysaccharide and a study of its inhibition by sucrose derivatives, Carbohydr. Res. 181:163.

Torii, M., Kabat, E. A., and Weigel, H., 1966, Immuno-logical studies of dextrans: VI. Further characterization of the determinant groups on various dextrans involved in their reactions with the homologous human antidextrans, J. Immunol. 96:797.

Van Cleve, J. W., Schaefer, W. C., and Rist, C. E., 1956, The structure of NRRL B-512 dextran. Methylation studies, J. Amer. Chem. Soc. 78:4435.

Vorlop, K. D., and Steinert, H. J., 1987, Cell immobilization within coated alginate beads or hollow fibers formed by ionotropic gelation, Ann. New York Acad. Sci. (US), 501:339.

Wallenfels, K., Bender, H., Keilich, G., and Bechtler, G., 1961, Pullulan, the slime capsule of *Pullularia pullulans*, Angew. Chem. 73:245.

Wallenfels, K., Keilich, G., Bechtler, G., and Freudenberger, D., 1965, Pullulan. IV. Elucidation of the structure by physical, chemical, and enzymic methods, Biochem. Z. 341:433.

Whitaker, J. R., 1963, Determination of molecular weights of proteins by gel filtration on Sephadex, Anal, Chem. 35:1950.

Wilham, C. A., Alexander, B. H., and Jeanes, A., 1955, Heterogeneity in dextran preparations, Arch. Biochem. Biophys. 59:61.

Yuen, S., 1974a, Properties and applications of the polysaccharide, pullulan, Process Biochem. 9(11):7.

Yuen, S., 1974b, Pullulan and its applications, Hayashibara Biochemical Laroratories, Inc. Okayama, Japan.

ASPECTS OF THE CHEMISTRY AND NUTRITIONAL EFFECTS OF NON-STARCH POLYSACCHARIDES OF CEREALS

Peter J. Wood*

Centre For Food and Animal Research, Agriculture
Canada, Ottawa, ONT K1A 0C6

INTRODUCTION

In the last decade or so the non-starch polysaccharides (NSP) of cereals have received prominent attention as dietary fiber. It has been generally accepted that the diets of the majority of the population in western developed societies are deficient in dietary fiber and this may contribute to the development of a variety of diseases including coronary heart disease and colon cancer. A more health and nutrition conscious consumer has responded by increasing whole grain cereal consumption. The benefits are probably two-fold in that cereals substitute for less desirable constituents of the diet such as fat, but specific responses to fiber intake also occur. The chief of these are improved colonic function, improved glucose regulation and reduced serum cholesterol levels. The latter two properties seem associated more with soluble NSP whereas the former is a property of insoluble polysaccharide. Dietary fiber is thus increasingly distinguished as insoluble and soluble although the latter is difficult to measure and lacks a clear physicochemical basis.

Non-starch polysaccharides in cereals used for animal feed reduce the available metabolizable energy of the feed for non-ruminants. Additional undesirable effects are associated specifically with soluble, viscous polysaccharides of rye and barley. Thus similar properties of NSP result in opposite desirabilities in feed and food.

This article will identify the main NSP of some of the common cereals and specific nutritional roles will be discussed. In view of the recent controversy over the value, or otherwise, of oat bran in reducing serum cholesterol levels, particular emphasis will be placed on the metabolic effects of viscous soluble fiber and the potential of oat bran as a source of such fiber.

*For the Department of Agriculture, Centre for Food and Animal Research, Government of Canada on the metabolic effects of viscous soluble fiber and the potential of oat bran as a source of such fiber.

ANALYSIS AND CHEMICAL NATURE OF CEREAL NON-STARCH POLYSACCHARIDES

General analytical methods
The most general methods of analysis of NSP use acid hydrolysis followed by chromatographic separation and estimation of the constituent monosaccharides.

Sulfuric acid (0.5-1M) is usually chosen for hydrolysis, typically at 100°C for two to four hours. Acid is neutralized with barium carbonate and monosaccharide residues converted to alditol acetates for analysis by gas liquid chromatography (GLC) (Sawardeker et al 1965, Blakeney et al 1983, Englyst and Cummings 1984). Full recovery of the monosaccharides is difficult because of potential hydrolytic, isolation and derivatization losses. Other schemes such as methanolysis and silylation (Roberts et al 1987) have potential but do not appear to have been developed for cereal NSP analysis. For analysis of fructans, since reduction of fructose leads to ambiguity, an alternative approach, for example silylation of butaneboronate esters (Wood et al 1975) is required.

HPLC analysis has the advantage of not requiring derivatization but conventional methods have lacked adequate separating power and specificity in detection to allow general application. However, the major monosaccharide residues of interest, arabinose, xylose and glucose, separate well with aqueous elution from cation based carbohydrate analysis columns (Neilson and Marlett 1983).

Essential prerequisites for the analysis are removal of low molecular weight sugars (which may also remove the small amounts of fructans), deactivation of endogenous enzymes and removal of starch. These are achieved by extraction with hot (refluxing) aqueous ethanol and treatment with amylolytic enzymes. Complete liquefaction of starch followed by conversion to glucose or low molecular weight, aqueous ethanol soluble maltodextrins is essential. Heat stable, liquefying amylases generally achieve the necessary solubilization, "Termamyl" (Novo BioLabs, 33 Turner Rd., Danbury, CT

06810-5101) being a popular choice in dietary fiber analysis. Amylolytic enzymes used must be free from glycanases capable of significantly degrading NSP. For example, Termamyl contains β-glucanase, which is considerably less stable than the amylase at the temperature (90-100°C) used to liquefy the starch. However, if samples are not pre-heated to this temperature significant degradation of β-glucan may occur. Although this may be analytically recoverable, the β-glucan may have suffered depolymerization and viscosity loss.

In some cooked and processed products there remains a "resistant" starch (Englyst and Cummings 1984) which requires special consideration if not to be included as NSP.

The hot aqueous treatment used in starch hydrolysis solubilizes a portion of NSP, water-soluble NSP or soluble dietary fiber. This may be precipitated by ethanol and either separately analyzed or included with the digested residue. Literature values for soluble NSP may, however, be based on very different protocols. For example, soluble barley β-glucan usually refers to 65°C water soluble, based on brewing practice. Less obvious protocol differences may also produce significantly different solubility characteristics.

Specific analytical methods

A number of specific methods of analysis have been developed for the (1→3)(1→4)-β-D-glucan of cereals, initially to service the brewing industry. The most widely used is the McCleary and Glennie-Holmes assay (1985) in which specific enzymes convert the β-glucan (but not cellulose or starch) to glucose, then measured by a glucose oxidase method. The assay is available in kit form (Megazyme, Aust. Pty. Ltd., 6 Altona Place, North Rocks, Sydney, NSW 2151). Specific calcofluor binding to β-glucans in solution (Wood 1980) was developed by Jørgensen (1988) as a non-enzymic assay. Bound dye shows an increased fluorescence which may be used to quantitate the β-glucan. However, photosensitivity hinders manual measurement and automated equipment (flow injection) as well as a fluorescence detector are required (Jørgensen 1988). Although specific under the conditions described, the method should not be used uncritically since other food ingredients, such as substituted celluloses (e.g., xyloglucan) and proteins with hydrophobic binding sites (e.g., casein) may respond in the assay and inhibition of binding and fluorescence enhancement may also occur (Wood 1980 and unpublished).

Specific methods are sometimes used for pentosan, based on the different acid dehydration products of pentose and hexose. Distillation and estimation of furfural has been used (Saastamoinen et al 1989) or direct colorimetry (Douglas 1981, Henry 1985, Hashimoto et al 1987a). Acid hydrolysis and chromatographic analysis may give lower values (Henry 1985) but whether this reflects hydrolytic losses or overestimation due to interference in the colorimetry is uncertain.

Cellulose can be distinguished on the basis of its resistance to hydrolysis unless subject to prior solubilization in 72% sulfuric acid. The estimation assumes that glucose released by hydrolysis with dilute sulfuric acid, without prior solubilization, is non-cellulosic which may not be true. These cellulose values may therefore be underestimates. An alternative is subtraction of $(1\to3)(1\to4)$-β-D-glucan from total anhydroglucose. This may overestimate by including glucose from other sources such as glucomannan and $(1\to3)$-β-D-glucan.

Uronic acids create resistance to hydrolysis and are difficult to analyze by hydrolysis and GLC or HPLC. They are determined by colorimetry (Blumenkrantz and Absoe-Hansen 1973) or decarboxylation (Theander and Åman 1982).

Identity and concentrations of cereal NSP
The origins of analyzed monosaccharide residues can, in general, be inferred from structures previously established in more detailed studies (Fincher and Stone 1986, Juliano 1985). The major constituents are pentosan, or arabinoxylan to which some glucuronic and 4-O-methylglucuronic acid may be attached, $(1\to3)(1\to4)$-β-D-glucan and cellulose. There may be, in addition some $(1\to3)$-β-D-glucan and mannan or glucomannan. These are all cell wall constituents. An additional component is protein associated arabinogalactan. Fructans are also NSP, and may exert potentially beneficial physiological

effects (Nilsson et al 1988, Hidaka et al 1990). Generally these are present in low amounts but may be as much as 1.7% in rye (Åman and Hesselman 1984). Rice is different from the other major food cereals having both pectin and xyloglucan (Juliano 1985) as part of its cell walls.

This article will focus mainly on wheat, rye, barley and oats. Values for pentosan, $(1\rightarrow3)(1\rightarrow4)$-$\beta$-D-glucan and cellulose, compiled or calculated from the literature are summarized in Table I and the relevant references listed in Table II.

Table I
Pentosan, β-glucan and cellulose concentration in cereals and cereal fractions (% dry weight basis)[a]

Sample	Pentosan	β-Glucan	Cellulose
Barley	5.7 - 11.0	3.0 - 10.6	4.3
Hulless barley	5.1 - 6.2	5.1 - 10.7	1.5 - 3.9
Corn	4.3	0.1	2.0 - 2.3
Oats	3.2 - 12.6	2.2 - 5.4	6.0 - 12.9
Oat groats	2.2 - 4.1	3.9 - 6.8	0.4 - 1.0
Oat bran	3.2 - 6.1	5.8 - 8.9	1.0
Rice	1.2 - 4.0	0.1	1.7 - 2.0
Rye	5.4 - 12.2	1.0 - 2.1	1.5
Rye flour	3.1 - 9.3	nr[b]	1.4
Rye bran	12.5 - 19.9	nr	nr
Wheat	4.3 - 8.1	0.5 - 1.0	2.0
Wheat flour	1.0 - 2.7	nr	0.1 - 0.2
Wheat bran	10.7 - 27.9	nr	7.4 - 8.2

[a]Data compiled from references in Table II
[b]nr, not reported

297

Table II
References for Table I

Reference	Analyses[a]	Cereals[a,b]
Åman 1987	Pentosan (GLC), cellulose, β-glucan	O
Åman and Graham 1987	β-Glucan	B,O
Åman and Hesselman 1984	Pentosan (GLC)	B,O,R,W
Åman and Hesselman 1985	β-Glucan	B,O,R,W
Anderson et al 1978	β-Glucan	B,O,R,Ri
Bengtsson et al 1990	Pentosan (GLC), β-glucan cellulose	B
Beresford and Stone 1983	β-Glucan	W
Delcours et al 1989	Pentosan (orcinol)	R
Englyst et al 1989	Pentosan (GLC), cellulose	B,O,R,W
Frφlich and Nyman 1988	Pentosan (GLC)	O
Hashimoto et al 1987a	Pentosan (orcinol)	W
Hashimoto et al 1987b	Pentosan (orcinol)	B,O,Ob
Henry 1985	Pentosan (Phloroglucinol), β-glucan	B,O,R,Ri,W
Lehtonen and Aikasalo 1987	Pentosan (furfural)	B
Marlett 1991	Pentosan (GLC), cellulose, β-glucan	B
McCleary and Glennie-Holmes 1985	β-Glucan	B,O,R,W,C
Michniewicz et al 1990	Pentosan (Phloroglucinol)	W
Nyman et al 1984	Pentosan (GLC)	B,C,R,Ri,W
Saastamoinen et al 1989	Pentosan (furfural), β-glucan	R
Saini and Henry 1989	Pentosan (GLC)	W,R
Salomonsson et al 1984	Pentosan (GLC)	B,O,R,W
Shinnick et al 1988	Pentosan (GLC), β-glucan	O
Ulrich et al 1991	β-Glucan	B
Varo et al 1984	Cellulose	C,Ri
Wood et al 1991a	β-Glucan	O

[a]Lists data from reference used to compile Table I: other analyses and cereals may have been included in the study.
[b]B, barley; C, corn; O, oats; Ob, oat bran; R, rye; Ri, rice; W, wheat

Caution is required in comparing data as listed in Table I since cultivar, sample source and preparation, and analytical methods may affect values. This is especially true of oats and barley which may have hulls or be hulless. With oats, for example, analysis might be of whole grain which may contain 25-30% hull, a rich (85-90%) source of insoluble fiber. Oats for food are always de-hulled and analysis then refers only to the groat. The literature may, however, simply refer to "oats" and it is only from high pentosan, cellulose and insoluble fiber values that the inclusion of hulls can be recognized. Regardless of hull content, cultivar variation can be large, for example the β-glucan concentration quoted for oats and barley. Critical evaluation of the wide range of values noted (Table I) requires reference to the original studies. Nevertheless, some general conclusions are possible. The cellulose content of the different cereals is similar if on a hulless or dehulled basis. It is chiefly associated with the outer layers of the kernel, or bran. The microstructure of the cereals - the thickness of the aleurone cell-walls and the crushed, or empty nature of the cells of the pericarp and seed coat - leads to increased cellulose levels compared to the endosperm with its starch-packed cells. Insoluble pentosan is also associated with the bran layers, so brans normally have both an increased total and increased proportion of insoluble NSP relative to the whole grain. On the other hand, refined flours decrease in total NSP, but the amount of soluble NSP remains constant consequently the proportion of soluble NSP increases in the flour (Nyman et al 1984).

Bran usually includes some sub-aleurone endosperm layers with thicker cell walls and/or smaller cells than the inner endosperm. This is especially true of oat bran, leading to considerable potential for β-glucan enrichment in the bran (Wood et al 1991a). Wheat and especially rye are good sources of pentosan. Rye is also a good source of soluble NSP, provided by both β-glucan and pentosan. Oats and barley are the richest sources of β-glucan, but the best commercially available food source is oat bran. There are some structural differences between oat and barley β-glucan (Wood et al

1991b), but the possible functional effects of this, if any, are unknown.

Table III shows the monosaccharide residues present in the NSP of cereals. It is not always clear in the literature whether values reported refer to anhydro-sugar, as they should if intended as polysaccharide analysis. Arabinose, xylose and glucose are the major constituents.

Table III
Monosaccharide constituents (as anhydrosugar) of NSP of cereals
(% dry weight basis)

Sample	Ara[a]	Xyl	Man	Gal	Glu	UA	Ref[b]
Barley (flake)	2.8	5.1	0.2	0.2	8.2	0.2	1
Barley	2.1	3.9	0.5	0.1	7.7	0.8	2
Barley flour	0.9	1.1	0.3	0.1	5.5	nd[c]	2
Hulless barley[d]	1.7	1.9	0.8	0.5	7.2	0.2	3
Hulless barley[e]	2.0	2.5	0.7	0.5	13.4	0.2	3
Corn	1.9	2.4	0.2	0.4	2.6	0.6	2
Corn flour	1.1	1.0	0.1	0.2	2.3	0.4	2
Corn meal	0.6	0.5	t[f]	0.1	0.7	0.1	1
Oats	1.5	5.4	0.3	0.7	12.2	1.1	4
Oat groat	1.0	2.3	0.3	0.2	5.5	0.5	5
Rolled oats	1.1	1.4	0.1	0.2	5.5	0.2	1
Oat bran	1.5	2.1	t	0.2	9.4	0.3	1
Oat bran	1.4	1.8	0.3	0.2	8.2	0.5	5
Oat hull	3.5	22.2	t	1.9	29.5	1.9	5
Rice	0.8	3.2	0.1	0.3	7.4	1.2	2
Rice flour	t	t	0.1	0.1	0.8	0.1	2
Rye	3.5	5.4	0.3	0.3	3.5	0.2	1
Rye	3.1	4.6	0.4	0.4	3.5	0.5	2
Rye flour	1.7	2.1	0.3	0.2	2.1	0.3	2
Wheat	3.3	4.8	t	0.3	2.8	0.2	1
Wheat	2.3	3.4	0.1	0.3	3.1	0.3	2
Wheat flour	1.1	1.4	0.1	0.2	0.6	t	1
Wheat flour	0.6	0.8	0.1	0.2	0.7	nd	2
Wheat bran	10.0	18.2	0.2	0.6	10.7	2.1	1

[a]Ara, arabinose; Xyl, xylose; Man, mannose; Gal, galactose; Glu, glucose; UA, uronic acid
[b]References: 1, Englyst et al. 1989; 2, Nyman et al. 1984; 3, Bengtsson et al. 1990; 4, Åman 1987; 5, Frφlich and Nyman 1988
[c]nd, not detected
[d]Data from cultivar with 5.2% β-glucan
[e]Data from cultivar with 10.7% β-glucan
[f]t, trace

The focus of this article is on NSP. Since total and insoluble DF determined gravimetrically may include lignin and other non-carbohydrate components it is tempting to omit gravimetric data. However, the physiological effects of the insoluble fiber of cereals include potential effects of lignin and other phenolic components of the cell-wall, such as ferulic acid, since it is these non-carbohydrate components which confer, in part, the insolubility. Nutritional effects therefore cannot be attributed, or apportioned, to one chemical component or another. This is true of most functional characteristic of foods – that microstructure and micro-chemical organization are crucially important. Soluble dietary fiber is somewhat simpler, since there is no intricately associated lignin, and nutritional effects can be observed with isolated, purified polysaccharides. In real foods, however, microstructure might again modify these effects. Dietary fiber values determined by GLC with and without Klason lignin are compared with gravimetric data in Table IV.

NUTRITIONAL EFFECTS OF CEREAL NON-STARCH POLYSACCHARIDES

Bioavailability of Nutrients of Feed

The true metabolizable energy (TME) of cereals used as feed for non-ruminants is reduced by fiber. The TME of different cereals fed to adult poultry is listed in Table V. Not all differences in this bio-assay are due to the fiber; cellulose, for example, has zero TME. The difference between regular and hulless oats therefore can be attributed mostly to the hulls. Starch has a similar TME to whole corn. The ratio between the TME of starch and glucose (1.1:1.0) reflects the conversion of anhydroglucose to glucose. This assay was used to determine the effect of addition of oat gum (80% β-glucan, Wood et al 1989) to the TME of corn (Table VI). The reduction in TME was linearly related to the amount of oat gum added to the corn feed and is essentially a dilution effect. Thus oat gum, fed at 10.5% and 20.9% of the corn diet, reduced the TME by 9% and 18% respectively. Starch was essentially fully digested although the presence of gum at 20.9% significantly ($p < 0.01$) increased the small amount recoverable in the feces. Approximately one third of the β-glucan was recovered in the feces.

Table IV
A comparison of dietary fibre values determined by GLC, with
and without Klason lignin and gravimetrically determined values
(% dry weight basis)

Sample	NSP (GLC)			Total Dietary Fiber (Gravimetric)	Ref[b]
	Soluble	Total	Total + KL[a]		
Barley	3.9	15.1	18.6	18.8	1
Barley flour	3.9	7.9	7.9	8.2	1
Corn	<1	8.0	9.4	9.3	1
Corn flour	<1	4.1	4.9	3.9	1
Oats	3.8[c]	nr	nr	31.2[c]	3
	nr	21.2	29.6	nr	4
Oat groats	3.0	9.0	11.0	11.5	2
Oat bran	5.3	12.6	15.6	15.2	2
Oat hulls	0.4	58.9	78.9	83.9	2
Rice	<1	13.0	16.9	19.2	1
Rice flour	<1	1.1	1.1	0.7	1
Rye	3.8	12.5	14.6	16.1	1
Rye flour	3.8	6.8	7.1	7.5	1
Rye bran	2.5	20.6	23.5	nr	5
Wheat	1.3	9.5	11.5	12.1	1
Wheat flour	1.3	2.4	2.4	2.8	1
Wheat bran	1.2[d]	22.5[d]	26.6[d]	nr	5
	3.3	36.6	nr	nr	6
	1.2	32.9	37.5	nr	7
	nr	nr	nr	42.2	8

[a]Abbr: KL, Klason lignin; nr, not reported
[b]References: 1, Nyman et al 1984; 2, Frølich and Nyman 1988;
 3, Mongeau and Brassard 1990; 4, Åman 1987; 5, Salomonsson 1984;
 6, Englyst et al 1989; 7, Varo et al 1984; 8, Prosky et al 1984
[c]Average of 4 samples
[d]Average of 2 winter wheat brans

Table V
True metabolizable energy (TME) of different cereals in adult poultry[a]

Sample (N[b])		TME (MJ/kg)
Barley	(9)	14.03 ± 0.68
Hulless barley	(23)	14.54 ± 0.74
Corn	(10)	17.17 ± 0.42
Whole oats	(10)	13.62 ± 0.63
Hulless oats	(9)	16.71 ± 0.30
Oat groats	(2)	17.3 ± 0.59
Rye	(8)	14.46 ± 0.47
Wheat	(12)	15.74 ± 0.47
Wheat bran	(2)	8.54 ± 0.66
Cornstarch	(2)	17.25 ± 0.42
Glucose		15.6
Cellulose	(1)	0

[a]From Sibbald 1977 and 1983
[b]N = number of samples

Table VI
Effect of oat gum on True Metabolizable Energy (TME) of a corn diet in adult poultry (n=8 at each level)[a]

Oat Gum % of diet (dwb)	Mean weight of excreta (g/bird)	Mean TME MJ/kg	Mean Fecal Starch % (dwb)	Mean Fecal β-Glucan % (dwb)
0	9.1	16.86 ± 0.34	1.3 ± 0.6	t[b]
10.5	11.7	15.29 ± 0.79	1.7 ± 0.9	6.7 ± 1.1
20.5	14.0	13.62 ± 0.51	3.2 ± 1.2	13.1 ± 2.3

[a]From Wood and Sibbald (unpublished data)
[b]t, trace

When assayed with young chicks, addition of oat gum decreased TME to a greater extent than due to a dilution effect alone (Cave et al 1990). In addition, bioavailabilities of amino acids and fat were decreased by 6 to 10%.

Effects of Viscosity on Growth and Feed Efficiency in Chicks

Soluble viscous fibers in feed have additional undesirable chronic effects. The poor feed value of barley and rye for young chicks has been attributed to viscous non-starch polysaccharide, β-glucan in barley but mainly pentosan in rye (Antoniou et al 1981). The effect of rye pentosan was directly demonstrated by extraction and addition to a wheat diet (Antoniou et al 1981) but the effect of β-glucan has generally been established indirectly by reducing the viscosity with enzyme, which improves feed:gain ratios, and by correlations with extract viscosities (Campbell et al 1989).

The reduction in feed efficiency caused by oat β-glucan was recently directly demonstrated by addition to feed containing hulless oats (Table VII). The effect of oat or barley β-glucan on chick growth is not entirely predictable on the basis of β-glucan concentration as seen in Table VII where the feed:gain ratio with the oat diet was not significantly different from the corn diet whereas in other experiments (Cave et al 1990) feed efficiency declined significantly with increasing levels of naked oats in the diet. As in barley (Hesselmann et al 1981) this variability probably relates to endogenous enzymes from contaminating microorganisms. Such enzyme activity might increase solubilization, and hence viscosity, but subsequent depolymerization would decrease viscosity. Furthermore, microorganism load will vary with sample and storage time. Effects are therefore difficult to assess or predict.

Nutritional effects of insoluble non-starch polysaccharides in humans

Insoluble non-starch polysaccharides of cereals, as exemplified in wheat bran, improve laxation in humans, producing an increased stool bulk and frequency, and decreased transit time (Jenkins et al 1985, Eastwood

Table VII
Effects of dietary oat β-glucan on performance of male broiler chicks, 8-20 days of age[a,b]

	Corn-Soy	Naked Oats[c]	Naked Oats[d] + Oat Gum	Oat Bran[e]
			Diet	
Feed intake (g)	478a	435a	377b	307c
Weight gain (g)	342a	295b	211c	149d
Feed:gain	1.42a	1.49a	1.79b	2.38c
Mortality (%)	1	4.8	0	9.5
Excreta score[f]	1.5a	1.4a	1.7a	3.7b

[a] Data from Cave et al, 1990
[b] Means followed by different letters are significantly different (p<0.05) from corn-soy control
[c] 25 g/kg β-glucan
[d] 65 g/kg β-glucan
[e] 77 g/kg β-glucan; enzyme-deactivated bran
[f] Visual assessment at 14 days on scale 1-5; 1 dry and firm, and 5 wet and sloppy

1990, Stephen 1991). Fiber reaches the colon undigested where it is largely fermented by bacteria. The degree to which this occurs varies with fiber source. Soluble fiber is readily fermented. The insoluble fiber of wheat bran, however, partially resists fermentation and is one of the most effective fecal bulking agents known. Increased fecal bulk occurs from both undigested fiber residues (and associated water) and increased bacterial cell mass. When fermentation occurs, there may be further consequences from alterations in bacterial metabolites such as short chain fatty acids (Eastwood 1990, Stephen 1991). This is more a characteristic of soluble fiber (e.g. β-glucan) and may play a role in serum cholesterol reduction (Anderson et al 1990). Most studies indicate that insoluble dietary fiber does not influence serum cholesterol levels. Colonic effects, however, may provide protection against various diseases including colon cancer.

Nutritional effects of soluble non-starch polysaccharides in humans

Viscous polysaccharides such as guar and xanthan gum and pectin find many applications in the food industry, for example as thickeners, emulsion stabilizers and gelling agents. The commercial availability of these polysaccharides has enabled studies of the effects of the isolated products on glucose regulation (Jenkins et al 1978 and 1985) and serum cholesterol levels (Anderson et al 1990). Generally, a reduction in postprandial blood glucose and insulin levels and improvement in longer-term glucose regulation in diabetics has been reported. There appear to be associated improvements in serum cholesterol levels. Doses used in most of these studies are high (15-30 g/day) and not readily achieved with real foods.

Oat bran as a source of β-glucan - physiological effects

Inclusion of oat bran or rolled oats in significant amounts in the daily diet has generally been shown to lower serum cholesterol (Anderson et al 1990). A study by Swain et al (1990) made very public a controversy over whether this is a specific effect attributable to oats or simply a consequence of accompanying dietary modification, most particularly altered fat intake. Although to the consumer the basis for effect may not be of concern, this question is of crucial importance scientifically. Resolution of the controversy requires a better understanding of mechanism.

It seems to be more generally accepted, even by those who suggest that oat bran has no specific cholesterol lowering effect (Sacks 1991), that soluble viscous polysaccharide supplements can reduce serum cholesterol levels significantly. Since the effect, if any, of oat bran has been attributed to the viscous polysaccharide, β-glucan, (Anderson and Siesel 1990) it is worthwhile considering whether oat products are capable of delivering significant doses of this component. Unfortunately, many studies have been done without reporting the β-glucan or soluble fiber concentration. Good quality commercial bran normally contains 7-9% β-glucan and the minimum suggested by the American Association of Cereal Chemists was 5.5% (Anon 1989). In our experience some unfractionated oat cultivars exceed this level and with

Table VIII
Summary of data from dry and wet-milling (Burrows et al 1984) of 11 oat cultivars (From Wood et al 1991a and unpublished data)

Analysis	Mean ± SD[a]	Range
β-Glucan (%)[b]		
Groat	5.1 ± 0.8	3.9 – 6.8
Dry Mill, Bran	7.4 ± 1.0	5.8 – 8.7
Wet Mill, Bran[c]	16.7 ± 1.6	14.7 – 19.1
Bran Yield (%)		
Dry Mill	53 ± 4	48 – 58
Wet Mill	18.5 ± 2.8	13.2 – 22.8
Enrichment Factor		
Dry Mill	1.5 ± 0.1	1.3 – 1.6
Wet Mill	3.3 ± 0.3	2.8 – 3.5

[a]Standard deviation
[b]Dry weight basis, determined by method of McCleary and Glennie Holmes (1985)
[c]Wet-milled bran provided by D. Paton

about 50% extraction by sieving, brans containing 5.8%-8.7% β-glucan were readily obtained (Table VIII). Wet milling techniques achieved levels of 15-19% β-glucan. Oat bran is thus a potentially very rich source of β-glucan.

Oat gum, containing 80% β-glucan (Wood et al 1989) reduced serum cholesterol in rats 30-48% relative to a cellulose or wheat bran control. The effect was linearly dose dependant (Jennings et al 1988, Anderson and Siesel 1990). Oat β-glucan also reduced serum cholesterol in chicks (Welch et al 1988).

In clinical studies, oat gum lowered the postprandial glucose and insulin response to an oral glucose load similarly to guar gum (Braaten et al 1991). Reducing the viscosity either by lowering the dose or by partial acid hydrolysis reduced the effectiveness of the gum (Wood et al unpublished). Studies to determine the effect of oat β-glucan on serum cholesterol in humans are in progress.

CONCLUSIONS

Cereals are potentially rich sources of soluble and insoluble NSP. This reduces the value of the cereal as animal feed but has nutritional benefits for humans. Wheat bran is an excellent source of insoluble NSP, mainly pentosan and cellulose, and improves laxation, with attendant health benefits. Oats and barley, on the other hand, are sources of β-glucan, a soluble dietary fiber component. The β-glucan concentration in some barley cultivars exceeds 10%, and in experimental oat brans reaches 19%. Oat β-glucan showed effects on glucose metabolism similar to other soluble fibers such as guar gum and may contribute to oat bran's ability to reduce serum cholesterol. Rye is a good source of soluble pentosan and also contains β-glucan.

LITERATURE CITED

Åman, P. 1987. The variation in chemical composition of Swedish oats. Acta. Agric. Scand. 37:347-352.

Åman, P. and Graham, H. 1987. Analysis of total and insoluble mixed-linked (1→3)(1→4)-β-D-glucans in barley and oats. J. Agric. Food Chem. 35: 704-709.

Åman, P. and Hesselman, K. 1984. Analysis of starch and other main components of cereal grains. Swedish J. Agric. Res. 14:135-139.

Åman, P. and Hesselman, K. 1985. An enzymic method for analysis of total mixed-linkage β-glucans in cereal grains. J. Cereal Sci. 3:231-237.

Anderson, J.W. and Siesel, A.E. 1990. Hypocholesterol-emic effects of oat products. Pages 17-36 in: New Developments in Dietary Fiber. Advances in Experimental Medicine and Biology, Vol. 270. I. Furda and C.J. Brine, eds. Plenum Press, New York and London.

Anderson, J.W., Deakins, D.A., Floore, T.L., Smith, B.M. and Whitis, S.E. 1990. Dietary fiber and coronary heart disease. Crit. Rev. Food Sci. Nutr. 29:95-147.

Anderson, M.A., Cook, J.A., and Stone, B.A. 1978. Enzymatic determination of 1,3:1,4-β-glucans in barley grain and other cereals. J. Inst. Brew. 84:233-239.

Anon. 1989. AACC committee adopts oat bran definition. Cereal Foods World 34:1033.

Antoniou, T., Marquardt, R.R. and Cansfield, P.E. 1981. Isolation, partial characterization and antinutritional activity of a factor (pentosans) in rye grain. J. Agric. Food Chem. 29:1240-1247.

Bengtsson, S., Åman, P., Graham, H., Newman, C.W. and Newman, R.K. 1990. Chemical studies on mixed linked β-glucans in hulless barley cultivars giving different hypocholesterolemic responses in chickens. J. Sci. Food Agric. 52:435-445.

Beresford, G., and Stone, B.A. 1983. (1→3)(1→4)-β-D-Glucan content of Triticum grains. J. Cereal Sci. 1:111-114.

Blakeney, A.B., Harris, P.J., Henry, R.J. and Stone, B.A. 1983. A simple and rapid preparation of alditol acetates for monosaccharide analysis. Carbohydr. Res. 113:291-299.

Blumenkrantz, N. and Absoe-Hansen, G. 1973. New method for quantitative determination of uronic acids. Anal. Biochem. 54:484-489.

Braaten, J.T., Wood, P.J., Scott, F.W., Riedel, K.D., Poste, L.M. and Collins, M.W. 1991. Oat gum, a soluble fibre which lowers glucose and insulin in normal individuals after an oral glucose load: comparison with guar gum. Am. J. Clin. Nutr. 53:1425-1430.

Burrows, V.D., Fulcher, R.G. and Paton, D. 1984. Processing aqueous treated cereals. U.S. Patent 4,345,429.

Campbell, G.L., Rossnagel, B.G., Classen, H.L. and Thacker, P.A. 1989. Genotypic and environmental differences in extract viscosity of barley and their relationship to its nutritive value for broiler chickens. Anim. Feed Sci. Technol. 26:221-230.

Cave, N.A., Wood, P.J. and Burrows, V.D. 1990. The nutritive value of naked oats for broiler chicks as affected by dietary additions of oat gum, enzyme, antibiotic, bile salt and fat-soluble vitamins. Can. J. Anim. Sci. 70:623-633.

Delcour, J.A., Vanhamel, S. and De Geest, C. 1989. Physicochemical and functional properties of rye nonstarch polysaccharides. I. Colorimetric analysis of pentosans and their relative monosaccharide compositions in fractionated (milled) rye products. Cereal Chem. 66:107-111.

Douglas, S.G. 1981. A rapid method for the determination of pentosan in wheat flour. Food Chem. 7:139-145.

Eastwood, M. 1990. Function of dietary fibre in the large intestine. Pages 211-221 in: Dietary Fibre: Chemical and Biological Aspects. Southgate, D.A.T., Waldron K., Johnson, I.T. and Fenwick, G.R., eds. Royal Soc. Chem., Cambridge, UK.

Englyst, H.N. and Cummings, J.H. 1984. Simplified method for the measurement of total non-starch polysaccharides by gas-liquid chromatography of constituent sugars as alditol acetates. Analyst 109:937-942.

Englyst, H.N., Bingham, S.A., Runswick, S.A., Collinson, E. and Cummings, J.H. 1989. Dietary fibre (non-starch polysaccharides) in cereal products. J. Hum. Nutr. Diet. 2:253-271.

Fincher, G.B. and Stone, B.A. 1986. Cell walls and their components in Cereal Grain Technology. Pages 207-295 in: Advances in Cereal Science and Technology, VIII. Y. Pomeranz, eds. Am. Assoc. Cereal Chem., St. Paul, MN.

Frølich, W. and Nyman, M. 1988. Minerals, phytate and dietary fibre in different fractions of oat grain. J. Cereal Sci. 7:73-82.

Hashimoto, S., Shogren, M.D. and Pomeranz, Y. 1987a. Cereal pentosans: their estimation and significance. I. Pentosans in wheat and wheat milled products. Cereal Chem. 64:30-34.

Hashimoto, S., Shogren, M.D., Bolte, L.C. and Pomeranz, Y. 1987b. Cereal pentosans: their estimation and significance. III. Pentosans in abraded grains and milling by-products. Cereal Chem. 64:39-41.

Henry, R.J. 1985. A comparison of the non-starch carbohydrates in cereal grains. J. Sci. Food Agric. 36:1243-1253.

Hesselman, K., Elwinger, K., Nilsson, M. and Thomke, S. 1981. The effect of β-glucanase supplementation, stage of ripeness and storage treatment of barley in diets fed to broiler chickens. Poultry Sci. 60:2664-2671.

Hidaka, H., Hirayama, M., Tokunaga, T., and Eida, T. 1990. The effects of undigestible fructooligosaccharides on intestinal microflora and various physiological functions on human health. Pages 105-177 in: New Developments in Dietary Fiber. Advances in Experimental Medicine and Biology, Vol. 270. I. Furda and C.J. Brine, eds. Plenum Press, New York and London.

Jenkins, D.J.A., Bright-See, E., Gibson, R., Josse, R.G., Kritchevsky, D., Peterson, R.D. and Rasper, V.F. 1985. Report of the Expert Advisory Committee on Dietary Fibre, published by Health Protection Branch, Health and Welfare Canada.

Jenkins, D.J.A., Wolever, T.M.S., Leeds, A.R., Gassull, M.A., Haisman, P., Dilawari, J. Goff, D.V., Metz, G.L. and Alberti, K.G.M.M. 1978. Dietary fibers, fiber analogues, and glucose tolerance: Importance of viscosity. Br. Med. J. 1:1392-1394.

Jennings, C.D., Boleyn, K., Bridges, S.R., Wood, P.J. and Anderson, J.W. 1988. A comparison of the lipid-lowering and intestinal morphological effects of cholestyramine, chitosan and oat gum in rats. Proc. Soc. Exp. Biol. Med. 189:13-20.

Jørgensen, K.G. 1988. Quantification of high molecular weight $(1\to3)(1\to4)$-β-D-glucan using calcofluor complex formation and flow injection analysis. 1. Analytical principle and its standardisation. Carlsberg Res. Commun. 53:277-285.

Juliano, B.O. 1985. Polysaccharides, protein and lipids of rice. Pages 59-174 in: Rice Chemistry and Technology (2nd Ed.). B.O. Juliano, ed. Am. Assoc. Cereal Chem., St. Paul, MN.

Lehtonen, M., AND Aikasalo, R. 1987. Pentosan in barley varieties. Cereal Chem. 64:133-134.

Marlett, J.A. 1991. Dietary fibre content and effect of processing on two barley varieties. Cereal Foods World 36:576-578.

McCleary, B.V. and Glennie-Holmes, M. 1985. Enzymic quantification of $(1\to3)(1\to4)$-β-D-glucan in barley and malt. J. Inst. Brew. 91:285-295.

Michniewicz, J., Biliaderis, C.G. and Bushuk, W. 1990. Water-insoluble pentosans of wheat: composition and some physical properties. Cereal Chem. 67:434-439.

Mongeau, R. and Brassard, R. 1990. Determination of soluble, insoluble and total dietary fiber in oats, groats, oat flour and oat bran by different methods. 75th Annual Am. Assoc. Cereal Chem. Meeting, Oct. 14-18. Abstract 83.

Neilson, M.J. and Marlett, J.A. 1983. A comparison between detergent and non-detergent analyses of dietary fiber in human foodstuffs, using high performance liquid chromatography to measure neutral sugar composition. J. Agric. Food Chem. 31:1342-1347.

Nilsson, U., Öste, R., Jagerstad, M. and Birkhed, D. 1988. Cereal fructans: in vitro and in vivo studies on availability in rats and humans. J. Nutr. 118:1325-1330.

Nyman, M., Siljeström, M., Pedersen, B., Bach-Knudsen, K.E., Asp, N.-G., Johansson, C.-G. and Eggum, B.O. 1984. Dietary fiber content and composition in six cereals at different extraction rates. Cereal Chem. 61:14-19.

Prosky, L., Asp, N-G., Furda, I., Devries, J., Schweizer, T.F. and Harland, B.F. 1984. Determination of total dietary fiber in foods, food products, and total diets: interlaboratory study. J. Assoc. Off. Anal. Chem. 67:1044-1052.

Roberts, E.J., Godshall, M.A., Clarke, M.A., Tsang, W.S.C. and Parrish, F.W. 1987. Methanolysis of polysaccharides: a new method. Carbohydr. Res. 168:103-109.

Saastamoinen, M., Plaami, S. and Kumpulaimen, J. 1989. Pentosan and β-glucan content of Finnish Winter rye varieties as compared with rye of six other countries. J. Cereal Sci. 10:199-207.

Sacks, F.M. 1991. The role of cereals, fats, and fibers in preventing coronary heart disease. Cereal Foods World 36:822-826.

Saini, H.S. and Henry, R.J. 1989. Fractionation and evaluation of triticale pentosans: comparison with wheat and rye. Cereal Chem. 66:11-14.

Salomonsson, A.-C., Theander, O. and Westerlund, E. 1984. Chemical characterization of some Swedish cereal whole meal and bran fractions. Swedish J. Agric. Res. 14:111-117.

Sawardeker, J.S., Sloneker, J.H. and Jeanes, A. 1965. Quantitative determination of monosaccharides as their alditol acetates by gas liquid chromatography. Anal. Chem. 37:1602-1604.

Shinnick, F.L., Longacre, M.J., Ink, S.L. and Marlett, J.A. 1988. Oat Fiber: Composition versus Physiological Function in Rats. J. Nutr. 118:144-151.

Sibbald, I.R. 1977. The true metabolisable energy values of some feedingstuffs. Poultry Sci. 56:380-382.

Sibbald, I.R. 1986. The T.M.E. system of feed evaluation: methodology, feed composition data and bibliography. Technical Bulletin 1986-4E, Ministry of Supply and Services Canada. Available from Centre for Food and Animal Research, Agriculture Canada, Ottawa, ON K1A 0C6.

Stephen, A.M. 1991. Starch and dietary fibre: their physiological and epidemiological interrelationships. Can. J. Physiol. Pharmacol. 69:116-120.

Swain, J.F., Rouse, I.L., Curley, C.B. and Sacks, F.M. 1990. Comparison of the effects of oat bran and low-fiber wheat on serum lipoprotein levels and blood pressure. N. Engl. J. Med. 322:147-152.

Theander, O. and Åman, P. 1982. Studies on dietary fibre. A method for the analysis and chemical characterisation of total dietary fibre. J. Sci. Food Agric. 33:340-344.

Ulrich, S.E. and Clancy, T.A. 1991. Analysis of β-glucans in barley and malt: a comparison of four methods. J. Am. Soc. Brew Chem. 49(3):110-115.

Varo, P., Laine, R., Veijalainen, K., Pero, K. and Koivistoinen, P. 1984. Dietary fibre and available carbohydrates in Finnish cereal products. J. Agric. Sci. Fin. 56:39-48.

Welch, R.W., Peterson, D.M. and Schramka, B. 1988. Hypocholesterolemic and gastrointestinal effects of oat bran fractions in chicks. Nutr. Rep. Int. 38:551-561.

Wood, P.J. 1980. Specificity in the interaction of direct dyes with polysaccharides. Carbohydr. Res. 85:271-287.

Wood, P.J. 1985. Dye-polysaccharide interactions - recent research and applications. Page 267 in: New

Approaches to Research on Cereal Carbohydrates. R.D. Hill and L. Munck, eds. Elsevier, Amsterdam.

Wood, P.J., Siddiqui, I.R. and Weisz, J. 1975. The use of butaneboronic esters in the gas-liquid chromatography of some carbohydrates. Carbohydr. Res. 42:1-13.

Wood, P.J., Weisz, J. and Blackwell, B.A. 1991b. Molecular characterization of cereal β-D-glucans. Structural analysis of oat β-D-glucan and rapid structural evaluation of β-D-glucans from different sources by high performance liquid chromatography of oligosaccharides released by lichenase. Cereal Chem. 68:31-39.

Wood, P.J., Weisz, J., and Fedec, P. 1991a. Potential for β-glucan enrichment in brans derived from oat (<u>Avena sativa</u> L.) cultivars of different (1→3)(1→4)-β-D-glucan concentration. Cereal Chem. 68:48-51.

Wood, P.J., Weisz, J., Fedec, P. and Burrows, V.D. 1989. Large scale preparation and properties of oat fractions enriched in (1→3)(1→4)-β-D-glucan. Cereal Chem. 66:97-103.

STRUCTURE AND FUNCTIONALITY
OF CARBOHYDRATE HYDROCOLLOIDS
IN FOOD SYSTEMS

Donald B. Thompson

Department of Food Science
Penn State University
University Park, PA 16802

INTRODUCTION

One approach to the topic of carbohydrate hydrocolloid functionality is to list the various ingredients of interest, and for each to describe the properties that the ingredient can impart to various types of food systems. A second approach is to list the various types of food systems of interest, and for each to describe the ingredients that can impart desired properties. Such descriptive approaches are justified based on what has traditionally been empirical knowledge concerning the practical uses of carbohydrate hydrocolloids. Although much specific knowledge may be gained from a descriptive listing, it is difficult for the reader to grasp the body of the information; a good index to allow ready access to specific details may be the most important attribute of such material. In the present chapter, emphasis is on presentation of concepts to increase the reader's fundamental understanding of the behavior of carbohydrate hydrocolloid ingredients. No apology is made for lack of specific detail regarding ingredient usage in foods; the reader is urged to seek such information from the industrial technical literature or from published reviews, including those contained in Glicksman (1982; 1983; and 1986). Excellent overviews of hydrocolloid use in foods are given by Glicksman (1984) and Walker (1984). Information regarding the use of hydrocolloids as found in specific commercial retail products is presented in Morley (1984).

CARBOHYDRATE HYDROCOLLOIDS

A colloidal particle was first distinguished from a molecule in true solution using a parchment membrane; only the molecules in true solution could pass through the membrane. An arbitrary and traditional dividing line according to particle size is approximately 1 nm (Dickinson and Stainsby, 1982). On this basis, glucose molecules (~0.5 nm) dissolve in water to produce a true solution. The upper size limit for a colloidal particle is based on the notion that the particle is much larger than the molecules in which it is dispersed, and yet it is still small enough that Brownian motion can overcome the force of gravity. An arbitrary upper limit on size of 1000 nm is sometimes described (Dickinson and Stainsby, 1982; Everett, 1988). The colloidal particle is large enough to be considered a separate solid phase, at least in some respects.

A hydrocolloid may simply be defined as a colloidal particle capable of being dispersed in an aqueous system. Many globular proteins are in the range of 4-7 nm. Although it is common usage to refer to a solution of protein, the mixture is more appropriately considered a colloidal dispersion due to the size of these molecules.

The most common non-protein hydrocolloids are polysaccharides, which for the most part may be considered carbohydrates. In nature these molecules tend to function eithet as energy storage compounds or as structural or protective elements. Starch and galactomannans are examples of energy storage molecules, and cellulose and pectins are well-known examples of structurally important polysaccharides. It is likely that we presently understand only a small fraction of the complex interactions among polymers in nature, interactions that might well be useful in food systems. The behavior of individual carbohydrate polymers and their interactions in food systems is the focus of this chapter.

In a dispersion there is more than one phase. Many foods are complex mixtures involving more than one type of dispersion. A discontinuous (or dispersed) solid phase in a continuous liquid (often aqueous) phase is a sol. A simple example of a sol is a globular protein dispersed in water. A dispersed carbohydrate hydrocolloid molecule may constitute all or part of the solid phase of a food sol, as for example in unclarified apple juice. A discontinuous liquid phase in a second liquid phase is an emulsion. In food systems one of these phases is lipid and the other aqueous. A carbohydrate hydrocolloid molecule suspended in the aqueous phase can often stabilize the emulsion, as for xanthan or alginate in pourable salad dressings. A discontinuous gas phase dispersed in a liquid phase is a foam, and again a carbohydrate hydrocolloid molecule in the aqueous phase can stabilize the foam; in fact, the hydrocolloid may even be necessary to generate the foam, as when guar gum is added to physically damaged egg white to allow foaming.

Another type of dispersion of importance in foods is a gel, for which two continuous phases may be described: usually an aqueous phase and a solid phase. The continuous solid phase is produced when hydrocolloid molecules interact to form a network. The mechanisms for formation of junction zones to link the hydrocolloid molecules vary. An understanding of these mechanisms is important to controlling the potential sol-gel-sol transformations in food systems.

POLYSACCHARIDE STRUCTURAL FEATURES

Excellent detailed overviews of polysaccharide structure are available (see Rees, 1977; Blanshard and Mitchell (eds.), 1979; MacGregor and Greenwood, 1980; Rees et al., 1982; and Dea, 1982).

Since a polysaccharide is composed of saccharide units, the first consideration in describing the structure of the polysaccharide is the component unit(s). If all units are identical, the molecule is a homopolysaccharide; otherwise it is a heteropolysaccharide. Other than the amylose and amylopectin of starch, most carbohydrate hydrocolloids in commercial use are heteropolysaccharides.

The ring conformation of the constituent saccharide units makes a critical contribution to polysaccharide structure. For glucose units, the ring conformation is thought to be most stable as the $^4C_1$, or C1 (Rees et al., 1982), for which the OH groups at positions 2, 3, and 4 and the CH_3OH at position 5 are all attached equatorially (see Figure 1). For glucose units the anomeric carbon is carbon 1. The original anomeric OH group may have been either α or β. The β anomer of D-glucose has an equatorial OH. Since the saccharide units are always linked to others through the anomeric OH (except for the single saccharide unit at the reducing terminus), the anomeric designation is always important to describe the backbone structure of the polysaccharide. Based on the assumption that the ring conformation is fixed (an assumption adjusted by French et al., 1991), one need only determine the carbon to which the other glycosidic O is attached to define the structure of a linear homopolymer with identical linkages. (If more than one non-anomeric OH groups are involved, the monosaccharide unit represents a branch point.) Because the ring conformation is considered fixed, of the bonds involved in the polysaccharide backbone, only the two glycosidic C-O bonds may rotate. Calculation of potential energy as these two bonds are rotated leads to

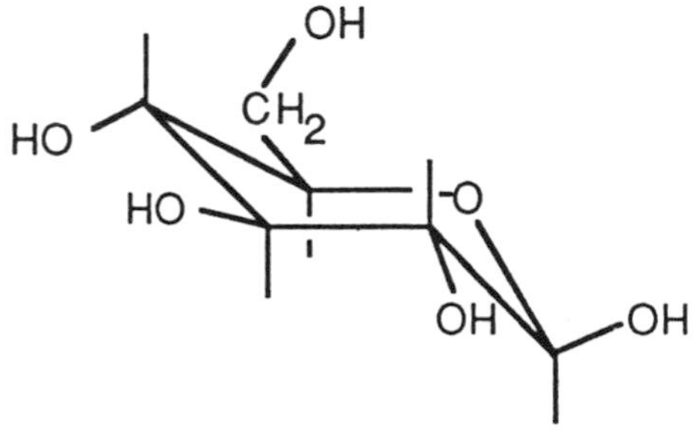

Figure 1. The $^4C_1$ conformation of β-D-glucopyranose.

conformational energy maps, which have been used to predict favored relative orientations of adjacent rings (Brant and Crist, 1991). These favored orientations have led to reasonable inferences regarding conformations of the analogous polysaccharide molecules (Brant and Crist, 1991).

Variation in the above structural features is a particularly vexing problem. For example, if the molecule is a heteropolysaccharide, how are the constituent units arranged? They may alternate in a regular manner; they may be found in homopolymeric blocks; or they may be arranged totally randomly. This arrangement is a critical determinant of the structure and consequently of the properties of galactomannans (McCleary et al., 1985), pectins (de Vries et al., 1984; Roumbouts and Thibault, 1986), and alginates (Gacesa, 1988). To further complicate the situation, there is no reason to assume that all the molecules in a "purified" sample follow the same pattern. Variation in the carbons involved in the glycosidic linkages can be of several types: the linkages may alternate regularly, they may be arranged in repeating blocks, or they may be entirely random. Determining the location of branch points relative to each other can be an especially subtle problem, as for amylopectin fine structure (Manners, 1989).

As if these complications were not enough, polysaccharide molecular weights also vary. For a high-molecular weight linear molecule there is no mechanism for completing synthesis at a precise molecular weight. Unlike a specific protein, which is made from a template molecule which in effect defines the molecular weight precisely, polysaccharide molecular weight will always be best described in statistical terms as an average for the particular sample. Even if all molecules were identical, determination of the molecular weight would be non-trivial. For higher molecular weight samples, chemical analysis is generally not an effective way to determine molecular weight, which by chemical analysis would have been in terms of the most commonly appreciated expression, the "number average" molecular weight. Instead, methods such as light scattering and viscosity are often used to obtain "weight average" and "viscosity average" molecular weights, respectively . Unless the sample is mondisperse, these values for molecular weight will not agree (MacGregor and Greenwood, 1980).

The conformation of a polysacharide in solution will be determined by the above structural features as well as by the characteristics of the aqueous phase. Since the behavior of the polysaccharide in a food system will often reflect its conformation in solution, solvent considerations may profoundly influence the behavior of the food system. For a charged polysaccharide such as alginate, the pH, ionic strength, and presence of specific cations all can be critical to the behavior of the system containing this polysaccharide. Furthermore, the presence of other charged polymers may be important as well.

The conformation and shape of a polysaccharide may result from varying degrees of short-term or long-term order. At one extreme is the random coil arrangement leading to a more-or-less spherical shape if the polymer is sufficiently large. At the other extreme is a fixed helical molecule, which may approach being an extended rod. The physical behavior of molecules with these conformations is vastly different. Thus structural features and solvent considerations are both important to understand the behavior of food ingredients.

HOMOPOLYMERS AND POLYMERS WITH HOMOPOLYMERIC REGIONS

For linear homopolymers, ordered structures, if they occur, will be as some form of helix. The most energetically favorable ordered conformation of the molecule will be determined by the most favored rotation about the two glycosidic C-O bonds, ϕ and ψ. For some molecules the helix will be fully extended, and for others the helix will be more coiled. Cellulose is an example of a molecule capable of forming a highly extended ribbon structure in which the adjacent glucose moieties are in identical ring conformations, but rotation around the glycosidic bonds is such that each ring is flipped 180 degrees about the axis of the extended chain (see Figure 2). An unsubstituted mannan polymer has a nearly identical structure to

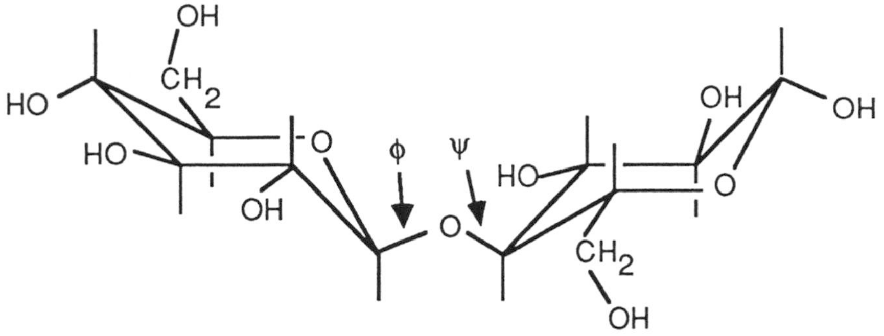

Figure 2. The structure of cellobiose, the repeating disaccharide of cellulose. The glycosidic bonds with allowed rotation are indicated.

cellulose except that the equatorial OH at C2 of the glucose moieties of cellulose is axial for the mannan. The behavior of these two polysaccharides is very similar, based on the similarity of the backbone structure. Both form highly ordered aggregates not dispersible in water. Chitin, with the identical backbone as cellulose, has similar physical properties despite the N-acetyl group at C2 (Muzzarelli, 1985).

Cellulose occurs in nature as partially crystalline fibers, in which crystalline micellar regions alternate with amorphous regions (Marchessault and Sundararajan, 1983). One type of cellulosic ingredient is made from simple purification of this material and grinding to the appropriate mesh size (Solka Floc®). Another type is made by homogenizing the purified material (Nutracel®). When the purified cellulose is subjected to acid, the glycosidic linkages in the amorphous regions are hydrolyzed, leaving the crystalline regions, producing a third type of cellulose ingredient (Avicel®). In each of these cases the cellulose ingredient functions to provide particles in a coarse suspension, and the interaction of these particles with water and other ingredients may provide interesting functional properties to a food. Furthermore, the cellulose is not hydrolyzed by mammalian enzymes, and thus is essentially non-caloric. As suspensions, these materials are often used at much higher levels than are the readily dispersible hydrocolloids.

Because the glycosidic linkages of amylose are α, one of the glycosidic bonds is axial (see Figure 3). Consequently, a fully extended ribbon structure is not energetically favorable, and a more compact helix

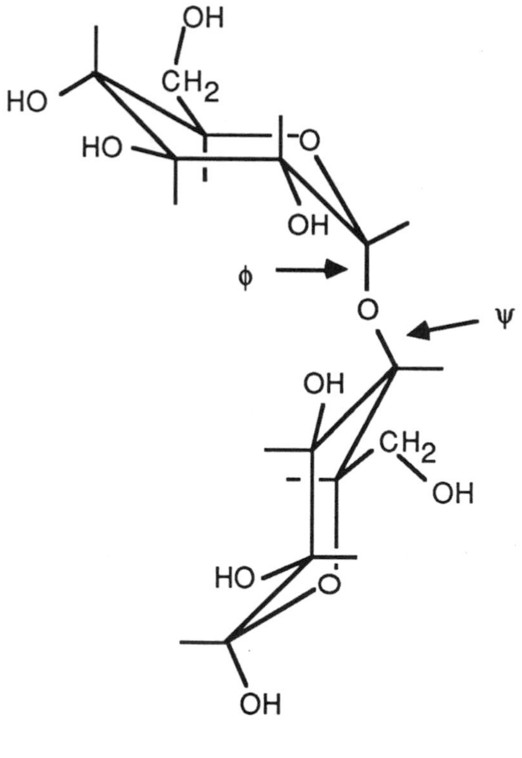

Figure 3. The structure of maltose, the repeating unit of amylose.

describes the energetically favored ordered structure. One type of helix is termed the V-helix, after the characteristic x-ray diffraction pattern. This helix forms as an inclusion complex, in which the helical amylose surrounds an appropriate guest species. The familiar blue color of starch and iodine reagent is understood as an amylose inclusion complex. The same sort of complex can form with monoacyl lipid molecules in food, with profound effect on starch physical behavior (Galloway et al., 1989). A second favored helical structure for amylose is a double helix involving two molecules of amylose. This sort of structure is important in retrogradation phenomena, both for amylose and for amylopectin (Morris, 1990).

Homopolymeric regions of β-L-guluronate in alginate and of α-D-galacturonate in demethoxylated pectin result in an even more constrained type of helix, termed a "buckled ribbon" (Rees et al., 1982). The anomeric C-O linkage of β-L-guluronate is axial (the β anomer of an L sugar being analogous to the α anomer of a D sugar), as is the C-O bond at C4. When an ordered structure occurs, the buckled structure is stabilized by calcium ions, which connect the negative charges in two folded helices to form an "egg box" complex. (This analogy may be more correctly considered as calcium "eggs" which have cracked and leaked, the dried egg cementing the box together.) The ability of alginate to gel on addition of calcium ions may be explained by this complex. Although α-D-galacturonate of pectin and β-L-guluronate of alginate would seem very different constituent monomer units, Rees et al. (1982) have pointed out that the homopolymer units are very nearly mirror images (see Figure 4). As a result, the ordered structures induced by calcium are very similar. Because alginate polymannuronate homopolymeric regions have the same backbone structure as cellulose, they are unable to form the stable egg box structure. The calcium-induced gelation of low-methoxyl pectin has been used to produce low-sugar jams and jellies.

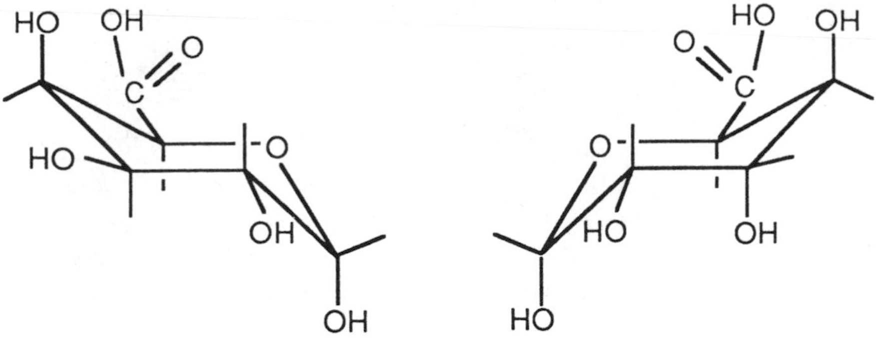

Figure 4. The structures of α-D-galacturonic acid (left) and β-L-guluronic acid (right), monomer units responsible for the homopolymeric egg box regions in pectin and alginate, respectively.

SUBSTITUTED HOMOPOLYSACCHARIDES

The properties of a homopolysaccharide mannan (mentioned above as an analogue to cellulose) may be altered by substitution of galactose moieties at C6 of the mannan backbone, forming galactomannans. The two best known examples of galactomannans are locust bean gum (carob) and guar gum. Both represent the primary storage polysaccharides of the respective seed endosperms. These molecules differ according to the extent to which galactose moieties are attached to the mannan chain (McCleary et al., 1985). Although an ordered structure is still possible for the chain backbone, the association of adjacent chains is less favored as the degree of substitution increases. The less-substituted molecule, locust bean gum, is not dispersible in cold water but is dispersible in hot water, whereas the more substituted molecule, guar gum, is dispersible in hot or cold water. These differences have been rationalized on the basis of a greater tendency for the less substituted mannan to form the ordered mannan backbone. These ordered regions are thought to associate, creating a kinetic barrier to water dispersibility (Rees et al., 1982). Whether the galactose substitution is random or according to a pattern can also influence galactomannan physical behavior, and this question has been explored in detail for these molecules (McCleary et al., 1985). Once dispersed, both tend to be used to contribute viscosity to aqueous food systems.

A more complex substituted homopolysaccharide is xanthan, which has the cellulosic backbone structure. In this case, however, the substitution pattern is so extensive that it normally dominates the steric configuration of the cellulosic backbone. The pentasaccharide substituent is attached to C6 of every glucose unit; the result is a stable helix with bulky side chains projecting outward. The extended helix may result in a molecule with a rigid, rod-shaped nature, or it may be part of a double helix (Rees et al., 1982). At rest, a dilute suspension of xanthan sets up a weak gel structure due to the associated rod-like segments or double helical regions (Shatwell et al., 1991). However, when shear forces are applied, such as pumping or shaking or pouring, the structure is quickly lost, and extreme shear-thinning behavior results. When the molecule is heated to above the helix-coil transition temperature, a random coil structure results, and the extreme shear-thinning behavior is lost (Dea, 1982).

DERIVATIZED HOMOPOLYSACCHARIDES

Analogous to the substituted homopolysaccharides are the derivatized homopolysaccharides, the most important types in food use being the modified celluloses. Just as the properties of the cellulosic-type backbone can be altered by saccharide substitution, they may also be altered by derivatization of the glucose moieties. After dispersion of cellulose chains in NaOH to allow reagents access to the individual chains and to prepare the OH groups for reaction, the methyl group of acetate may be ether-linked to the glucose moieties of the chain (Batdorf and Rossman, 1973). The resulting carboxymethylcellulose is a highly charged polymer (usually provided in the sodium salt form) at neutral pH due to the ionized carboxylic acid groups. The physical properties of the molecule will depend upon the degree of carboxymethyl substitution and the length of the cellulosic chains. The chain will have an extended (but not rigid) random coil conformation generating high viscosity at low usage levels. The charged nature of the molecule also dictates that the viscosity will be sensitive to pH and ionic strength.

Cellulose may be derivatized to produce uncharged polymers as well: for example, methylcellulose and hydroxypropylmethylcellulose. These materials are capable of forming gels with heat (by a poorly understood mechanism). What is remarkable is that the initial sol is recovered upon cooling (Glicksman, 1969). This unusual behavior has been suggested to be of use in reducing fat uptake during frying. Differences in degree of substitution and chain length can result in differences in viscosity behavior, and in differences in thermal gelation temperature and gel properties.

COPOLYSACCHARIDES

A regular, ordered structure may be achieved in a copolysaccharide when the arrangement of the saccharide units is itself regular. The simplest case is of two alternating saccharide units, which together constitute the repeating unit of the polymer. The seaweed polysaccharides agar and carrageenan are both based on the predominately regular repeating structure of a β-galactose derivative linked 1-4 to an α-galactose derivative: the disaccharide units are linked 1-3 (see Painter,1983, for full details of the structural complexity). The energetically favored ordered structure which forms under appropriate conditions is a helix, which can associate with another chain to form a double helix. The α-galactose derivative of agar is peculiar in two respects: it is an L-sugar, and the ring contains a 3,6-anhydro linkage. The anhydro linkage forces the following ring conformation: the OH's at positions 2 and 3 are axial, as is the CH_3OH attached at C5, yet the OH's at positions 1 and 4 are equatorial (see Figure 5). Consequently, all glycosidic linkages in this disaccharide repeating structure are equatorial relative to the ring conformation.

Unlike in agar, the α-galactose derivative of kappa-
and iota-carrageenan is in the D form (Rees et al., 1982).
Nevertheless, a 3,6-anhydro linkage again forces the OH's
at positions 2 and 3 to be axial, as well as the CH_3OH at
C5, and the OH's at positions 1 and 4 are equatorial (see
Figure 5). Thus the same helical backbone structures form

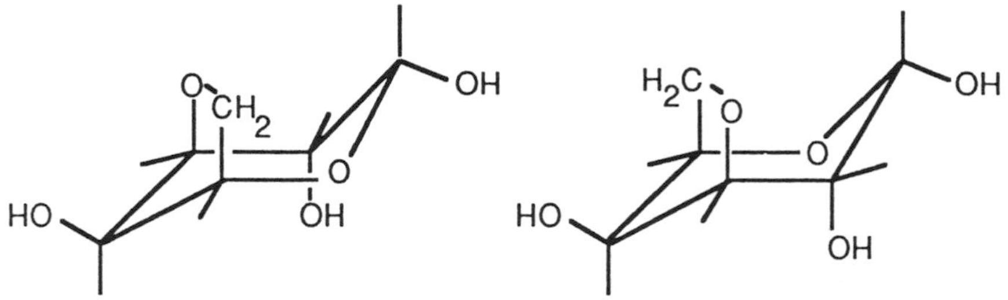

Figure 5. The structures of 3,6-anhydro-α-L-galactose
(left) and 3,6-anhydro α-D-galactose (right), derivatives
of which occur in agar and carrageenans, respectively.
For clarity, sulfate esters are not shown. Note that all
OH's at positions 1 and 4 are equatorial.

in agar and in kappa- and iota-carrageenans when an
ordered structure is energetically favorable. These helical
structures are a prerequisite to gel formation with these
polymers. The difference in the behaviors of these
polysaccharides is largely the extent to which the OH
groups have been converted to sulfate esters. Agar is
essentially uncharged, kappa-carrageenan has about one
sulfate group per two-saccharide repeating unit, and iota-
carrageenan has about two sulfate units per two-saccharide
repeating unit (Rees et al., 1982). (It is this difference in
sulfate levels that accounts for the cation-specific behavior
of the carrageenans.) Lambda-carrageenan differs from the
other carrageenans in that the α-galactose moieties contain
few anhydro linkages (Painter, 1983). Consequently, the
OH's at positions 1 and 4 are axial not equatorial, and the
ordered helical structure is not possible for the copolymer.
As a result, lambda-carrageenan will not gel, but simply
adds to viscosity (Rees et al., 1982).

The ability to form helices or double helices is not sufficient for stable gel formation. Rees et al. (1982) have pointed out that although gel formation requires sufficient lengths of ordered structures to allow association of these structures into junction zones, the order must be interrupted periodically to prevent excessive association of ordered chains and an eventual loss of gel structure. Disordered regions allow the junction zones to be connected in a stable way. In agar and the gelling carrageenans about 1 in 10 of the α-galactose units does not have an anhydro linkage, leading to an interruption in the ordered structure.

As negatively charged polymers, the carrageenans also have the ability to ionically interact with proteins, a property exploited in use with dairy products. The gel-forming ability of carrageenans has been recently exploited in a reduced-fat formulated ground beef used by a fast-food purveyor to make a reduced-fat hamburger.

A more complex case of a regular, repeating heteropolysaccharide structure is gellan gum, in which the repeating unit is a tetrasaccharide composed of glucose, glucuronate, glucose, and rhamnose. The double-helical ordered structure which can result accounts for its gelling ability (Morris, 1991).

STRUCTURALLY COMPLEX POLYSACCHARIDE GUMS

The polymers described above have generally been either linear or based on a well-defined linear backbone. Many of the exudate gums are much more complex in structure. In fact, structural details are largely lacking for gum ghatti and gum karaya, and are still incomplete for gum tragacanth (Glicksman, 1982). The latter exudate appears to be a mixture of at least two polymer types, an acidic polymer with a backbone of largely galacturonate and containing short side chains, and a neutral galactan polymer heavily substituted with arabinose. Although these polymers have a history of food use, they are expensive and have been replaced in most uses.

Gum arabic is an exudate that is still used extensively in foods as a result of some unusual physical properties. The exudate is harvested from several species of *Acacia*, with the chemical structure varying with species. The structure is based on a linear 1-3-linked galactan backbone, to which other linear 1-3-linked galactan chains are joined periodically at C6. The side chains also contain considerable arabinose as well as glucuronic acid and rhamnose (Clarke et al., 1979). In addition, there appears to be a small proportion of protein associated with the material.

The physical behavior of gum arabic is unusual in that highly concentrated (to ~45%, w/w) aqueous dispersions are possible. The viscosity of many other hydrocolloids increases dramatically around 1%, showing pronounced shear-thinning. Gum arabic viscosity is approximately Newtonian even at the higher concentrations (Glicksman, 1982). It appears that the highly branched structure allows dispersion with minimal viscosity, and yet the complex nature of the branches precludes their self-association. Gum arabic can function as an emulsifier for oil-in-water emulsions. It also has been used in flavor encapsulation, where its high dispersibility allows a more extensive barrier upon drying of the gum arabic/flavor mixture (Glicksman, 1982). Although its emulsifying activity is not understood, this behavior is probably related to its ability to interact with lipophilic flavor compounds in the manufacture of flavor emulsion concentrates.

Another polysaccharide having extensive dispersibility is larch gum, a arabinogalactan similar to gum arabic in terms of the galactan backbone and branching scheme. Clarke et al. (1979) have described larch gum, gum arabic, and the arabinogalactan of gum tragacanth as arabinogalactans with underlying similarity of structure.

Polydextrose® is another example of a structurally complex polysacharide gum. It is manufactured from glucose, citric acid, and sorbitol. The resulting structure is composed of glucose moieties linked through a variety of glycosidic bonds, with 1-6 links being most common. Citrate and sorbitol are bound to this structure as esters and ethers, respectively. As might be anticipated from the irregular structure, the polymer is amorphous and has little tendency to self-associate or to gel. Dispersions of up to 80% (w/w) in water can be prepared (Torres and Thomas, 1981).

OVERVIEW OF POLYSACCHARIDE MOLECULES IN FOOD SYSTEMS

Interactions with other Similar Polysaccharide Molecules

Viscosity
The most common reason for using hydrocolloid ingredients is to generate viscosity. Carbohydrate hydrocolloid dispersions rarely exhibit Newtonian behavior; instead viscosity is affected by shear stress. To describe viscosity behavior, flow curves, for which shear stress is plotted against shear rate, may be generated for several polymer concentrations. Most hydrocolloids show shear-thinning behavior, to an extent which varies with the hydrocolloid. The varying degrees of shear-thinning makes comparison of apparent viscosities (each determined from a single shear rate/shear stress point, assuming Newtonian behavior) inappropriate unless the particular shear stress is the condition of interest. However, at very low shear there is usually a Newtonian region of the flow curve. The zero-shear viscosity of a hydrocolloid dispersion pertains to this Newtonian region of its flow curve. A plot of this zero-shear viscosity versus concentration is shown in Figure 6.
For most polymers zero-shear viscosity increases with concentration relatively gradually to a point, above which the concentration has a more pronounced effect. Dilute and concentrated solutions have been distinguished according to the two regions of the curve. When the solution becomes sufficiently concentrated that molecules begin to compete for solvent and occupied volume, significant entanglement of polymers occurs, and the dependence of zero-shear viscosity on concentration increases, accounting for the discontinuity in the curve in Figure 6.
The ability of several types of polymers to generate viscosity varies widely, as shown in Table 1. In the absence

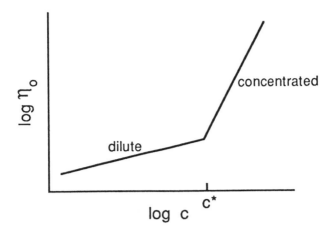

Figure 6. The effect of concentration on zero-shear
viscosity for a particular hydrocolloid. The critical
concentration separating dilute and concentrated
solution behavior is given by c*. When the abscissa is
expressed as log c[η], the behaviors of various
hydrocolloids are superimposable. Adapted from Morris
(1985).

Table 1. Comparative Carbohydrate Hydrocolloid
Apparent Viscosities.[a]

% (w/w)	guar gum	locust bean gum	gum tragacanth	gum arabic
1	4,500	100	54	
1.5	15,000	1,600		
2			906	
5			111,000	7
10				17
20				41
40				936

a. Viscosities in centipoise. Data from Sanford and Baird,
1983.

of other considerations, apparent viscosity information together with the price per pound may be useful in deciding which hydrocolloid ingredient to include in a formulation. However, one must use caution because other aspects of the physical behavior of the molecule may be important in a formulation (Szczesniak, 1986).

In mechanical spectroscopy, the effect of oscillatory shearing forces of differing frequencies is monitored . Using this approach the viscoelastic nature of a hydrocolloid dispersion may be separated into its viscous and elastic components, given by the loss modulus (G") and storage modulus (G'), respectively. Using such an approach, the behaviors of a concentrated dispersion and of a gel may be similar at high frequencies. Due to the difficulty of reorienting the dispersed hydrocolloid over short times, the concentrated dispersion takes on characteristics of a gel, in that the storage modulus may begin to exceed the loss modulus (Morris, 1989).

Gelation

The mechanism of gelation has been alluded to above as reflecting a balance between the order needed to allow polymer segments to associate in junction zones and the disorder needed to prevent excessive association. The mechanisms of interchain association involve association of helical segments resulting from polymer order (see Oakenful, 1987; Edwards et al., 1987). These helical segments may associate by stacking of helices, as for those of amylose inclusion complexes, and by more intimate interactions such as double helix formation, as seen with amylose, agar, carrageenans, and gellan. When double helices form, it is possible to generate additional interactions in junction zones through stacking of the double helices. The stacking of double helices formed from negatively charged polymers, such as low-methoxyl pectins, carrageenans and gellan, is possible by interaction of the charged helices with specific cations of a charge density and size suitable for helix packing.

The initial stages of gelation involve the nucleation necessary to generate junction zones. Once nucleated, the junction zone generally is formed by cooperative interaction of the two chains or helices through numerous energetically small interactions. The cooperative nature of these interactions implies that interactions will continue to occur along the polymer with time when sterically feasible. This tendency explains the importance of disordered regions, which terminate the cooperative process. Without such disordered regions, cooperative changes will continue and gel properties will change accordingly. Amylose and amylopectin are two molecules which, from a food technology perspective, suffer from excessive order. Amylose may self-associate excessively, ultimately leading to increasingly perfect junction zones and increasing resistance to enzymatic hydrolysis. A gel which forms from unmodified amylose will ultimately show syneresis.

Complete hydrolysis of native starch produces glucose, and is of great importance in the sweetener industry. Less complete hydrolysis can be used to make corn syrups, which vary in composition according to the method of hydrolysis. Even less complete hydrolysis is employed to make maltodextrins, which may vary in DE from 2-19. Some of the low DE maltodextrins are marketed as fat substitutes: e.g., Paselli SA-2 ® (from potato starch), N-Oil® (from tapioca starch), and Maltrin 040® (from corn). (See Alexander, "Carbohydrates Used as Fat Replacers," in this volume.) Stellar® (from corn) is a hydrolyzed starch marketed as a modified starch instead of a maltodextrin, even though its DE is around 10 (after cooking, based on company data). However, due to chain association in the form of crystallites, the molecular weight of the crystallite units as they occur in an uncooked gel will be higher than the DE would suggest. At high concentration these starch-derived materials are capable of substituting for fat by forming soft gels, thus mimicking some of the creamy properties of fat in certain formulated food products.

Crystallinity

The logical extreme of the associative behavior necessary for gelation is the extensive, ordered packing of molecules in crystalline structures. For starch, an understanding of its potential extensive crystallinity is important to an understanding of its functionality. The starch granule is the functional unit of starch as an ingredient. (See Zobel, "Starch Granular Structure," in this volume.) The granule is an immense (on a molecular scale) highly organized structure. A typical 15 um corn starch granule is composed of over a billion molecules. Although amylose is only about 25% of the granule weight, there are many more amylose molecules due to its relatively small size (DP of about 1000) compared to amylopectin (DP>100,000). In the native form, the granule exhibits considerable crystallinity (about 40% by x-ray diffraction, due primarily to packing of short, linear branches of the amylopectin fraction), but unlike the crystallinity of native cellulose, the order is readily lost by heating in water below 100° C (except for certain mutant genotype starches, such as *ae* starch). Starch molecules differ from those of most other carbohydrate hydrocolloids because they are made functionally useful only by altering this granule package.

Commonly the granule is heated in the presence of water during processing, causing loss of crystallinity and a largely irreversible uptake of water and swelling of the granule, a process termed gelatinization (see Atwell et al., 1988, for discussion of the definition of this term). The individual molecules are hydrated, and some (especially amylose) are molecularly dispersed. If the treatment is harsh enough, granule integrity may be lost. Gelatinization is analogous to dispersion of other hydrocolloids, except that to the extent that granule integrity is preserved, the result is not molecular dispersion for starch. Dispersion of any hydrocolloid must be carefully accomplished if appropriate functionality is to be achieved, often requiring careful attention to conditions and treatments.

Unlike for many other relatively stable hydrocolloid dispersions, gelatinized starch begins to change as it is cooled. Whether inside or outside the swollen granule, the molecules have a strong tendency to associate, resulting in gelation when amylose is present at above 1.5% (w/w). In these gels crystallinity may be readily observed by x-ray diffraction (Morris, 1990). Because the native granule is partly crystalline, the regeneration of crystallinity is, in a sense, retrograde behavior, and the process is termed retrogradation (see Atwell et al., 1988, regarding use of this term).

There is no reason to assume that the crystallinity regenerated is entirely the same as that in the native granule, and there is much evidence to the contrary. Not initially present, "resistant starch," which is produced by certain physical treatments, is generated by the highly ordered association of amylose molecules. Although not involved in resistant starch, even highly branched molecules such as amylopectin can self-associate to a considerable extent. Even though the branch points do represent significant interruptions in the linear otherwise amylosic structure, they also appear to function as nucleation sites for association of the shorter amylosic chains into double helices, which can then pack into crystalline arrays. Any time starch is used in a formulation, the tendency toward retrogradation must be considered as a potential destabilizing influence on the physical properties generated by gelatinization. This concern is one of the primary reasons for using modified starch.

Interactions with Different Polysaccharide Molecules

Different polysaccharide molecules can interact to produce properties different than either molecule could produce alone, even at higher concentration, or they can interact synergistically to produce a similar effect but more efficiently. If different polysaccharides are used for viscosity generation and there is no specific interaction between them, the viscosity effect is likely to be additive. However, if the molecules are capable of some sort of specific interaction, an opportunity exists for synergism in generation of viscosity. The interaction between xanthan and galactomannans is used to advantage in several food products. The basis for this interaction has been ascribed to be an association of the xanthan helix and the relatively unsubstituted (termed "smooth") regions of a galactomannan (Dea, 1982). More recent evidence is consistent with the interpretation that under appropriate conditions the xanthan helix "denatures" allowing a specific ordered interaction between its cellulosic backbone and the highly similar mannan backbone of the galactomannan (Morris, 1991). Morris has noted that under normal solution conditions, the suitability of the xanthan backbone for interaction is disguised (Morris, 1991). It would appear that xanthan has the potential to have more than one ordered conformation in solution. It is interesting to speculate about what other polysaccharides might have a similar potential, and the synergistic interactions which might result.

Galactomannans have also been shown to interact with kappa-carrageenan. This interaction has been ascribed to the formation of junction zones between a carrageenan double helix and the relatively unsubstituted region of a galactomannan (Dea, 1982). However, an alternative explanation proposes an entirely different sort of mechanism. Morris (1991) has proposed that the galactomannan may serve to maintain the swollen carrageenan gel network by entangling with the portion of the carrageenan molecules exterior to the junction zones.

Interactions with Non-Polysaccharide Molecules

Interactions between polysaccharides and water have been discussed above in terms of viscosity and gelation. Water interactions with polymers are also important in two other respects; during freezing, storage, and thawing; and in controlling mobility of food components in foods containing hydrocolloids. These two seemingly disparate situations have both been dealt with using the food polymer science approach taken by Slade and Levine (1991). Polysaccharide interaction with water can prevent significant fractions of the water from crystallizing at temperatures well below the freezing point. However, the unfrozen water in the amorphous polysaccharide-water mixture can be kept immobilized if the polymer is held below the temperature for the transition from a glassy solid to a rubbery material. Consequently, undesirable changes mediated by mobile water can be minimized. The effect of different solutes and polymers on the glass transition temperature has been thoroughly studied (Slade and Levine, 1991).

These same authors have addressed the question of starch gelatinization and retrogradation from a food polymer science perspective, and have rationalized the behavior of starch in foods based on the plasticizing nature of water on amorphous glassy material, which occurs in the amorphous regions in a native starch granule, throughout the freshly gelatinized starch granule, and in a heat-treated granule under low-moisture conditions.

Polysaccharide interactions with protein occur primarily due to ionic interaction between negatively charged polysaccharides and the positive charges of protein (Ledward, 1979). Above the protein isoelectric point, the protein will have a net negative charge and the polysaccharide will also be negatively charged. As pH is decreased, there will be an increasing tendency for the protein and polysaccharide to associate due to simple considerations of charge. The first interactions are often dispersed complexes, but as pH is decreased the probability of precipitation increases. Such precipitation reactions may be of use in protein recovery from waste streams. At neutral pH an extensively positively charged region of kappa-casein may be responsible for the specific interaction with carrageenan (Ledward, 1979), an interaction useful for increasing viscosity and thus more efficiently suspending chocolate particles in chocolate milk. Above the protein isoelectric point, interactions between negatively charged protein and polysaccharide may be mediated by calcium ions.

Most interactions between polysaccharide and lipid have been indirect. For example, oil-in-water emulsions may be stabilized by, among other things, an increase in the viscosity of the aqueous phase, a property readily achieved by including a carbohydrate hydrocolloid.

Polysaccharide interactions with flavor molecules have been described above. The effect of polysaccharide use on flavor release and perception is another concern of the product developer using these ingredients. Cyclodextrin interaction with flavor molecules may be of use during processing or in finished products (Szejtli, 1990).

Because certain charged polysaccharides have strong and specific interaction with cations, the ability of these polymers to influence the chemistry of minerals in food is of interest. From a nutritional perspective, bioavailability might be either increased or decreased, depending upon the food and the mineral (Ink and Hurt, 1987). From a technological perspective, it might be an advantage to make certain minerals, especially iron and copper, less available to catalyze lipid oxidation in a food system.

LITERATURE CITED

Atwell, W. A., Hood, L. F., Lineback, D. R., Varriano-Marston, E., and Zobel, H. 1988. The terminology and methodology associated with basic starch phenomena. Cereal Foods World. 33: 306-311.

Batdorf, J. B. and Rossman, J. M. 1973. Sodium carboxymethylcellulose. Page 695 in: Industrial gums. polysaccharides and their derivatives. R. L. Whistler, ed.. Academic Press, New York.

Blanshard, J. M. V. and Mitchell, J. R., eds. 1979. Polysaccharides in food. Butterworths, London.

Brant; D. A. and Christ, M. D. 1991. Realistic conformational modeling of carbohydrates: applications and limitations in the context of carbohydrate-high polymers. Page 42 in: Computer modeling of carbohydrate molecules. A. D. French and J. W. Brady, eds. American Chemical Society, Washington.

Clarke, A. E., Anderson, R. L., and Stone, B. A. 1979. Form and function of arabinogalactans and arabinogalactan-proteins. Phytochem. 18: 521-540.

Dea, I. C. M. 1982. Polysaccharide conformation in solutions and gels. Page 420 in: Food carbohydrates. D. R. Lineback and G. E. Inglett, eds. AVI Publishing Co., Westport, CT.

de Vries, J. A., Rombouts, F. M., Voragen, A. G. J., and Pilnik, W. 1984. Comparison of the structural features of apple and citrus pectic substances. Carbohydr. Polymers. 4:89-101.

Dickinson, E. and Stainsby, G. 1982. Colloids in food. Applied Science Publishers, New York.

Edwards, S. F., Lillford, P. J., and Blanshard, J. M. V. 1987. Gels and networks in practice and theory. Page 1 in: Food structure and behavior. J. M. V. Blanshard and P. Lillford, eds. Academic Press, London.

Everett, D. H. 1988. Basic principles of colloid science. Royal Society of Chemistry, London.

French, A. D., Rowland, R. S., and Allinger, N. L. 1991. Modeling of glucopyranose: the flexible monomer of amylose. Page 120 in: Computer modeling of carbohydrate molecules. A. D. French and J. W. Brady, eds. American Chemical Society, Washington.

Gacesa, P. 1988. Alginates. Carbohydr. Polymers. 8:161-182.

Galloway, G. I., Biliaderis, C. G., and Stanley, D. W. 1989. Properties and structure of amylose-glyceryl monostearate complexes formed in solution or on extrusion of wheat flour. J. Food Sci. 54: 950-957.

Glicksman, M. 1969. Cellulose gums. Chapter 12 in: Gum technology in the food industry. Academic Press, New York.

Glicksman, M., ed. 1982. Food hydrocolloids, Volume I, CRC Press, Inc., Boca Raton, FL.

Glicksman, M., ed. 1983. Food hydrocolloids, Volume II, CRC Press, Inc., Boca Raton, FL.

Glicksman, M. 1984. The role of hydrocolloids in food processing -- cause and effect. Page 297 in: Gums and stabilisers for the food industry. 2. Applications of hydrocolloids. G. O. Phillips, D. J. Wedlock, and P. A. Williams, eds. Pergamon Press, Oxford.

Glicksman, M., ed. 1986. Food hydrocolloids, Volume III, CRC Press, Inc., Boca Raton, FL.

Ink, S. L. and Hurt, H. D. 1987. Nutritional implications of gums. Food Tech. 41(1): 77-82.

Ledward, D. A. 1979. Protein-polysaccharide interactions. Page 205 in: Polysaccharides in food. J. M. V. Blanshard and J. R. Mitchell, eds. Butterworths, London.

MacGregor, E. A. and Greenwood, C. T. 1980. Polymers in nature. John Wiley and Sons, New York.

Manners, D. J. 1989. Recent developments in our understanding of amylopectin structure. Carbohydr. Polymers. 11:87-112.

Marchessault, R. H. and Sundararajan, P. R. 1983. Cellulose. Page 12 in: The polysaccharides, Volume 2. G. O. Aspinall, ed. Academic Press, New York.

McCleary, B. V., Clark, A. H., Dea, I. C. M., and Rees, D. A. 1985. The fine structures of carob and guar galactomannans. Carbohydr. Res. 139: 237-260.

Morley, R. G. 1984. Utilisation of hydrocolloids in formulated foods. Page 211 in: Gums and stabilisers for the food industry. 2. Applications of hydrocolloids. G. O. Phillips, D. J. Wedlock, and P. A. Williams, eds. Pergamon Press, Oxford.

Morris, E. R. 1985. Rheology of hydrocolloids. Page 57 in: Gums and stabilisers for the food industry. 2. Applications of hydrocolloids. G. O. Phillips, D. J. Wedlock, and P. A. Williams, eds. Pergamon Press, Oxford.

Morris, E. R. 1989. Polysaccharide solution properties: origin, rheological characterisation and implications for food systems. Page 132 in: Frontiers in carbohydrate research -- 1: food applications. R. P. Millane, J. N. BeMiller, and R. Chandrasekaran, eds. Elsevier, New York.

Morris, V. J. 1990. Starch gelation and retrogradation. Trends Food Sci. & Tech. July, pp. 2-6.

Morris, V. J. 1991. Weak and strong polysaccharide gels. Page 310 in: Food polymers, gels, and colloids. E. Dickinson, ed. Royal Society of Chemistry, Cambridge.

Muzzarelli, R. A. 1985. Chitin. Page 418 in: The polysaccharides, Volume 3. G. O. Aspinall, ed. Academic Press, New York.

Oakenfull, D. 1987. Gelling agents. CRC Crit. Rev. Food Sci. Nutr. 26: 1-25.

Painter, T. J. 1983. Algal polysaccharides. Page 196 in: The polysaccharides, Volume 2. G.O. Aspinall, ed. Academic Press, New York.

Rees, D. A. 1977. Polysaccharide shapes. Chapman and Hall, London.

Rees, D. A., Morris, E. R., Thom, D., and Madden, J. K. 1982. Shapes and interactions of carbohydrate chains. Page 196 in: The polysaccharides, Volume 1. G. O. Aspinall, ed. Academic Press, New York.

Roumbouts, F. M. and Thibault, J.-F. 1986. Enzymic and chemical degradation and the fine structure of pectins from sugar beet pulp. Carbohydr. Res. 154:189-203.

Sanford, P. A. and Baird, J. 1983. Industrial utilization of polysaccharides. Page 411 in: The polysaccharides, Volume 2. G. O. Aspinall, ed. Academic Press, New York.

Shatwell, K. P., Sutherland, I. W., Ross-Murphy, S. B., and Dea, I. C. M. 1991. Influence of the acetyl substituent on the interaction of xanbthan with plant polysaccharides -- I. Xanthan-locust bean gum systems. Carbohydr. Polymers. 14: 29-51.

Szejtli, J. 1990. The cyclodextrins and their applications in biotechnology. Carbohydr. Polymers. 12: 375-392.

Slade, L. and Levine, H. 1991. Beyond water activity: recent advances based on an alternative approach to the assessment of food quality and safety. CRC Crit. Rev. Food Sci. Nutr. 30: 115-360.

Szczesniak, A. S. 1986. Rheological basis for selecting hydrocolloids for specific applications. Page 311 in: Gums and stabilisers for the food industry. 3. G. O. Phillips, D. J. Wedlock, and P. A. Williams, eds. Pergamon Press, Oxford.

Torres, A. and Thomas, R. D. 1981. Polydextrose ... and its applications in foods. Food Tech. 35(7): 44.

Walker, B. 1984. Gums and stabilisers in food formulations. Page 137 in: Gums and stabilisers for the food industry. 2. Applications of hydrocolloids. G. O. Phillips, D. J. Wedlock, and P. A. Williams, eds. Pergamon Press, Oxford.

CARBOHYDRATES USED AS FAT REPLACERS

Richard J. Alexander

Penwest Foods Company
Cedar Rapids, IA 52406

INTRODUCTION

Probably the earliest work on low-fat (or lower-fat) foods started with what is commonly known today as spoonable salad dressings. These were essentially derived from mayonnaise, which, based on its 'standard of identity' contains 80% fat. Food processors found that they could make a fairly comparable product using only 30 to 50% fat with 4 to 5% of a modified food starch to provide viscosity and emulsion stability.

Fat content was also kept to a minimum in the pourable, french-style salad dressings, where a number of gums, such as xanthan or alginate, or modified food starches were employed as the viscosifier and emulsion stabilizer.

The current technology seems to stem from the early work of Richter and co-workers (1976A and 1976B), who first described the preparation and properties of a low DE potato-based maltodextrin. This product forms a white, glossy, thermoreversible gel when dispersed in water. The gel has a neutral taste and is freeze-thaw stable.

As discussed by Hannigan (1981) and Schierbaum (1978), this maltodextrin gel is miscible with fats and oils, and provides stable emulsions. It produces a fat-like mouthfeel, and can be used to replace fat in normally high fat foods, such as mayonnaise, salad dressings, frozen desserts, and fillings for baked goods.

CLIMATE FOR FAT REPLACERS

If it weren't for several fat-related health issues, raised during the early to mid-1980's, the fat replacement and fat substitute technologies developed prior to and after that time period would not have skyrocketed like they have done.

Health Issues

As indicated in several articles in the literature (Anon., 1991A; Duxbury, 1991; Egbert, 1991; Meyer, 1991; Przybyla, 1990) the Surgeon General recommended that fat intake of Americans be reduced to 30% of the total calories consumed, with one-third (10%) coming from saturated fat, one-third from unsaturated fat and one-third from polyunsaturated fat. Various health-related organizations, such as the American Cancer Society and the American Heart Assc., have supported the Surgeon General's recommendations (Egbert, 1991) because of the effects of lowering fat and cholesterol on reducing incidents of cancer and coronary heart disease (Przybyla, 1990).

More than 50% of all adults in the U.S. have increased risk of heart attack because of high blood cholesterol (Inglett, et al, 1991). Consequently, fat reduction has become our society's healthiest obsession (Przybyla, 1990; Best, 1991B), although consumer concern for 'high fat' foods have at least partially replaced dietary emphasis on cholesterol (Duxbury, 1991). Based on data reported by Hegenbart (1991), fat and cholesterol are almost equivalent (at about 40-45% of the population) as America's top health-related food concerns.

Solutions to these health problems have been provided by a variety of product modifications. The products described in this chapter have obviously been successful in reducing both fat and calories in many food systems. The incorporation of fiber has been effective in reducing problems associated with cancer (Przybyla, 1990), as well as caloric reduction, since fiber is non-digestible or only partially digestible (Schierioth, 1991). Soluble fiber, such as the beta-glucan in oat-based products, has also been associated with lowering of blood cholesterol (Inglett, et al, 1991). Products low in fat, saturated fat and cholesterol have also been helpful with factors relative to coronary heart disease.

Trends in Production and Consumption

Because of these health issues, a variety of related trends have been observed through several recent surveys.

According to a survey conducted for the Calorie Control Council, 124 million adult Americans consume reduced- or low-fat foods on a regular basis, and 2/3 of the same group believe there is a need for food ingredients as fat replacers (Anon., 1991A; Hegenbart, 1991). Another pole indicated that 141 million people (Anon, 1991G) used either low-cal/low-fat foods (45%), low-fat foods (22%), or low-cal foods (9%). This represented about 76% of the adult population.

Another survey conducted in 1990 showed 76% of 37 baking companies were expected to increase efforts in the low-cal food area over the next two years (Anon., 1991C). Also in 1990, 10% of all new food products were low-fat or reduced-calorie products (Anon., 1991B). In the salad dressing area, however, 24% of 201 products were low-fat or reduced-cal (Anon., 1991B), while in 1989 over 450 new dairy products were labeled low-fat or no-fat (Anon., 1990C).

The current market for all fat replacement products is estimated at $ 1 billion. In the baked goods area, the current market is estimated at $ 250 million and will grow to $ 800 million by 1995 (Swientek, 1991).

Most of what people in the food processing area hear about today has to do with fat and fat reduction (Duxbury, 1991). Also, fat and calorie reduction remains the leading R&D priority slated for increased investment by 76% of 352 food processors polled in 1991 (Best, 1991B).

CARBOHYDRATES AS FAT REPLACERS

In the last 5-6 years, a number of authors have written general articles covering the use of carbohydrates as fat replacers. Each article describes a few of the products that are being used, but none of them includes all of the varied materials that have found utility as fat replacers. This chapter will attempt to include all of the carbohydrate-based products.

Perhaps one of the key points made by several authors is that there is no ideal product; one that will provide all the benefits needed in a complex food system (Anon., 1991A; Miller, 1991; Penichter, 1991). As other authors phrase it, "effective fat replacement requires a systems approach" (Anon., 1991C; Best, 1991B). This can be accomplished by certain combinations of gums or gums and starches (Ballard, 1986; Best, 1991A; Dillon, 1990) or gums and maltodextrins (Cain, 1989).

Another important point made in several of the articles is that its important to match or imitate fat's sensory appeal when replacing fat (Anon., 1991E) with one or more of the carbohydrate products. Several of the properties discussed throughout the literature as being important in matching the sensory appeal include texture or mouthfeel (Best, 1991B; Hegenbart, 1991), moisture retention (Anon., 1991C), flavor (Hegenbart, 1991), and how and when flavor is released. More details of how specific products function and how they help mimic the properties of fat will be discussed within specific product categories.

SPECIFIC CARBOHYDRATES USED AS FAT REPLACERS

For specific brand names of products and their manufacturers see Tables I through V.

Cellulose-Based Products

Refined Cellulose
Ang and Miller (1991A) described the use of powered cellulose produced by James River Corp. (Solka Floc).

Table I.

Commercial Celluloses and Microcrystalline Celluloses Used as Fat Replacers[1]

Product	Manufacturer	Product Composition
SOLKA FLOC®	James River	100% Cellulose
Avicel® RCN-30	FMC Corp.	75% Microcrystalline cellulose, 20% maltodextrin, 5% xanthan gum
Avicel® RCN-15	FMC Corp.	85% Microcrystalline cellulose, 15% guar gum
Avicel® RCN-10	FMC Corp.	90% Microcrystalline cellulose, 10% guar gum
Avicel® RC-501	FMC Corp.	91% Microcrystalline cellulose, 9% CMC
Avicel® RC-581	FMC Corp.	89% Microcrystalline cellulose, 11% CMC
Avicel® RC-591	FMC Corp.	88% Microcrystalline cellulose, 12% CMC
Avicel® CL-611	FMC Corp.	85% Microcrystalline cellulose, 15% CMC

[1] Many of the product names are registered trademarks of the manufacturer cited. The ® or ™ notations will be used, but only in the tables.

Table II.

Commercial Gums Used as Fat Replacers

Product	Gum Type	Supplier or Manufacturer(s)
[2]	Algin (Alginates)	Kelco
Viscarin® SD 389	Carrageenan (Iota)	FMC, Sanofi
Kelcogel™	Gellan Gum	Kelco
Supercol F	Guar Gum	Henkel, National Starch
Konjac Flour	Glucomannan Gum	Kyowa Shokuhin, FMC
Keltrol™	Xanthan Gum	Kelco, Rhone-Poulenc
Methocel®	Hydroxypropyl methyl cellulose	Dow Chemical

[2] No commercial trade name available in the literature reviewed.

Table III.
Commercial Maltodextrins Used
as Fat Replacers

Product	Manufacturer	Starting Starch	Approx. DE
Instant N-Oil® II	National Starch	Tapioca	[4]
Maltrin® M-040	GPC	Corn	4
Maltrin® M-100 or M-150	GPC	Corn	9-13 14-18
Oatrim[3]	U.S.D.A.	Oats	<5
Paselli SA2	Avebe	Potato	2
STAR DRI® 1	A.E. Staley	Corn	1

[3] Material is a development product of the USDA
   Labs in Peoria, IL., and is not yet commercial.

[4] DE was not reported in the literature.

Table IV.
Special Starch-Derived Products
Used as Fat Replacers

Product	Manufacturer	Product Type
N-Oil®	National Starch	Tapioca Dextrin
Instant N-Oil®	National Starch	Tapioca Dextrin
Litesse™	Pfizer, Inc.	Polydextrose[5]

[5] Product made by recombining dextrose with 10%
   sorbitol under acidic conditions to yield a low
   mol. wt. dextrose-based polymer. First product to
   market place was labeled just Polydextrose. Second
   generation product is labeled Litesse™.

Table V

Commercial Starches Used as Fat Replacers

Product	Manufacturer	Starch Base
Amalean™ I	American Maize	HP high amylose corn
Amalean™ II	American Maize	Instant HP high amylose corn
STA-SLIM™ 142	A.E. Staley	Instant potato
STA-SLIM™ 143	A.E. Staley	Potato
STA-SLIM™ 150	A.E. Staley	Instant tapioca
STA-SLIM™ 151	A.E. Staley	Tapioca
STA-SLIM™ 171	A.E. Staley	[a]
STELLAR™	A.E. Staley	Acid modified corn
Remyline	Remy Industries (A&B Ingredients)	Rice
Remygel	Remy Industries	Rice
Starch Plus Regular Type SPR	California Natural Products	Rice
Starch Plus Waxy Type SPW	California Natural Products	Waxy Rice
Mira-Thik® 468	A.E. Staley	[a]
Instant Pure-Flo	National Starch	HP waxy corn
Paselli BC-EX-Tunnel	Avebe	Waxy corn acetate
Instant W-11	American Maize	Crosslinked waxy corn
Frigex	National Starch	[a]
Gel 'N' Creamy	National Starch	[a]

[a] Starch base was not available from the literature reviewed for this manuscript.

In addition to having utility in baked goods to increase cake volumes, improve cell structure, etc., it can also reduce calories by allowing less fat pick-up in those products. Batters containing the fiber are stronger, and it allows the batter to stay in tact more readily during and after frying (Ang, et al, 1991B). It is particularly useful with batter coatings for items such as meat, fish and mushrooms.

It is also useful with donuts where less oily, lighter colored donuts are produced with 10-20% lower fat content (Anon, 1991D). In all cases the cellulose caused an increase in the moisture content of the baked products in which it was used.

In these latter applications the cellulose did not really function as a fat replacer, but it did reduce fat and calories by preventing fat absorption.

In other areas Miller and Werstok (1980) defined a low fat spread (20-40% fat) in which a cellulose ether was incorporated as the thickening agent. Examples were given of hydroxypropyl methyl cellulose, plus micro-crystalline cellulose with a small amount of CMC. Barnett, et al (1990) claimed the use of a mixture of hemicelluloses to replace fat and sucrose in ice cream-type desserts and as a fat replacer in cookies.

Microcrystalline Cellulose
This variety of cellulose has also solved some of the problems associated with the removal of fat in low-cal foods (Penichter, 1991). The Avicel products from FMC supposedly contribute fat-like organoleptic properties to a variety of food systems (Anon., 1990C). They provide the mouthfeel of high-fat emulsions in low-fat products, they lend a glossy, fat-like appearance to food products and contribute zero calories.

When dispersed in aqueous systems, the RC and CL types of Avicel form soft, creamy gels, which give the mouthfeel and texture desired in high-fat foods (Penichter, 1991). In low pH systems, xanthan gum or CMC are added as protective colloids. New 'all natural' products, introduced in 1991, included RCN-30: 75% cellulose, 20% maltodextrin and 5% xanthan gum; RCN-15: 85% cellulose and 15% guar gum; and RCN-10: 90% cellulose and 10% guar gum.

In specific product areas the microcrystalline celluloses are particularly useful in low-fat or no-fat salad dressings (Baer, et al, 1991). They are described in low-cal baked products (Robbins, 1984), where they are used in combination with fat substitutes such as sucrose polyester. One patent (Howard, 1991) claims their use in preventing the laxative effect in low-cal fat substitutes by incorporation of a small percentage in the polyol polyester.

Gums

A number of different gums have been cited as being useful as fat replacers. A recent patent (Dartey, et al, 1990) for preparing a low cholesterol mayonnaise claims at least one gum selected from the group of xanthan, guar, gum tragacanth, gum arabic , locust bean gum, alginates, pectins or mixtures thereof. A preferred mixture is a blend of xanthan, locust bean and guar.

A relatively new product is based on Konjac flour, which is used because of the high viscosity gels obtained on heating (Tye, 1991). The flour is composed of sacs that swell and rupture to release a high molecular weight, soluble glucomannan gum. Konjac is particularly useful in sausage-like products, in low-fat spreads, and in pasta products. It interacts, or is synergistic, with kappa carrageenan, xanthan gum and certain starches.

A new use for iota carrageenan was recently found with the development of low-fat ground beef (Egbert, et al, 1991). Based on a project at Auburn University for McDonalds Hamburgers, a low-fat (8%) hamburger was developed which contains 0.5% iota carrageenan as the fat replacer. In addition to retaining water in the system, the gum appears to form particles similar in size to lipid or fat droplets, and thus provides similar organoleptic properties to fat.

Gellan gum, a fermentation product somewhat similar to xanthan gum, is reported to be particularly useful as a fat replacer in jams and jellies (Anon., 1991J), in fruit fillings for baked goods, and in pie fillings (Duxbury, 1990). It can also be used in icings, frostings and glazes. The product can be used to produce a mayonnaise with as little as 20% oil. Unfortunately, the gum is still being reviewed by FDA, and it is not yet available to the food industry.

The product is a gelling agent as opposed to being a thickener/stabilizer as are most gums and many starches. It can be modified with other gums, and starch/gellan blends can improve gel structure and flavor release.

Maltodextrins

As indicated before, the early fat replacer technology of recent history seems to have originated with the starch hydrolysis products (SHP) claimed by Richter, et al (1976A, 1976B). These products were low DE (about 5) maltodextrins derived from potato starch and were useful in a variety of normally high-fat products (Hannigan, 1981; Schierbaum, 1978), because they formed a thermoreversible gel. This gel mimicked the mouthfeel and texture of high-fat foods.

Since then a number of similar maltodextrins have been produced commercially from corn, tapioca, and potato starches, and at least one in research quantities from oats.

Instant N-Oil II

This product is a low DE tapioca maltodextrin made by procedures described by Lenchin, et al (1985). It is part of series of tapioca-based materials starting with N-Oil and Instant N-Oil, which are dextrins. Instant N-Oil II is a maltodextrin of DE less than 5, and usually about 3 or less. As described by Light (1990), the maltodextrin gives a gel with fat-like texture, probably similar to the SHP product of Richter. The product is particularly useful in low-fat foodstuffs, such as frozen desserts, margarine, salad dressings and sauces

Maltrin M-040

Low DE corn-based maltodextrins claimed by Morehouse and Sander (1987A, 1987B) appear to be the basis for Maltrin M-040. This product has been characterized as having a bland flavor, smooth mouthfeel, and short texture (Dziezak, 1989). It can be used to replace fats or oils in salad dressings, dry beverage mixes and coffee whiteners,(Morris, 1984); in frozen desserts and extruded cereals,(Dziezak, 1989); and in margarine or butter-like spreads (Morehouse and Lewis, 1985).

According to GPC data (Anon., 1990A), Maltrin M-100 and M-150 can also be used in certain formulas to replace vegetable oil, such as salad dressings, sauces and gravies. The higher DE products, however, are not discussed in connection with the usual gel forming ability of the lower DE materials.

Oatrim

This product was developed in the laboratory by scientists at the Northern Regional Research Labs, USDA, Peoria, IL (Inglett, 1990; Worthy, 1990). The material has been referred to as a mixture of beta-glucan and amylodextrins. The latter is a term synonymous with maltodextrin. This portion of Oatrim was reported to be a low DE product, 5 DE or less (private communication with G. E. Inglett).

Oatrim is produced by an alpha-amylase hydrolysis of either oat bran or oat flour (Anon., 1991I; Anon., 1990E; Duxbury, 1990; Inglett, 1990). After removal of the insoluble fractions, the remaining product is a water soluble mixture of maltodextrin (derived from the oat starch) and beta-glucan. There are actually three potential products, varying primarily in beta-glucan content. Oatrim-1 is made from debranned oat flour and contains 1-2% beta-glucan; Oatrim-5 is made from whole oat flour and contains 5-6% beta-glucan; while Oatrim-10 is made from oat bran and contains about 10% beta-glucan (Anon., 1990E; Inglett, 1991).

The materials are of particular interest because of their two-pronged effect on reducing calories and cholesterol (Anon., 1990D; Worthy, 1990). Similar to other maltodextrins the products form gels, which allow for the reduction of fat and cholesterol in many high-fat foods. They further reduce cholesterol because of the properties of beta-glucan.

354

The products have found utility in a number of foods including frozen desserts, beverages, salad dressings and sauces (Worthy, 1990), as well as in baked goods (Anon., 1990D), yogurts and sour cream (Duxbury, 1991), instant breakfast drinks, spreads, gravies, soups, peanut butter, cheeses and cereal products (Anon., 1990E).

As yet, Oatrim is not a commercial product. The USDA has acquired three licensees for the technology, who are interested in further development and possible manufacture and sale of the product (Inglett, 1991).

Paselli SA2

Paselli SA2 is a 2 DE potato-based maltodextrin, probably somewhat similar to the original SHP, also based on potato starch. The product has been described as possessing a thermoreversible gel with smooth, fat-like texture and neutral taste (Anon., 1989A).

The material has been found useful as a fat replacer in a number of food products including baked goods (Anon., 1989B), as well as in dips, salad dressings and dessert toppings (Dziezak, 1989). One patent (Cain, et al, 1989) claimed its use as a gelling maltodextrin along with a higher DE non-gelling maltodextrin or other natural gums in a fat-reduced spread.

Other Products

Although not mentioned in the literature, other products listed in Table III, such as Maltrin M-100 and Star-Dri 1 have been advertised by their suppliers and are considered as fat replacers. Low DE maltodextrins have been described as being useful in low-fat food products in general (Vorwerg, 1988) and in reduced caloric spreads (Lowery, 1989). In addition, reduced-fat foodstuffs have recently been reported to be produced using a low DE maltodextrin made from high amylose starch (Furcsik, et al, 1991A).

Cain, et al (1989) claimed a margarine-like spread in which a gelling agent and viscosity enhancer were combined to provide a fat-reduced product. The gelling agent could be a low DE maltodextrin, such as Paselli SA2, Instant N-Oil II, or gum, such as guar gum or carrageenan. The viscosifier could be a non-gelling maltodextrin, such as Maltrin M-150, or a gum, such as xanthan gum or locust bean gum.

Special Starch Derived Products

Two tapioca-based dextrins, N-Oil and Instant N-Oil, and Polydextrose make up this category of carbohydrate fat replacers.

N-Oil and Instant N-Oil have been mentioned in several articles over the past few years, particularly in 1991. The products have been used to replace fat in cakes and muffins (Anon., 1991C), processed meats (Anon., 1991C), frozen desserts and salad dressings ((Anon., 1990B; Duxbury, 1991; Dziezak, 1989; Haumann, 1986), puddings (Duxbury, 1991), spreads (Anon., 1990B, Duxbury, 1991), and sour cream (Duxbury, 1991; Dziezak, 1989).

Polydextrose has also been mentioned in a number of articles since 1980. It was described quite extensively by La Bell (1991B) as serving primarily as a bulking agent in systems where an artificial sweetener is used, followed by a secondary use of replacing a portion of the fat. This idea was repeated in a number of articles (Anon., 1990B; Anon., 1991C; Anon., 1991G; Best, 1991B; Haumann, 1986; La Bell and Duxbury, 1991). The product supposedly adds texture and mouthfeel, humectancy and, in some cases, flavor and appearance to certain foods. It can not be used alone to replace all of the fat (Schierioth, 1991)

Polydextrose has been reported as an ingredient in the following low-fat foods: chocolate confections (Anon., 1991G), cakes, cookies, frostings and whipped toppings (Best, 1991B), baked goods and frozen desserts (Haumann, 1986). It is also claimed as an ingredient, along with microcrystalline cellulose, in dietetic desserts (Wolkstein, 1986).

Starches

This is by far the largest group of products with nearly 20 different starches being names in the technical and/or patent literature as fat replacers.

There are four groups of starches that have received considerable coverage in both company brochures and data sheets as well as the technical literature, and these will be discussed separately, There is a fifth group in which a starch is mentioned at least once in the technical literature, and these will be described collectively.

Amalean Starches

The first product in the series, Amalean I, was introduced by American Maize-Products and reported by Duxbury (1990C). It is defined as a fat sparing starch that can replace up to 100% of the fat or oil in a food product. It is targeted for salad dressings, sauces, cream fillings, dips and spreads. It can also be used in cakes, cookies and frozen novelties. A pregelled or instant version of the product, Instant Amalean II, was introduced in 1991.

Amalean was defined by Furcsik, et al (1991B, 1991C) in the recent patent literature. Here a method for making a reduced fat foodstuff is claimed in which at least a portion of the fat is replaced with an hydroxypropyl starch made from high amylose starch. The preferred starch is made from Amaizo 5, a product containing 50-65% amylose, and the preferred substitution level is 0.05 to about 0.3 (9.7% HP groups).

Rice Starches

As recently described by La Bell (1991A), as well as in the product literature, rice starches have been found to be useful as fat replacers because of the following properties:

a. They have very small starch granules (2-8 microns), which helps provide a smooth texture. This is also true when used in ungelatinized form, such as an unheated cookie filling.
b. The gelatinized rice starch has a smooth, creamy, spreadable gel, which has fat-like characteristics.
c. The products have excellent heat and freeze-thaw stability, and water absorption capacity.

The products are particularly considered as fat replacers in ice cream, cheesecake, sour cream, mayonnaise and salad dressings, spreads and cookies. It has also been claimed as a partial fat replacement in a reduced calorie sausage (Johnson, et al, 1988). Potential suppliers of rice starches include Remy Industries and California Natural Products.

STA-SLIM Starches

This is a group of five starches recently introduced by the A. E. Staley Co., including two potato-based products (an instant starch-142, and a cook-up starch-143), two tapioca-based products (an instant starch-150 and a cook-up starch-151), and Sta-Slim 171, a new starch designed for meat and sausage patties.

Most of these products have recently been described in the technical literature. Sta-Slim 143 was discussed by Dillon (1990) as one of the starches being considered for texture and mouthfeel of no-fat, no-cal foods. The tapioca products, Sta-Slim 150 and 151, were mentioned as being useful as fat mimetics (Anon., 1991B), particularly in salad dressings. A combination of Sta-Slim 150 and Mira-Thik 468 was also described for use in a no-oil pourable dressing. Sta-Slim 143 (Anon., 1991H) was reported as being used in frankfurters and bologna products with as much as a 50% reduction in fat and 33% calorie reduction.

STELLAR Starches

Stellar is probably the newest starch-based fat replacer to hit the food industry. It was introduced on June 11, 1991 by A. E. Staley Mfg. Co. (Freedman, 1991) as a product designed to replace as much as 100% of the fat in a variety of food products. The product is reportedly smoother and better tasting than other products on the market (Ramirez, 1991). Staley states that Stellar is already approved for food use having met FDA regulations for food starch, modified (Anon., 1991F).

The product is classified as an acid hydrolyzed starch (Meyer, 1991). When blended or sheared in water a creme-like dispersion is produced. This creme consists of a loosely associated network of aggregated submicron starch crystallites in immobilized water. (Swientek, 1991). A particle gel of 3 to 5 microns in size is formed (Anon., 1991K), which is in contrast to other starches and gums, that form polymer gels without distinct particles. This gel simulates the mouthfeel of fat because it can be deformed, and because the particles are about the same size as fat crystals or globules.

Stellar is claimed to be particularly useful as an anti-staling agent in baked foods; as a stabilizer and toughening inhibitor in microwave foods; and as a fat mimetic in salad dressings, spreads, cheeses, cheesecakes, meats and cake frostings (Anon., 1991K).

Other Starches

One starch claimed by Ballard, et al (1986) for use in a low-oil salad dressing is Instant W-11 Starch, a crosslinked waxy maize starch from American Maize. The preferred fat replacer is a mixture of the starch with a combination of xanthan and guar gums. An instant hydroxypropyl starch from National known as Instant Pure Flo is the preferred starch in a 1988 British patent application (Platt, 1988) for a low-fat spread. An instant acetylated potato-based starch from Avebe, known as Paselli BC-EX-Tunnel, was also claimed.

Three starches were mentioned by Dillon (1990) as having become successful as fat replacers in the new generation of no-fat, no-cal food products. These included Sta-Slim 143 (already discussed) from Staley, and Frigex and Gel 'N' Creamy from National. Also, as described earlier (Anon., 1991B), the mixture of Mira-Thik 468 and Sta-Slim 150 (both products from Staley), provided both a creamy mouthfeel and short texture in a no-oil pourable salad dressing.

In addition to carrageenan being used to produce a low-fat hamburger, a combination of potato and tapioca starches (Best, 1991A) has been used to produce a 15% fat hot dog duplicating those made with 30% fat.

FUNCTIONALITIES OF VARIOUS FAT REPLACERS

Gums

Most of the gums are cold water soluble and disperse in water to form very viscous colloidal dispersions at low concentrations (1.0% or less). Some of the gums interact with multivalent metal ions to form rigid gels, that have special applications in the food area. However, most of the products form relatively smooth pastes or gels, similar to chemically modified starches.

Regular Starches

In most cases we are dealing with starches that function by virtue of their thickening properties. They gelatinize in hot water to form viscous, translucent pastes that have smooth, creamy texture. If the product has been chemically treated and is crosslinked and/or substituted with acetate or hydroxypropyl groups, for example, the paste would tend to be short in texture, but smooth and creamy. If the starch is unmodified it would tend to form more of a gel, particularly on cooling, and have a firmer texture. In general, these starches have been said to form a polymer gel.

Some of the newer, specific products will be discussed in more detail below.

Amalean

In the case of the Amalean products, we are dealing with fairly high viscosity starches that have not been reduced in molecular weight by acid or other means. According to the Amaizo patents (Furcsik, et al, 1991B and 1991C) the starch pastes are translucent with thick, smooth consistency. As indicated by their product data sheets, the Amalean starches enhance the body and mouthfeel of reduced-fat foods, and provide high viscosity at low solids.

Rice Starches

According to their data sheets, Remy starches were selectively grown for their fat replacer and thickener

properties. They are extremely stable for freeze-thaw
and heat sterilization. They have smooth, creamy
texture achieving a greater fat-like mouthfeel in fat-
reduced foods than other products. One gram of rice
starch will replace 4-5 grams of fat. Their small
granular size (2-8 microns) definitely has an effect on
the smoothness of a rice starch slurry, and probably has
some effect on the mouthfeel of a cooked starch paste.

STA-SLIM Starches
        Based on Staley's Tech. Data Sheet 507-096060, the
Sta-Slim starches provide the smooth, creamy texture,
appearance and eating qualities of products much higher
in lipid content. Based on Brabender viscosity curves
it appears that the starches have been acid modified,
giving them somewhat lower viscosity than their
unmodified counterparts. However, their setback
properties should be greater, providing starch gels with
more structure and firmness.
        According to Staley, the tapioca starches (Sta-Slim
150 and 151) provide a creamier, smoother texture with
stable viscosity. The potato-based products (Sta-Slim
142 and 143) provide more texture and setback or
firmness.

Specialty Starches

        The one product in this category is Stellar. As
indicated in Staley's Tech. Data Sheet TIB 29-195060.
Stellar is produced by controlled acid hydrolysis to
give an acid modified starch. The functionality of the
product occurs when it is sheared in water to create a
firm, deformable gel or creme.
        This creme is made up of submicron size
crystalline particles loosely associated to form large
aggregates. The large surface area of the particles
causes a high level of water immobilization, allowing
the aggregates to behave like fat particles. The creme
shows gel rheology similar to hydrogenated vegetable
shortenings. It provides the texture, tenderness,
lubricity and fat-like mouthfeel to a variety of food
products.

Although this theory has not been discussed in the literature, it is believed that the low DE (1-5) maltodextrins function in the following manner:

These low DE products are mid-way between acid modified starches and the water soluble corn syrup solids. As such, they contain a certain number of intermediate molecular weight carbohydrate molecules. These molecules include a certain fraction of short, straight-chain amylose-like molecules.

These molecules tend to line up and become associated in aqueous dispersion and form a gel. The molecules are too short to form a rigid, irreversible gel like the higher molecular weight starches. Plus, they contain a certain percentage of lower molecular weight oligosaccharides that remain in solution and inhibit gel formation.

The result is a soft, thermoreversible gel. Under cold to ambient temperatures the gel mimics the appearance, texture and mouthfeel of the kind of fat emulsion that exists in an ice cream, salad dressing, margarine-type spread of other related high-fat food.

Specific maltodextrins and their properties will be described below.

Instant N-Oil II

According to National's Tech. Service Bulletin 15889-238, at concentrations in excess of 20% Instant N-Oil II develops texture similar to a hydrogenated shortening. In replacing the fat in normally high-fat foods, it provides the texture, creamy mouthfeel, body and sometimes the gel character of higher fat products.

Maltrin M-040

As indicated in GPC's Tech. Bulletin on maltodextrins, their products, particularly their low DE materials, form solutions that are characterized by a bland flavor, smooth mouthfeel, and short texture, and can partially or totally replace fat in a variety of formulations. According to Morris (1984), Maltrin M-040 forms a heat-reversible gel and provides texture and mouthfeel, particularly as a fat replacer in low-cal salad dressings and margarines.

Paselli SA2

Based on Avebe's Technical Bulletin on Paselli SA2 (05.12.32.167EF), gels made from the product show a fat-like texture; they are smooth and bland. Products derived from potato starch are particularly well suited for this type of application because of their purity and molecular composition.

Oatrim

As indicated in some of the references on Oatrim, the product, like other low DE maltodextrins form a gel that has many of the characteristics of shortenings (Anon., 1990D; Inglett, 1990; Inglett, 1991).

Dextrins and Polydextrose

Dextrins

As indicated in National's Tech. Service Bulletins (34888-50 and 34388-508), N-Oil and Instant N-Oil function as thickeners, providing organoleptic and textural properties similar to high-fat foods. At concentrations above 20% the products develop textures similar to a hydrogenated shortening.

Polydextrose

Based on Pfizer's Technical Brochure on Litesse, as well as the technical literature, polydextrose has about the same viscosity as sucrose at equal concentrations. This enables the product to provide the desirable mouthfeel and body of sucrose when employed as a bulking agent or replacement for sucrose. When used as a bulking agent it can also help provide a fat or oily-like mouthfeel in certain fat-reduced foods, but it apparently can not function alone as a fat replacer in most systems. This is not surprising since it is too low in molecular weight to provide the viscosity or thickening properties of a starch or gum, nor does it have the necessary characteristics to form a gel like maltodextrins or microcrystalline cellulose.

Microcrystalline Cellulose

As indicated in FMC's Tech. Bulletin C-96, the colloidal RC or CL grades of Avicel are used in fat replacement systems. These grades are mixtures of microcrystalline cellulose with a small amount of CMC as stabilizer. When dispersed in water with shear, the colloidal particles form an insoluble network of cellulose crystallites.

The average size of these particles in dispersion is 0.2 microns, and as such they simulate the sensation of fat in an oil-in-water emulsion. This crystallite network imparts a short, creamy mouthfeel that gives the sensation of products with high-fat content.

Summary of Functionalities

There appears to be three main types of mechanisms by which carbohydrates function as fat replacers.

1. The first is a polymer gel formation which includes most starches and gums. The polysaccharides are basically thickeners, and function as fat replacers because of the large amount of water they imbibe in forming a high viscosity paste or gel.
2. The second mechanism is the thermoreversible gel formation which occurs with all low DE (1-5) maltodextrins made from common amylose-containing varieties of starch.
3. The third type of functionality occurs with microcrystalline cellulose and Stellar starch. They are from different carbohydrate sources, but they both are based on particle or crystallite gel formation. The gel is formed when Stellar or Avicel is sheared in a aqueous system. In the case of Stellar submucron particles form aggregates with diameters of about 3-15 microns. With Avicel particles of about 0.2 microns are formed. Such gels bind water and behave rheologically like fat, thus mimicking the mouthfeel of high fat foods.

LITERATURE CITED

Ang, J. F. and Miller, W. B. 1991A. Multiple functions of powdered cellulose as a food ingredient. Cereal Foods World. 36:558.

Ang, J. F., Miller, W. B. and Blais, I. M. 1991B. Fiber additives for frying batters. U.S. patent 5,019,406.

Anonymous. 1989A. Caloric control commentary. Calorie Control Council. Atlanta, GA.

Anonymous. 1989B. Alternatives for fat/oils lower...calories. Food Engineering. July. p.40.

Anonymous. 1990A. G.P.C. Tech. Bulletin # 10. Feb.

Anonymous. 1990B. Fat substitute update. Food Tech. March. pp. 92-97.

Anonymous. 1990C. Lose the fat, retain the taste: cellulose gel helps do both. Prepared Foods. May. p.157.

Anonymous. 1990D. A.R.S. seeks help in development of its "double whammy" Oatrim. Milling & Baking News. June 12. pp. 47-48.

Anonymous. 1990E. U.S.D.A.'s Oatrim replaces fat in many food products. Food Tech. Oct. p. 100.

Anonymous. 1991A. Fat replacers: food ingredients for healthy eating. Calorie Control Council. Atlanta, GA.

Anonymous. 1991B. Low-fat salad dressings demand a "total systems" solution. Prepared Foods. Feb. p.73.

Anonymous. 1991C. Development of low-fat, no-fat products targeted by bakers. Milling & Baking News. March 19. pp. 17-18.

Anonymous. 1991D. Fat-blocking fiber aids deep-fried foods. Prepared Foods. April. p. 71.

Anonymous. 1991E. Reduced fat and cholesterol conference. Food Engineering. May. pp. 41-58.

Anonymous. 1991F. Staley manufacturing sets starch-based fat replacer. Chem. Market. Reporter. June 17.

Anonymous. 1991G. 75% of Americans consume lite foods. Food Processing. July. pp. 14-16.

Anonymous. 1991H. Modified potato starch partially replaces fats and oils. Food Processing. July. p.74.

Anonymous. 1991I. Oatrim: fat reducer, cholesterol fighter. Food Processing. July. p. 80.

Anonymous. 1991J. Gums replace fat. Prepared Foods. July. p. 82.

Anonymous. 1991K. Anti-staling fat mimetic: too good to be true? Prepared Foods. August. pp. 133-34.

Baer, C. C., Buliga, G. S., Hassenheutti, G.L., Henry, G. A., Heth, A.A., Jackson, L. K., Kennedy-Tolstedt, J. M., Kerwin, P. J., Miller, M. S., Parker, E. M., Paul, N. K., Pechak, D. G., Smith, G. F. and Witte, V. C. 1991. Low calorie food products having smooth, creamy, organoleptic characteristics. U.S. patent 5, 011, 701.

Ballard, B. F., Schureid, J. M. and Dec, A. F. 1986. Dry mix for low oil salad dressings. U.S. patent 4, 596, 715.

Barnett, R. E., Dikeman, R., Liao, S-Y, Gill, J. and Pantaleone, D. P. 1990. Water soluble bulking agents. U.S. patent 4, 927, 654.

Best, D. 1991A. Technology fights the fat factor. Prepared Foods, Feb. pp. 48-9.

Best, D. 1991B. The challenges of FAT substitution. Prepared Foods. May. pp. 72-77.

Cain, F. W., Jones, M. G. and Norton, I. T. 1989. Spread. European patent 0-237, 120-B1.

Dartey, C. K., Trainor, T, M., and Evans, R. 1990. Low cholesterol mayonnaise substitute and process for its preparation. U.S. patent 4, 948, 617.

Dillon, P. M. 1990. Gums and starches bulk up low-cal foods. Food Engineering. Jan. pp. 87-90.

Duxbury, D. D. 1990A. Gellan gum-new gelling agent enhances many good attributes. Food Processing. June. PP. 54-57.

Duxbury, D. D. 1990B. Oatrim: fat reducer, cholesterol fighter. Food Processing. Aug. pp. 48-52.

Duxbury, D. D. 1990C. Fat sparing starch can replace 100% fat/oil for 96% caloric reduction. Food Processing. Dec. p. 38.

Duxbury, D. D. 1991. Special report: diet, nutrition and health. Food Processing. March. pp. 58-72.

Dziezak, J. D. 1989. Fats, oil and fat substitutes. Food Tech. July. pp. 66-74.

Egbert, W. R., Huffman, D. L., Chen. C. and Dylewski, D. P. 1991. Development of low-fat ground beef. Food Tech. June. pp. 64-73.

Freedman, A. M. 1991. Tate and Lyle unit set to weigh in with fake fat. Wall St. Journal. June 11. pp. B-1, B-6.

Furcsik, S. L., Mauro, D. J., Kornacki, L., Faron, E.J., Turnak, F. L. and Owen, R. 1991A. Method for making a reduced fat foodstuff, W.O. patent 91/01092.

Furcsik, S. L., Mauro, D. J., De Boer, E., Yahl, K. and Delgado, G. 1991B. Method for making a reduced fat foodstuff, W.O. patent 91/01092.

Furcsik, S. L., Mauro, D. J., De Boer, E., Yahl, K. and Delgado, G. 1991C. Method for making a reduced fat foodstuff, U.S. patent 4, 981, 709.

Hannigan, K. J. 1981. Fat replacer cuts calories in fatty foods. Food Engineering. Sept. p. 105.

Haumann, B. F. 1986. Getting the fat out: researchers seek substitute for full-fat fat. JAOCS. 63:278.

Hegenbart, S. 1991. Beyond fat replacement: making the system work. Food Product Design. Aug. pp. 28-40.

Howard, N. C. and Kleinschmidt, D. C. 1991. Low calorie fat substitute resistant to laxative side effect. U.S. patent 5, 006, 360.

Inglett, G. E. 1990. Hypocholesterolemic beta-glucan/ amyodextrins from oats as dietary fat replacers. Symposium on beta-glucan: Biotechnology and Nutrition. 199th ACS Meeting. Boston, MA.

Inglett, G. E. and Grisamore, S. B. 1991. Maltodextrin fat subst. lowers cholesterol. Food Tech. June.p.104.

Johnson, G. R., Jones, E. C., Jr. and Jones, M. C. 1988. Reduced calorie sausage containing cooked rice. U.S. patent 4, 735, 819.

La Bell, F. 1991A. Rice starch reduces fat, adds creamy texture without imparting flavor. Food Processing. March. pp. 95-96.

La Bell, F. 1991B. New ingredients for low-fat, sugar-free and microwave foods , Food Processing. April. pp. 52-54.

La Bell, F. and Duxbury, D. D. 1991. Less fat, more healthy ingredients. Food Processing. Aug. pp.98-107.

Lenchin, J. M., Trubiano, P. C. and Hoffman, S. 1985. Converted starches for use as a fat- or oil-replacement in foodstuffs. U.S. patent 4, 510, 166.

Light, J. M. 1990. Modified food starches: why, what, where and how. Cereal Foods World. 35:1081.

Lowery, A. N. 1989. Meltable spread composition: low-fat substitute for butter or margarine. U.S. patent 4, 869, 919.

Meyer. P. A. 1991. Stellar, corn-based fat replacer from Staley enters $1-billion derby. Milling & Baking News. June 25. p. 44.

Miller, D. E. and Werstok, C. E. 1980. Low-fat comestible spread substitutes. U.S. patent 4, 238, 520.

Miller, M. 1991. Fat replacement technologies. Paper presented to Midwest Section, AACC. April 16. Elmwood Park, IL.

Morehouse, A. L. and Lewis, C. J. 1985. Low-fat spread. U.S. patent 4, 536, 408.

Morehouse, A. L. and Sander, P. A. 1987A. Low DE starch hydrolyzates. U.S. patent 4, 699, 669.

Morehouse, A. L. and Sander, P. A. 1987B. Low DE starch hydrolyzates. U.S. patent 4, 699, 670.

Morris, C. E. 1984. New applications of maltodextrins. Food Engineering. June. p. 105.

Penichter, K. A. and McGinley, E. J. 1991. Cellulose gel for fat-free food applications. Food Tech. June. p. 105.

Platt, B. L. and Gupta, B. B. 1988. Low fat spread. U.K. patent GB2193221 A.

Przybyla, A. E. 1990. Formulating healthy foods. Food Engineering. Feb. pp.49–59.

Ramirez, A. 1991. Staley develops starch substitute for fat. New York Times. June 12. pp. 1 and D–8.

Richter, M., Schierbaum, F., Augustat, S. and Knock, K.-D. 1976A. Method of producing starch hydrolysis products for use as a food additive. U.S. patent 3, 962, 465.

Richter, M., Schierbaum, F., Augustat, S. and Knock, K.-D. 1976B. Method of producing starch hydrolysis products for use as a food additive. U.S. patent 3, 986, 890.

Robbins, M. D. and Rodriguez, S. S. 1984. Low calorie baked products. U.S. patent 4, 461, 782.

Schierbaum, F. R., Richter, M. and Augustat, S. 1978. Central Inst. of Nutrition Academy of Sciences. Potsdam-Rehbruecke, G.D.R. International Congress of Food Science & Tech. Abstracts. p. 84.

Schierioth, W. 1991. Nutritionally modified bakery foods. Presented at AACC short course on low calorie food product development. San Diego, CA.

Swientek, R. J. 1991. New fat replacer-creme of the crop? Food Processing. Aug. pp. 48–50.

Tye, R. J. 1991. Konjac flour: properties and applications. Food Tech. March. pp. 82–92.

Vorwerg, W., Schierbaum, F., Reimer, G., and Gringmuth, B. 1988. Process for manufacture of food preparations. D.D. patent 256, 645.

Wolkstein, M. 1986. Dietetic frozen desserts containing aspartame. U.S. patent 4, 626, 441.

Worthy, W. 1990. Evidence mounts for dietary soluble fiber benefits. C&EN. May 28. pp. 23–24.

NON-FOOD USES OF STARCH

Kenneth W. Kirby

Penford Products Co.
Cedar Rapids, IA

INTRODUCTION

The versatility of starch allows its use in a multitude of non-food uses. In these applications, the physical properties of aqueous dispersions, its bonding ability and the nature of its films forms the basis for its value.

As versatile as starch is, its utility can further be enhanced by reacting it with a variety of chemicals which contribute new properties to the native starch. Generally, derivatives of starch will correct or minimize some of the undesirable paste properties of native starch. These are related to retrogradation or re-association of hydrated starch chains and the ultimate effect on viscosity, filming, water holding and other physical characteristics.

While a study of the properties of starch and its derivatives is best understood by examining its chemical composition, molecular architecture and granular organization, these subjects are addressed elsewhere in this volume and will not be repeated here. The biochemist has learned that starch is comprised of alpha-D-glucopyranose units, regardless of its source. Despite this fact, there are major differences in the performance of starch from different biological sources.

These differences arise from the varying amounts of amylose and amylopectin, the frequency of branching that occurs in the amylopectin and the length of the chains in both amylose and amylopectin.

Plant sources of starch for non-food uses derive mainly from corn (maize), potato, tapioca and wheat. Barley, rice and sago palm provide minimal amounts except in isolated areas. Genetic selections of waxy maize (amylopectin starch) and high amylose (corn starch) have found increasing non-food markets in recent times.

Corn provides over 95% of the raw material for starch in the USA and probably about 77% worldwide (Wurzburg, 1986 and Jones, S.F., 1983). In the USA, corn starch production was about six billion pounds in 1990, of which about 61% went into the paper industry, its principal market. Another 19% went into food products, with the remaining 20% going for textiles, adhesives, chemicals, flocculants, building materials and other miscellaneous uses. The six billion pounds of corn starch produced in the USA represents only about 17% of all of the corn ground by wet-millers (Corn Annual, 1991). Of the six billion pounds of corn starch produced, approximately 1/6 or about one billion pounds may be considered as specialty starch. The discussion in this chapter will center on specialty non-food uses and the directions that new developments are going.

USES OF STARCH IN PAPER

Starch use in paper may be separated into three general application areas. These are the wet end or forming section of a paper machine, the size press and calendar stack section, and the coating section, which may be on machine or off-machine. Admittedly, there are some variations within these categories because there is a growing trend to perform some degree of coating at the size press to produce various coated grades of paper. Coated paper differs from other paper in that a pigment in the form of clay, calcium carbonate, titanium dioxide or other mineral is added to provide an improved surface for printing.

While this discussion will focus on the uses of

specialty starches in paper, it should be recognized that the largest volume of starch used in paper is untreated common corn starch for enzyme conversion, thermochemical conversion and for corrugated box board. Although common corn starch performs many functions, it has limitations became of its tendency for cooked pastes to retrograde, has limited water holding properties, forms non-continuous films and results in variable viscosity during storage and application. The choice of common corn starch is usually an economic decision and as higher quality paper is required, specialty starches are used.

Wet End Starches

Cationic starches are added to the raw pulp in a paper machine to bond the fibers prepared by chemical digestion, thermomechanical or simple ground wood methods. All wood fibers are anionic, whether formed chemically by oxidation or resulting from the residual lignin, hemicelluloses and other components intermixed with the cellulose fibers. Wood fibers develop bonds to hold them together in a sheet when the hydroxyl groups are hydrated so that hydrogen bonding is effective. Energy is required to break up the fibers and cationic starch can provide the adhesive to bond the fibers, thereby saving significant energy costs. Other positive features from adding cationic starch include drainage increases, ease of drying the sheet and increased machine speeds.

The cationic charge on the starch will also attract other anionic pulp additives, such as clay and titanium dioxide, as well as some more detrimental substances carried along in the pulping operation. As ash retention increases, sometimes a strength loss results.

Cationic starch is an efficient flocculent and must be added properly to a paper furnish so that good formation results. Efficient mixing is required to uniformly contact all the fibers and form a mat that dewaters rapidly and evenly. Improved formation results in increased machine speed and greater production and thereby increased profitability. Cationic starches are used as native unmodified viscosity products.

Cationic starches commercially available are either tertiary amino or quaternary ammonium derivatives of corn, potato, tapioca or waxy starch. The first successful cationic product was made (Caldwell and Wurzburg 1957) by reacting 2-diethylaminoethyl chloride with starch in a highly alkaline suspension. The starch ether formed had to be quaternized with mineral acid to develop cationicity. Tertiary amino starches are most efficient at acidic pH and as the pH becomes alkaline the cationic charge diminishes.

Quaternary ammonium starches are prepared by reacting starch with 2,3 epoxypropyl-trimethylammonium chloride in aqueous slurry with an alkaline catalyst and a swelling inhibitor present. Although most reactions are made with the chlorohydrin form of the reagent, there is some more recent interest in using the epoxide form. The latter is useful for dry reaction processes.

Quaternary ammonium substituted starches retain their cationic charge over the pH range of use in paper applications and have advantages over tertiary amino starches, particularly at high pH.

Amphoteric starches contain both an anionic and cationic group and the balance of these groups is said to affect the performance of the starch (Caldwell et al., 1969). The most common anions are phosphate, carboxyl or sulfonate, while the cationic group may be tertiary amino or quaternary ammonium. The starch product remains cationic but will complex with both anionic and cationic components in a paper furnish.

Amphoteric starches may be made from any commercial starch base. The most common are corn and waxy, although potato starch when made cationic becomes an amphoteric starch because of the naturally occurring ortho-phosphate group in potato starch.

Cationic starches have established themselves as valuable additives in paper making and will be a part of the process for many years to come. They become even more important as alkaline paper making grows because of the need for emulsifying the size and to bind the added amount of ash as calcium carbonate use grows. Alkaline sizing, to impart a degree of water resistance to paper, is done by adding either (ASA) alkenyl succinic

374

anhydride or (AKD) alkyl ketene dimer. These materials must be emulsified and conducted to the anionic fiber to develop water resistance. Cationic starches serve this purpose. A preferred starch appears to be a potato cationic with approximately 0.3% nitrogen added. However, other cationic starches made from waxy, corn or tapioca also perform well in emulsification.

Surface Sizing Starches

Starches used for surface sizing are depolymerized in varying degrees to allow one to raise the solids of a starch paste applied to the sheet of paper. The most common depolymerization is done by enzyme or acid digestion at the mill site. Thus, alpha amylase and ammonium persulfate conversion consume the largest amount of starch for surface sizing. Surface sizing also consumes the largest single volume of starch in the paper process.

The most common physical characteristic to deal with in surface sizing is the tendency of a depolymerized starch to retrograde or re-associate. This phenomenon causes varying viscosities, poor films, and, in some extreme cases, can cause rejection at the size press. When this happens a poor discontinuous film results and off-quality paper is produced.

With increasing demands for quality, increased speed of machines and the competitive nature of the business, the trend toward use of specialty starches has increased in recent years and will apparently continue to grow.

Hydroxyethyl starch is the preferred and dominant specialty surface sizing starch. This product was invented in 1950 (Kesler and Hjermstad) and continues to grow in volume and importance. The reason for its high degree of acceptance in over forty years of use is its strong resistance to retrogradation. Because of this characteristic, films are clearer and more flexible, water holding is increased, viscosity variations are minimal and storage problems are eliminated, all due to the resistance to retrogradation. Starch is still a source of energy for bacteria and mold and precautions must be taken to prevent deterioration from this source.

Alkaline sizing of paper has demonstrated the need for an improved film forming starch. The ease of penetration of a size into a sheet because of increased porosity and hydrophobicity has shown that a hydroxyethyl starch will allow controlled penetration for bonding of fibers and ash as calcium carbonate, and at the same time yield a surface film suitable for printing without press picking, print picking, dusting and severely reduced runnability.

Low level oxidation of starch by sodium hypochlorite treatment provides a product intermediate in quality between enzyme or ammonium persulfate converted starch and hydroxyethyl starch. These products rank in production volume in the same relationship as their quality provides to the papermaker.

Growing in popularity and use are mixed derivative specialty starches made with cationic and hydroxyethyl substitution. When applied as a paper size the cationic charge results in substantivity to the fiber. On repulping, which occurs in all mills, the starch is not lost to the effluent and BOD and waste treatment costs are reduced.

Coating Starches

Coating starches are made to a lower unit viscosity than other starches. The need to have a significant amount of binder present to hold the pigment and still control the viscosity so that a film can be applied requires a thinner viscosity starch. Because of this, the potential for retrogradation is increased as the starch chain becomes shorter. Thus, the hydroxyethyl derivative again excels in this application.

Coated paper has grown faster than the market for paper in general. This area has contributed significantly to the increased production volume of hydroxyethyl starch.

Growth in coated paper has come principally from the No. 5 grade used in newspaper supplements and advertising forms where offset printing or rotogravure is the printing vehicle.

Styrene-butadiene latex is a common additive to a coating formula to lend gloss and water resistance. Latexes are compatible with, and are frequently used in combination with hydroxyethyl starch. Recent innovations in starch chemistry have allowed the grafting of styrene and butadiene to starch to give a product which combines the properties of both (Nguyen et al., 1991) Additional benefits are contributed by the starch portion to give stiffness and the opportunity for higher solids coatings without exceeding viscosity limits on a high speed coater.

USE OF STARCH IN TEXTILES

Starch volume usage in the textile industry has roughly paralleled the trend in the industry to off shore production. Southeast Asia and other Asian countries now provide the bulk of textile products through imports. Since 1980 it is estimated that about 400,000 jobs have been lost in the US textile industry. Although the trend is continuing, starch usage has held steady through the past six years, partly due to the increased use of cotton fibers in the overall textile production.

Warp Sizing

It is estimated that about 80% of the starch used in textiles is consumed in warp sizing. A thorough discussion of warp sizing has been presented (Kirby, 1986).

The majority of starches used for warp sizing are acid modified or low oxidized products. These starches are provided in a number of fluidity (viscosity) ranges to allow for higher solids solutions. The films formed satisfy the needs of forming the warp so that the warp can be woven without shedding and forming fuzz balls. Characteristics required in a starch are abrasion resistance, flexibility, bonding sites and water holding.

Specialty starches provide improved flexibility, water holding, abrasion resistance and easier cooking and desizing operations. Starch acetates and hydroxyethyl starch are used to size cotton and cotton blends with synthetics. They are particularly suited to low temperature sizing at 140-160 °F of worsted yarns, orlon-wool mixtures and fibers that shrink on drying.

Carboxymethyl starches, depolymerized by oxidation (Hjermstad and Coughlin, 1972) are shown to be compatible with polyvinyl alcohol and may be desized with hot water. A preferred application is sizing polyester-cotton blends.

High amylose starch containing more than 50% amylose has found a market in sizing glass fiber.

With the exception of denims, all starch sized fibers are desized. Most commonly, desizing is with the enzyme alpha amylase. In some cases alkalis, acids, detergents or possibly, in the case of specialty starches, hot water is used.

Finishing

Finishing usually requires a soft hand, an indication of quality and specialty starches such as acetate and hydroxyethyl derivatives are used.

USE OF STARCH IN
FERMENTATION AND BIOCHEMICAL PRODUCTION

Starch, or its basic building block alpha-D-glucose, figures into the preparation of a multitude of products through chemical conversion or biochemical transformation. Many of these have industrial and food significance. As biotechnology grows in importance, the list will become longer.

The chemical use of alpha-D-glucose generally is not as dependent on the alpha-form, but rather, its polyol structure. On the other hand, the biochemical use relies mainly on the alpha-form of glucose in starch to be used as an energy source for organisms.

Fuel Alcohol

Fermentation derived ethanol is by far the principal chemical made from starch. Fuel ethanol is 100% ethyl alcohol and is usually blended with gasoline at 10% volume. In 1990, fuel ethanol production reached 900 million gallons, primarily from corn wet milling (Corn Annual, 1991). Estimates for expanded production are based on the 1990 Clean Air Act mandating the use of oxygenated fuels to control carbon monoxide emissions from the burning of gasoline. As other toxic and volatile organic compound control is addressed, the potential expansion of the industry could be as much as 75% above 1990 levels within five years. A catalyst for the growth of fuel ethanol has been legislation whereby the federal excise tax and a blender's tax credit have supported the economics of using corn as a raw material. These are in addition to several individual state tax forgiveness actions to assist corn growers and promote the industry.

Ethyl alcohol fermentation has been improved so that it yields a net positive 23,000 BTU/gallon energy balance. A 10% blend of ethanol in gasoline lowers the carbon dioxide emissions by 25 to 30% and increases the octane rating by two points. Gasoline, however, provides 120,000 BTU/gallon, while ethanol provides 76,000 BTU/gallon.

Other Fermentation Products

Because starch yields pure dextrose on hydrolysis, it is being preferred over molasses in a growing number of fermentation processes. Likewise, industry suppliers of dextrose are becoming directly involved in the use of dextrose as a raw material.

Citric acid, a food acid, may be the major chemical produced by fermentation outside of fuel alcohol and high fructose corn syrup. Following a close second may be lysine, which finds use as an additive in swine and poultry feed.

Vitamin C (or ascorbic acid) is a product that combines both chemical reaction and a biochemical process in its manufacture. Biotin is also being produced.

Other acids made by fermentation include lactic, gluconic and itaconic.

In addition to enzyme production for making phosphate free detergents, there is growth in feed grade penicillin and bacitracin.

Natural gums continue to be impacted by new synthetically formed gums. Xanthan gum has been established as a food and industrial product, partially because of its low sensitivity to salts and pH adjustment. Another synthetic gum recently introduced is gellan.

Possibly one of the more interesting developments is the introduction of cyclodextrins (Friedman and Hedges, 1989). These six, seven or eight membered, non-reducing ring compounds are said to be chemically and physically stable, offer solubility control and can be used for controlled release of entrapped molecules.

CHEMICALS MADE FROM STARCH

Aside from the many derivatives of starch which could be considered chemicals, possibly the highest volume product is sorbitol. Produced as the reduction product of glucose it finds many uses, including a base for ascorbic acid, for making plasticizers, resins, as a humectant, in writing inks, candy, as a sequesterant and softener and many other practical applications.

Methyl glucoside has been an item of commerce for more than thirty years being used as an extender and for making intermediates in resin and foam manufacture.

Alkyl polyglycosides combining glucose and fatty alcohols are to be made at a new plant starting in 1992. Uses planned are for foaming, solubilizing and grease cutting agents.

BIODEGRADABLE PRODUCTS FROM STARCH

This section is not intended as a treatise on the biodegradability or degradability of synthetic materials compounded with or containing starch. Rather, a number of product applications that have commercial status will be covered.

In the general subject of degradability, there has been much confusion, misperception, miscommunication and inaccurate information. Much of this is spawned by the high desirability to control waste disposal, landfill limitations and in general to reduce environmental pollution in the world. For a reasonable discussion of the subject of polyethylene film degradation we suggest an article by Gage (1990).

Starch, because of its digestibility by microorganisms and animals (humans), has been a principal target for inclusion into degradable materials. Biodegradability begins at 100% with pure starch and decreases as non-digestible materials are blended or compounded with starch. The digestion and ease of degradation depends on the amount and the accessibility of the starch granule in the mixture. Apparently a threshold value of about 59% starch loading exists when compounded with polyethylene (Studt, 1990). Below 59% starch only 16% of starch particles are accessible to each other. Above 59% starch, fully 77% of starch particles are accessible to each other.

Starch based blown films (Otey, Westhoff and Doane, 1980; Otey and Westhoff, 1982) have resulted from compounding with ethylene acrylic acid or polyethylene and are said to be biodegradable. About 40% starch was incorporated and gelatinized before extrusion or blowing.

Graft polymers of starch are said to be formed using water soluble acrylates and containing up to about 40% starch. Likewise, graft polymers with starch can be made with styrene (Otey and Doane, 1984). Commercialization of such films has resulted in mulching films and other extended uses.

Starch in the granule form may be incorporated into plastics in a variety of ways using catalysts for accelerating decomposition of the physical mixture (Griffin, 1977, 1978, 1980, 1983). These have been commercialized and use from 6-20% starch (Roper and Koch, 1990). An alternate but similar method uses a silane derivative of starch with a vegetable oil incorporated into the mixture. Large granule starches are a detriment and form a bumpy film where thin films are desired.

Biodegradability to where the starch becomes 100% soluble is claimed in another process based on extrusion of a destructured (acid modified) starch (Stepto, et al., 1988). Extenders may be added to enhance the performance in the form of hydrogenated triglyceride, lecithin, titanium dioxide and water. Potato starch is preferred became it can contain free electrolytes, although corn starch is claimed (Sachetto, et al., 1990).

Another application of extrusion to produce biodegradability is the preparation of an expanded high amylose starch to serve as a low density packing material, (Lacourse, 1989), thereby replacing the petrochemical based products presently used.

Future uses of starch may be possible with the development of biopolyesters such as poly-D-beta hydroxybutyrate of polylactic acid. Such materials are expected to be 100% biodegradable.

GENETIC SELECTIONS

Successful development of commercial lines of waxy corn and high amylose corn have allowed industrial product applications outside of food uses.

Waxy corn was introduced into the USA in 1908 from China where it had been grown since being transferred from the Americas (Whistler, 1984). Following a successful breeding program, the starch found use in foods because of its all branched structure and general resistance to retrogradation. Through chemical treatment its use expanded. In recent years it has found an industrial market as a cationic starch for wet end application both as the cationic form and as the amphoteric form.

High amylose varieties of corn starch have been developed (Shannon, 1984) and have become successful products for food and industrial purposes. Much of the genetic work was done by Bear Hybrid Seed Co. (Vineyard and Bear, 1952). Two principal levels of amylose used are Amylomaize V and Amylomaize VII containing about 55% and 70% amylose, respectively. Both of these starch bases require special cooking conditions to gelatinize the granules since a normal cook at 195 °F for 20 minutes will not cause granule dispersion. Once the starches are dispersed they show rapid retrogradation and, because of the high degree of linearity, they form stronger films.

Two principal volume uses of high amylose starch

are as a carrier starch in making corrugated box pastes and, as a size for fiber glass in the textile industry.

A newer use, previously mentioned, is as a derivatized extruded product to act as an expanded insulation material.

Miscellaneous Uses

Many miscellaneous uses of starch depend on its adhesiveness and its film forming nature. Starch acts as a flocculent in mining applications, as a binder in moldings and wall board as well as a dusting powder in surgeons gloves.

Possibly its newest future lies in the capability of the geneticist to alter the basic architecture of the chains, their distribution of branching and the ratio of branches to linear chains. Many genetic selections of corn are known but the hope of relating chemical structure to commercial applicability and performance still escapes the chemist.

LITERATURE CITED

Caldwell, C. G. and Wurzburg, O. B. 1957. Ungelatinized tertiary amino alkyl ethers of amylaceous materials. U. S. Patent 2,813,093.

Caldwell, C. G., Jarowenko, W. and Hodgkin, I, D. 1969 Method of making paper containing starch derivatives having both anionic and cationic groups, and the product produced thereby. U. S. Patent 3,459,632.

Corn Annual. 1991 Corn Refiners Association, Washington D. C. Friedman, R. B. and Hedges, A. R., 1989. Recent advances in the use of cyclodextrins in food systems. in: Frontiers in carbohydrate research -1. Food applications. Elsevier applied science. London and New York. (R. P. Millane, J. N. BeMiller and R. Chandrasekaran, Eds.)

Gage, P. 1990. Degradable polyethylene film - the facts. TAPPI. October 161-169.

Griffin, G. J. L. 1977. Biodegradable synthetic resin sheet material containing starch and a fatty material. U. S. Patent 4,016,177.

Griffin, G. J. L. 1978. Synthetic/resin based compositions. U. S. Patent 4,125,495.

Griffin, G. J. L. 1980. Shaped synthetic polymers containing a biodegradable substance, U. S. Patent 4,218,350.

Griffin, G. J. L. 1983. Plastics based composition containing a polyester resin and alkaline modified starch granules. U. S. Patent 4,420,576.

Griffin, G. J. L. 1991. Degradable plastics. U. S. Patent 4,983,651.

Hjermstad, E. T. and Coughlin, L. C. 1972. Etherified depolymerized starch product compatible with polyvinyl alcohol. U. S. Patent 3,652,541.

Hjermstad, E. T. and Coughlin, L. C. 1972. Process for preparing etherified depolymerized starch product compatible with polyvinyl alcohol. U. S. Patent 3,652,542.

Jones, S. F. 1983. The World-market for starch products. tropical development and research institute. G173, 56/62 Gray's Inn Road. London WC1X8LU.

Kesler, C. C. and Hjermstad, E. T. 1950. Preparation of starch ethers in original granule form. U. S. Patent 2,516,633.

Kirby, K. W. 1986. Textile industry in modified starches: properties and uses. O. B. Wurzburg, Ed. CRC Press. Boca Raton, FL.

Lacourse, N. L. and Slitieri, P. A. 1989. Biodegradable packing material and method of preparation thereof. U. S. Patent 4,863,655.

Nguyen, C. C., Martin, V. J. and Pauley, E. P. 1991. Starch graft polymers. U. S. Patent 5,003,022.

Otey, F. H., Westhoff, R. P. and Doane, W. J. 1980. Starch-based blown films. Industrial and Engineering Chemistry. December 592-595.

Otey, F. H. and Westhoff, R. P. 1982. Biodegradable starch-based blown films. U. S. Patent 4,337,181.

Otey, F. H. and Doane, W. J. 1984. Chemicals from starch. In: Starch chemistry and technology, 2nd Ed. 403-410. R. L. Whistler, J. N. BeMiller, E. F. Paschall, Eds. Academic Press, Orlando, FL.

Roper, H. and Koch, H. 1990. The role of starch in biodegradable thermoplastic materials. Starke, 42: No. 4, 123-130.

Shannon, J. C. and Garwood, D. L. 1984. Genetics and physiology of starch development: Starch Chemistry and technology, 2nd Ed., R. L. Whistler, J. N. BeMiller and E. F. Paschall, Eds, Academic press, Orlando, FL.

Sochetto, J. P., Stepto, R. F. T. and Zeller, H. 1990. Destructurized starch essentially containing no bridged phosphate groups and process for making same. U. S. Patent 4,900,361.

Stepto, R. F. T., Tomka, I. and Dobler, B. 1988. Destructurized starch and process for making same. E. P. Application 0,282,451.

Studet, T. 1990. New technologies for waste management. R & D Magazine. March pp 50-56.

Vineyard, M. L. and Bear R. P. 1952. Amylose content. Maize genetics cooperative newsletter. 26:5.

Whistler, R. L. 1984. History and future expectations of starch use. In: Starch chemistry and technology, 2nd Ed. P. 6. R. L. Whistler, J. N. BeMiller and E. F. Paschall, Eds. pp 6-7. Academic press, Orlando, FL.

Wurzburg, O. B. 1986. Modified starches: properties and uses. P 5. CRC Press, Boca Raton, FL.